Springer Geology

Series editors

The book series Springer Geology comprises a broad portfolio of scientific books, aiming at researchers, students, and everyone interested in geology. The series includes peer-reviewed monographs, edited volumes, textbooks, and conference proceedings. It covers the entire research area of geology including, but not limited to, economic geology, mineral resources, historical geology, quantitative geology, structural geology, geomorphology, paleontology, and sedimentology.

More information about this series at http://www.springer.com/series/10172

Mansour Ghorbani

Lithostratigraphy of Iran

Springer

Mansour Ghorbani
Shahid Beheshti University
Tehran, Iran

ISSN 2197-9545 ISSN 2197-9553 (electronic)
Springer Geology
ISBN 978-3-030-04962-1 ISBN 978-3-030-04963-8 (eBook)
https://doi.org/10.1007/978-3-030-04963-8

Library of Congress Control Number: 2018966438

This Springer imprint is published by the registered company Springer Nature Switzerland AG
The registered company address is: Gewerbestrasse 11, 6330 Cham, Switzerland

Foreword

While working recently with Prof. Mansour Ghorbani in Tehran, in connection with a review project on the biostratigraphy of Iran, I quickly came to appreciate his encyclopaedic knowledge of Iranian geology. I was also greatly impressed by the fact that he had visited so many of the localities that we were discussing. 'The Lithostratigraphy of Iran' is a masterful account that provides an excellent introduction to the stratigraphy of Iran for professional geologists but is written in such a way that it is also suitable for students with an interest in the geology of this region.

With a population of around 80 million and area of 1648 km^2, Iran is a fascinating country spanning the boundary between the Arabian and Eurasian plates. Although it is ranked fourth in the world in terms of proven oil reserves and has a history of hydrocarbon exploration dating back to the drilling of the first exploration well in the Middle East in 1908, relatively little of a summary nature has been published in the English language on its stratigraphy. 'The Lithostratigraphy of Iran' remedies this deficiency by presenting a comprehensive but concise account of the entire stratigraphic succession of the country, region by region. All aspects of the litho- and biostratigraphy of the succession are covered, together with the up-to-date chronostratigraphic interpretation of all of the rock units described. Depositional environments and paleogeographic and tectonic relationships are also discussed in some detail. In addition to the text, the stratigraphy of each geological period is summarized in table form. The descriptions of many of the rock units described are also augmented by clear diagrams and annotated field photographs. Where appropriate, comparisons are made to stratigraphic units in neighbouring countries, but this work deals primarily with Iran.

In Iran, as in most other countries, adherence to the established codes of stratigraphic practice has sometimes been poor in the past, with many formations being inappropriately named and defined, and/or having no clearly designated and described type sections. In other instances, conflicting interpretations of biostratigraphic work have led to confusion concerning correlation and dating. The author has dealt with these issues in a logical and systematic manner, carefully documenting the evidence and discussing how opinions have evolved through time. For example, changes to the postulated age of the Faraghan formation are traced from early suggestions that it was Devonian–Carboniferous, through to the publication of palynological evidence indicating that it is, in fact, Permian, and finally to the acceptance of the latter interpretation by the National Iranian Committee on Stratigraphy. The author's close links with the National Iranian Oil Company have enabled the inclusion of significant amounts of unpublished data on critical subsurface sections that considerably enhance the value of this book.

Professor Ghorbani is to be congratulated on, 'The Lithostratigraphy of Iran', his fifth book that follows 'The Economic Geology of Iran' (Springer, 2013). Few will read this book from cover to cover but that is not its purpose; it is a serious reference book that provides a wealth of up-to-date stratigraphic information on Iran.

Sheffield, UK
March 2018

Geoff Clayton
Visiting Professor of Palynology
Department of Animal and Plant Sciences
University of Sheffield

Preface

The mountains reared themselves, the streams gushed out,
While from the soil the herbs began to sprout
A farther step-man cometh into sight;
Locks had been made; he was the key of each
By nature first, in order last, art thou;
Hold not thyself then lightly. I have known
Shrewd men speak otherwise, but who shall know
The secrets that pertain to God alone?

Hakim Abu'l Qasem Ferdowsi, 977–1010 A.C.E.
Translated by Warner & Warner, 1905–25

This book is the results of three decades of fieldworks and laboratory works, intended to raise the knowledge of scholars, researchers and students about the geological knowledge of Iran that is one of the goals and missions of the 'Pars Geological Research Center' (Arianzamin). In addition to personal information and available resources, this book is intended to provide a wealth of reports from reputable geological organizations, academic dissertations and reports from companies and organizations related to geology, including the Geological Survey of Iran, National Company Iran Oil, Iran Steel Co. and the Iran Copper Co.

When writing of all parts of the book, I have been consulted by friends with the contents of certain sections. I am very grateful to all of them who encouraged and cooperated me, including Drs. Mehdi Zare, Abdul Rahim Houshmandzadeh, Bahauddin Hamdi, Sayed Ali Aghanabati, Anoshiravan Kani, Mohammad Hossein Motiei, Mozhgan Salehi Yazdi, Azin Ahifer, Asma Aftabi, Masoud Ovissi, Mohsen Ghorbani, Mohammad Ghasemi, Jafar Omrani, Reza Kohansal, Hamid Nazari, Reza Salamati and Jafar Sabouri.

I would like to express my gratitude to Keyvan Zandkarimi, Shahid Beheshti University, who translated the entire book, and as a stratigrapher checked the content and added some recent references and new researches under my authority. Masoud Oveisi is appreciated for editing this book. I am particularly indebted to Prof. Dr. Clayton for his linguistic and stratigraphic edition on the final draft of the book.

Tehran, Iran

Mansour Ghorbani
Associate Professor at Shahid Beheshti University
Manager of Pars Geological Research Center

Contents

About the Author

Dr. Mansour Ghorbani was born in Nanaj village, a rural district of Malayer County in the west of Iran on 1961. He completed his primary and secondary educations in his hometown by 1979.

He graduated from high school in 1983. He studied geology at the University of Shahid Beheshti and concurrently Chemistry at Islamic Azad University. He graduated from both courses in 1988 and 1989, respectively. He continued his academic studies in geology at the University of Shahid Beheshti and received his masters (M.Sc.) and Doctor of Philosophy (Ph.D.) degrees in 1993 and 1999, respectively.

Following his academic accomplishments, he joined the geology faculty at the Shahid Beheshti University and has been teaching undergraduates, postgraduates and Ph.D. students till now. He currently holds the associate professor position at the University.

From 1991 to 1996, he was involved in the treatise on the geology at geological survey of Iran. He wrote and compiled a lot of literatures on the geology and mineral deposits, such as economic deposits, soils, iron, antimony, arsenic, mercury, copper, lead and zinc in Iran.

Aside from teaching, he has been working on international and national research projects with mining, oil and gas companies.

The rewards and outcome of these years of studying and working are 38 books, more than 170 academic papers, over 120 scientific and technical reports in reference to natural and mineral resources in Iran and Iranology, as well as the compilation of international metallogeny and gem distribution maps of the Middle East.

He enjoys travelling around the country and abroad; he maintains that while visiting and working in various regions, he meets different ethnic groups with different cultures and traditions in Iran. He has learned how the habitat and the natural surroundings have a greater effect on the people's socio-economic aspects of life in some places than others.

Years of working experiences and personal beliefs in private research work compelled him to establish his own research centre called Pars Geological Research Centre (Arianzamin) in 2002. The staffs at the centre are all dedicated, diligent, experienced and qualified researchers in the fields of geosciences. The centre has performed successfully; it has compiled and published books, literature concerning different aspects of geosciences. The centre has its own website: http://arianzamin.com.

He, from a sociocultural standpoint, endeavours to help countries and people who speak the same language and have had the same or similar cultures, to establish a long-lasting sociocultural bond with one another.

Chapter 1
Precambrian

Being the longest Eon, the Precambrian is divided into the Archean and the Proterozoic which are further divisible into smaller units (Fig. 1.1). Although Archean and lower to middle Proterozoic rocks have not been previously described, Neoproterozoic rocks have been reported in some areas. Formerly, it was thought that Precambrian rocks with ages greater than 1 or even 1.5 Ga. are well developed in Iran. However, recent studies show that the these rocks are of less exposures and are of younger ages spanning the Neoproterozoic and especially Late Neoproterozoic rocks.

Some important outcrops of metamorphic rocks are described, and sedimentary rock units formerly assigned to the late Precambrian (Neoproterozoic)[1] are discussed.

The Neoproterozoic rocks of Iran are divided into three main groups.

(A) Metamorphic rocks and complexes (readers are referred to "The Magmatism and Metamorphism of Iran", Ghorbani 2018)
(B) Sedimentary rock units
(C) Igneous rocks (Ghorbani 2018).

(A) **Metamorphic rocks and complexes**

Some of the metamorphic complexes assigned to the Neoproterozoic are as follows:

The Poshtebadam, Chapdoni, Boneh-shorou and Naibaz complexes; complexes of the Anarak block such as Kaboudan, Chah-Gorbeh; Patyar and Morghab, Lakh marble unit and the Paryar Complex of Central Iran (Valeh and Haghipour 1970; Haghipour and Pelissier 1977; Haghipour 1978).

The Kuh-e sursat, Kheyrabad and Mahneshan Complexes, as well as the Amir Abad Complex of Takab area (Alavi-Naeini 1972).

Metamorphic rocks of Gasht and Shanderman, Talesh mountains (western Alborz) and Gorgan Schist of eastern Alborz have been assigned by recent studies to the Palaeozoic.

[1]Contrary to previous works, we use the term "Neoproterozoic" rather than "late Precambrian".

M. Ghorbani, *Lithostratigraphy of Iran*, Springer Geology,
https://doi.org/10.1007/978-3-030-04963-8_1

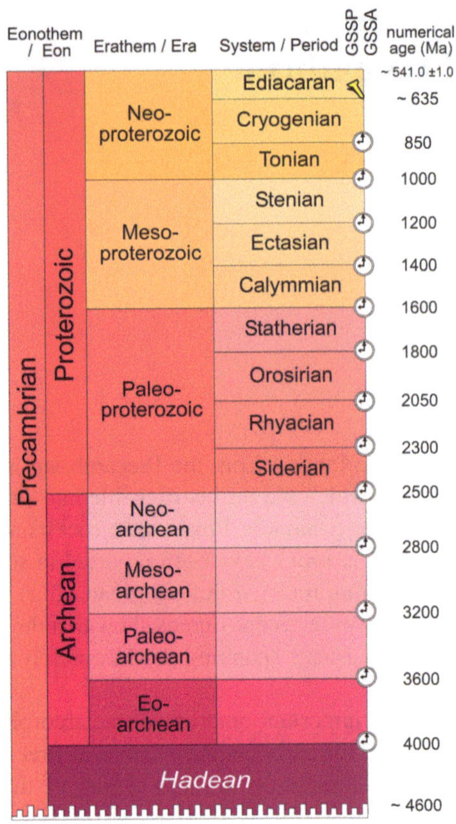

Fig. 1.1 International chronostratigraphic chart of Precambrian (ICS 2015)

It should be mentioned that the grade of metamorphism and absence of fossils lead to the assignment of some highly metamorphosed rocks without any age determination to the Neoproterozoic, for example the Gorgan and Shanderman Schists and the metamorphic rocks of Anarak, Muteh, Samnan, Aligodarz and Azna (Ghorbani 2007).

(B) **Sedimentary rock units of the Neoproterozoic**

The sedimentary rock units of the Neoproterozoic are as follows, although some of these are diachronous, and in some places may be of Neoproterozoic-early Cambrian age.

Alborz: The Kahar, Bayandor, Gharedash and Soltanieh Formations, the latter is definitely diachronous (Neoproterozoic-early Cambrian).

Central Iran: The Tashk I and Tashk II Formations; Kalmard, Morad, Natk, Saghand and Rizu 'Series' of Neoproterozoic Era. However, some evidence indicates the diachroneity of the Desu and Soltanieh formations. Several researchers (e.g. Ahmadzadeh-Heravi 1990) assigned the Hormuz 'Series' to the Neoproterozoic-early Cambrian but there is some evidence that implies that most of the 'Series' is of Neoproterozoic age and does not extend into the Cambrian. However, its upper parts have been published as Cambrian in the stratigraphic logs of NIOC (2016).

Recent paleontological and stratigraphic studies have revised the stratigraphic position of the lithostratigraphic units formerly assigned to the Neoproterozoic (e.g. Bagheri and Stampfli 2008). In the following account, both the traditional and more modern views are presented.

1.1 Neoproterozoic Rocks Based on Recent Studies

The new results show that the Neoproterozoic and lower Cambrian sedimentary facies of Northern Iran differ somewhat from their equivalents in Central Iran. Notably, the Kahar, Soltanieh, Gharedash and Barut formations differ in sedimentary facies from their correlative units, the Tashk Formation, Kushk, Kalmard, Morad, Rizu and Desu 'Series'.

Significantly, some of the metamorphosed rocks and formations were formerly assigned to the Neoproterozoic (Haghipour 1974), while the recent studies show that the protoliths are actually Neoproterozoic in age (e.g. Ghorbani 2013).

It is worth mentioning that metamorphic dating evidence confirms a late Neoproterozoic age for at least some of the units.

1.2 Northern Iran

Amongst the Neoproterozoic rocks of northern Iran, the Kahar, Gharedash, Bayandor and two constituent members of the Soltanieh Fm. (the Lower Dolomite and the Lower Shale) merit discussion. The Neoproterozoic-Cambrian transition coincides with the Middle Dolomite that represents continuous marine sedimentation without any hiatus (Hamdi 1989).

Ghorbani (1999), mainly based on lithological differences, considered the upper part of Bayandor Formation to be equivalent to the upper part of Kahar Formation. Additionally, the Neoproterozoic rocks of the area are metamorphic rocks of the Takab and Mahabad complexes as well as the igneous rocks of the Doran Granite. The results of field observations by the present author imply a Neoproterozoic age for the metamorphic complex in the Takab and Mahabad area, of which the latter is comparable to upper parts of the Kahar formation but with different sedimentary facies.

Kahar Formation

The Kahar Formation is the oldest formation of low metamorphic grade in Northern Iran. Being horizontally and vertically inhomogeneous, the Kahar Formation in the Alborz Mountains is mostly composed of slaty shale, sandy shale, phylite, quartzitic sandstone and, in some localities of carbonates (Dedual 1967). Additionally, in the

Mahabad and Takab regions, it contains some tuffaceous and volcanic beds showing increasing metamorphic grade. Its clastic content increases in its lower part.

In the Takab-Shahin-Dezh and Mahneshan area (1/250,000 geological map of Takab), there are several metamorphic complexes including the Kuh-e-sursat Complex (gneiss, migmatite, mica schist), the Amirabad Complex (schist and amphibolite), the Kheyrabad complex (gneiss, schist, amphibolite and migmatite) and the Mahneshan Complex, all of which have Neoproterozoic igneous-sedimentary protoliths, older than the Kahar Formation (Alavi-Naeini et al. 1976). These complexes have undergone two phases of metamorphism in the late Neoproterozoic and late Triassic.

Soltanieh Formation

As mentioned above, the Soltanieh Formation is a Neoproterozoic-lower Cambrian carbonate formation, the name of which is derived from the town of Soltanieh, Zanjan. Stöcklin et al. (1964) first proposed this name for a thick dolomite unit that forms conspicuous cliffs in the Soltanieh Mountain The presence of shale intercalations has led to subdivision of this formation into three members, namely "Lower Dolomite", "Chapoghlu Shale" and "Upper dolomite".

Previously, the formation was assigned to the Infracambrian due to its relative stratigraphic position and the presence of tube-like organism including forms referable to *Collenia soisser* Fenton and Fenton and *Hadrophycus immanis* Fenton and Fenton. Meyer (1967) correlated these beds to "*Collenia* Limestone Horizon" in the Central Alborz.

Stöcklin et al. (1964) also described *Eoredlichia* sp., Hyolithids, *Wutingaspis* sp. and *Salterella* sp. suggesting an early Cambrian for this formation. Salehi-Siavashani (1980) reported *Biconulites* sp., *Microgastropoda*, *Eoredlichia* sp., Brachiopoda, *Wutingaspis* sp., *Hyolithids* and Trilobites in the limestone intercalation of the Chapoghlu Shale, and suggested an early Cambrian age.

Ashkan (1986) described *Biconulites* sp., brachiopods, echinoid spines, microgastropds and trilobites, and assigned the formation to the early Cambrian.

Hamdi and Golshani (1983), based on the fossils *Hyolithes* sp., ostracods, brachiopoda, *Circotheca*? sp., *Hyolithellus* sp., *Lopochites* sp., *Sachites* sp., and Trilobite fragments proposed an early Cambrian age for Soltanieh Formation.

Later studies by Hamdi (1989) divided this formation into five members, in ascending order; "Lower Dolomite", "Lower Shale", "Middle Dolomite", "Upper Shale" and "Upper Dolomite", of which the lowest two members were assigned to late Precambrian and two highest members to the early Cambrian. The Precambrian-Cambrian transition coincides with the base of the middle Dolomite Member.

The Soltanieh Formation is 864 m thick in its type section but is of variable thickness elsewhere. It was revised by Hamdi (1989) as follows:

(1) **Lower Dolomite**: The Lower Dolomite Member (120 m-thick) is a yellow, well-bedded, recrystallized and cherty dolomite and limestone which, in the Valiabad section, contains thin phosphatic beds yielding phosphatic tubes of *Hyolithellus* sp., while others are resemble *Rugatotheca* sp., fragments of protoconodont *Protohertzina* sp., globomorphs of the *Olivooides multisulcatus* group, and indeter-

minate casts of ?monoplacophorans. In some localities such as the southeastern Soltanieh Mountains and the Hasanakdar section (central Alborz), this unit is thin or absent, so the Lower Shale rests directly upon the Kahar Formation.

(2) **Lower Shale**: the Lower, or Chapoghlu, Shale Member (130–247 m-thick) may intergrade with the underlying dolomite and consists of dark greenish-grey to dark grey shale with gypsum pseudomorphs, dark grey-green sandy shale and calcareous shale in the upper part. Supposedly Vendian acritarchs have been reported from north of Alamut (Seger 1977) while large discoidal algal vesicles assigned to *Chuaria* sp. are generally common. Boudinage structure may also be seen in this part of the succession.

(3) **Middle Dolomite**: the thickness of this member varies between sections. In the Dalir and Soltanieh sections it consists of 144 and 162 m carbonate rocks, respectively. Lower beds at some localities such as the Dalir Section contain grey siliceous limestone interbedded with shale, intergrading to light-color massive calcareous dolomite to dolomite. This member spans the Neoproterozoic-Cambrian boundary. The uppermost 7 m of this unit were assigned to Tommotian Stage (now Stage 2).

The Middle Dolomite Member also comprises approximately 78 m of strata including phosphatic beds and abundant and well-preserved small shelly fossils.

The Dalir section includes tubes of *Hyolithellus vladimirovae* and protoconodonts of the *Protohertzina anabarica* group. Less common elements include tubes of *Anabarites trisulcatus*, *Cambrotubulus decurvatus*, siphogonuchitids, *Palaeosulcachites* sp. and *Siphogonuchites* sp. and globomorphs.

(4) **Upper Shale**: Varying in thickness from place to place, this member is 87 m thick at its type section and 12 m-thick in the Barut section. The lithology of this unit is the same as that of the lower shale member but it tends to be coarser-grained and thicker-bedded in its upper parts. In the upper part of this member calcareous to conglomeratic sandstone grades into dolomitic shale in a transition into the Upper Dolomite Member.

The basal beds consist of a few meters of calcareous shale to black phosphatic limestone, with abundant tubular fossils and molluscs, comparable with those found in the upper part of the dolomite at Dalir but also containing *A.* cf. *tristichus,* hyoliths of the Allathecidae, and a pelagiellid.

The upper part of the Upper Shale Member contains abundant and diverse phosphatized molluscs in the Valiabad Section. Specimens of the *Latouchella korobkovi* group of monoplacophorans, including ‘close-coiled’ *Yangtzespira* sp., ‘lax coiled’ *Bemella* sp., and ‘uncoiled’ *Ceratoconus* sp. and *Obtusoconus* sp., appear c. 20 m from the top of the unit in Valiabad. Other typical elements at this level include *Purella tianzushanemis* and broad monoplacophorans resembling *Protowenella* sp. Pelagiellids of the *Pelagiellalorenzi* group appear in the top 10 m (Hamdi 1995).

(5) **Upper Dolomite Member**: this member is composed of cliff-forming, light grey to light recrystallized massive dolomite that is dark grey to brown weathering in color. The thickness ranges from 405 m in the Soltanieh area to 250–350 m in

the Dalir and Valiabad sections. *Collenia spissa* and *Hadrophycus immanis* are moderately common in the upper part of the member (Meyer 1967). Problematical records of *Salterella* have been reported (Assereto 1963; Stöcklin et al. 1964) but were not confirmed by Hamdi (1989).

In the Sarandon Section of the Central Alborz, this unit is represented by alternations of sandstones and shales. The contact with the overlying formation is transitional and its limit is drawn at the base of the first shale intercalation that initiates the regular shale/carbonate alterations characteristic of the Barut Formation. The Soltanieh Formation underlies the Bayandor Formation and the contact is conformable but marked by a sharp lithological break between purple shale at the top of the Bayandor Fm and the light-colored lower dolomite member of the Soltanieh Formation (Stöcklin and Setudehnia 1991). This member shows more homogeneity than other dolomitic members.

1.3 Central Iran

According to Hamdi (1985) the only Neoproterozoic rocks in Central Iran are the Kushk 'Series', consisting of clastics, acid volcanics, tuffs and carbonates (mainly dolomite).

Differentiation of the Neoproterozoic and Early Cambrian rocks of central Iran presents some difficulties. Other Neoproterozoic-early Cambrian formations include the Morad 'Series', the Rizu volcanic-sedimentary Formation, the Dezu and Tashk formations, the Aghda Limestone, the Kalmard 'Series', the Shorm Beds, and the Anarak metamorphic units.

Samani et al. (1993) established two formations namely the Natak and Saghand formations, considering these to be the oldest sedimentary rock units in Central Iran. Absolute age determinations of 874–750 Ma and 780–583 Ma have been established for the Natak Formation and the Saghand Formation respectively. These units are composed of clastics, sedimentary and volcanic rocks, and overlie the Rizu and Desu 'Series'.

According to the present author, in spite of slightly facies difference, these formations are lateral equivalents of the clastic-volcanic Tashk Formation and lower parts of the Rizu Formation, so the Natk and Saghand formations should not be accepted as new formations.

Field observations by the present author in the Bafgh region suggest a Neoproterozoic age for the volcanic rocks of the Rizu Formation as the overlying Desu Formation is dated as Neoproterozoic-early Cambrian. The Tashk Formation is overlain by Rizu Formation and consists of low-metamorphic clastic sediments, distinguished by their low grade of metamorphism (Fig. 1.2).

In contrast to the Desu and Rizu 'Series', the Tashk Formation tends to have a higher clastic contents and exhibits low regional metamorphic grade which are not seen in the Kushk and Rizu 'Series'. In the Behabab-Gandehdar road section, the

Fig. 1.2 A view of Tashk formation and Kushk 'Series' on the Kushk-Behabad road

Tashk Formation is overlain by the Rizu Fm. so one can conclude that the Tashk Formation is older than Rizu Fm.

Kalmard "Series"

The name of this formation is derived from the caravansary of Kalmard area on the Tabas-Yazd road. It is composed of arkosic sandstone and shale. There is some evidence from extensive outcrops of tuffaceous rocks of low grade greenschist facies metamorphism. These rocks were first described by Gansser (1955) who considered them to Cambro-Silurian in age. Later, Stöcklin (1971) assigned them a Precambrian age and stratigraphically correlated this unit with the Morad, Kahar, and Taknar formations. However, Nabavi (1976) regarded the Kalmard 'Series' as a lateral equivalent of the Heshem Formation and assigned it to the early Cambrian. Among other researchers who worked on the formation, Aghanabati (1975) considered a Neoproterozoic age based on relative stratigraphic position. Since rocks similar to the dolomites of Soltanieh overlie this succession in Tabas, the views of Stöcklin et al. (1964) and Aghanabati (2008) are correct.

It seems that Kalmard 'Series' is the equivalent to the Neoproterozoic Upper Tashk Member in the Behabad area. Because the rocks being equivalent to the Rizu 'Series' does not show any evidence of metamorphism. Hence because of the same lithology and grade of metamorphism, it is better to consider the Kalmard 'Series' to be equivalent to the Upper Tashk Member. However, the Kalmard 'Series' rocks exhibit greenschist facies regional metamorphism. From the above evidence, the Kalmard 'Series' is older than Rizu 'Series' and is considered equivalent to the Kahar Formation of Alborz and Tashk Formation of the Bafgh-Saghand area.

To show more details of Neoproterozoic-lower Cambrian rocks of Central Iran, a review of the Huckriede et al. (1962), Stöcklin (1971, 1986) as well as Hamdi and Zhion (1992) is presented in the following account.

Morad and Rizu 'Series'

Huckriede et al. (1962) established the name, "Morad 'Series'"[2] for a succession (more than 500 m-thick) of uniform clastic sediments including silty shale and sandy shale, fine-grained sandstone, arkosic mica-bearing sandstone and quartzite sandstone in the Abmorad, northern Kerman and northern Zarand (Rodshuro and Gogdel Mine area). The same authors, also described the Rizu 'Series' to include a Lower Cambrian succession of dolomite beds, sandstone, sandy and dolomitic shale, conglomerate as well as basic and acidic volcanic rocks.

According to Huckriede et al. (1962) the sediments of the Morad 'Series' were formed during the Neoproterozoic, and were affected by the Asiatic orogenic phase. The basal conglomerate of the Rizu 'Series' overlies the Morad 'Series' with an angular unconformity in the Abmorad area. However, in the Gogdel area, the Rizu Formation is overlain by the Desu 'Series' consisting of sandstones, evaporites and carbonate sediments together with acidic volcanic rocks.

Stöcklin (1971, 1986) correlated the Morad 'Series' with Kahar Formation, subsequently confirmed by field observations by the present author. He considered the Rizu and Desu 'series' as complex rather than 'Series', and finally correlated them with Soltanieh Formation of the Northern Iran. He, also, presumed that the Kushk 'Series' that contained Neoproterozoic Ediacaran fossils was a part of Rizu 'Series'. However, Hamdi and Zhion (1992) rejected the concept of the Morad and Rizu 'Series' of Huckriede et al. (1962) in the Kuhbanan area, northwestern Zarand, and instead considered it to be the lateral equivalent of the Heshem and Aghda formations. Additionally, the transition of the Heshem Formation to the Aghda Formation is continuous. The latter formation having been attributed by Hushmandzadeh et al. (1988) to the Cambrian.

An exactly similar sequence can be found at the Kushk mine to the west of Kharneghan Village. In this area, more than 200 m of shale of the Heshem Formation pass gradually into fetid, black limestone of Aghda Formation.

Consequently, the Cambrian Heshem and Aghda formations are underlain the Rizu 'Series', so the latter formation is certainly Neoproterozoic in age. The sequence at Kushk mine starts with shale, mineral-bearing unit of the Kushk 'Series', and is overlain by 200 m of tuffs and dolomites of the Rizu 'Series'. The latter unit is overlain by the Cambrian Heshem and Aghda formations, confirming the Neoproterozoic age of the Rizu 'Series'.

Shorm Beds

Ruttner et al. (1968) described a sequence exposed between Cheshmeh-e Shorm and Cheshmeh-e Kharmayu in the Derenjal Mountains, Shirgesht area. This consists of a

---

[2]Field observations by the present author in the Kushk Mine, Bafgh area, imply that it is better to rank the series as formation because the underlying (Kushk series) and overlaying (Heshem Formation) strata are well exposed.

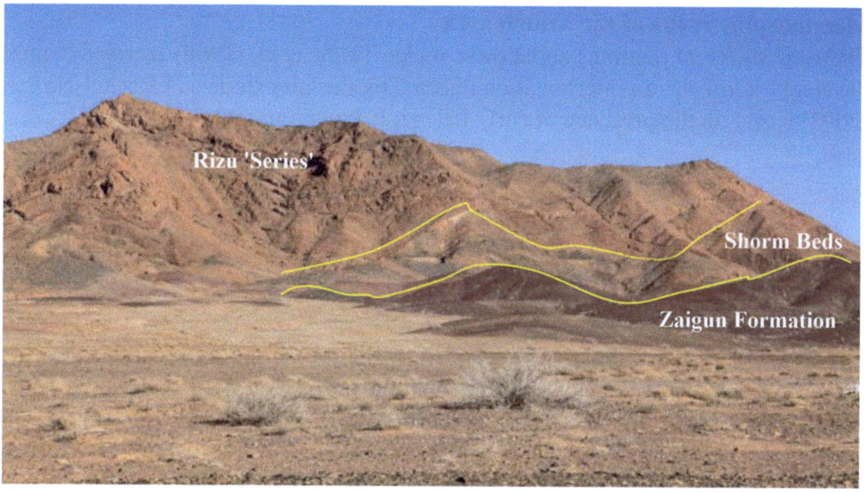

Fig. 1.3 A view of the Neoproterozoic-lower Cambrian rocks of Derenjal mountain

slightly metamorphosed complex of slaty limestone, calcite-vein shale, green slate, red phyllitic shale and siliceous dolomite assigned by them to Neoproterozoic. The same authors believed that these strata are the oldest rocks of the Shirgesht area and that they overlie rocks equivalent to Soltanieh with a paraconformity (Fig. 1.3).

In contrast, later studies by Hamdi (1995) implied an early Paleozoic age for the Shorm Beds and considered these to be the lateral equivalent of the Barut Formation.

Taknar Formation

Razzagh Manesh (1968) described a succession of slightly metamorphosed tuffaceous green schist with alternations of meta-rhyodacite and meta-rhyolites, exposed around the Taknar mine in the Kashmar area in northeastern Iran. This author assigned a Neoproterozoic age to the Taknar Formation that is disconformably overlain by the Paleozoic sediments.

Stöcklin (1972) and Müller and Walter (1983) both changed "Taknar Formation" to "Taknar 'Series'" and chronostratigraphically correlated this with the Neoproterozoic Kahar Formation.

However, Hamdi (1995) rejected their Neoproterozoic age assignment and attributed the Taknar Formation to the early–late Cambrian based on palynomorph and algae.

On the basis of regional field observations, Ghorbani (1999) correlated the Taknar 'Series' with the Ordovician Shanderman and Gorgan Schist. However, this relationship was disproved by further field observations by the same author that revealed that the supposed Ordovician strata in the Kashmar area are actually Cambro-Ordovician in age and are non-metamorphosed correlatives of the Shirgesht Formation. Hence, a Neoproterozoic age is correct.

Metamorphic rocks of the Muteh area

There are different opinions about these rocks. Thiele et al. (1968) ascribed them to the Neoproterozoic; an interpretation rejected by the later studies of Rashid-Nezhad (2002) based on the presence of early Paleozoic palynomorphs.

1.4 Zagros

Hormuz 'Series'

The description of the salts of Hormuz (Tavernier 1642) is one of the earliest recorded geological observations made in Iran. Blandford (1872) first introduced the name, Hormuz Salt Formation for the complex of salt and associated sedimentary and igneous rocks. Pilgrim (1908) changed the name, "Hormuz Formation" to "Hormuz 'Series'" comprising halite, gypsum, and blocks and contorted masses of many sizes and compositions, including black laminated fetid limestone, brown cherty dolomite, and red sandstone, and also igneous material (Figs. 1.4 and 1.5). He divided this 'Series' into four units as follows:

1. Purple coarse-grained sandstone
2. Volcanic rocks including tuff, agglomerate and gypsum
3. Dolomite, limestone, shale and rhyolite
4. Salt.

Stöcklin (1986) used the name "Hormuz Complex" and correlated this unit with the Bayandor, Soltanieh, Barut, Zaigun and Lalun formations, as well as the lowermost member of Mila Formation and the Punjab Saline 'Series' of the Salt Range.

Fig. 1.4 A view of the Hormuz 'Series'

Fig. 1.5 A view of rhyolites of the Hormuz 'Series'

Ahmadzadeh-Heravi (1990) despite the absence of the stratigraphic order renamed the "Hormuz 'Series'" the "Hormuz Formation" and divided it into 4 units; H1, H2, H3 and H4. These authors believed that, despite the lithological differences between domes, the overall lithological succession is similar. They also presented another description of the Hormuz Formation as follows:

H1 Unit, salt unit of Hormuz Formation

The base of this unit is not exposed in the studied sections. This unit is mainly composed of salt interbedded with thin-bedded tuffs, marls, laminated limestone and iron oxide and sulfide. It should be mentioned that the observed lineation is actually related to the dynamic and thermodynamic metamorphic structures, and so should be referred to as foliation.

One can conclude that this unit was derived from clastic sediments, based on the presence of fine- to coarse-grained carbonate and volcanic clasts. Nevertheless, the dominant brittle tectonism affecting the diapirs, has resulted in this feature. The H1 Unit is continuous with the H2 Unit Ahmadzadeh-Heravi (1990).

H2 Unit, red soil unit

This unit is mainly composed of red ochre rocks and is simply recognized by this feature. It is a succession of white to yellow marl, anhydrite deformed gradually to the gypsum, tuff, ignimbrite, black thin-bedded limestones as well as laminates of iron hydroxide and oxides. The composition of the ignimbrite is mainly rhyolite and dacite of alkalic origin. Because of the amalgamation during formation, these rocks include salts and gypsum observed in both thin section and hand samples. In some areas, there are a few beds of red ripple marked sandstone and fine-grained white tuff contaminated with salts and gypsum. The thickness of the unit differs from place to place.

H3 unit, black algal limestone

Fetid, black, thin-bedded limestone with algal fossil remnants are assigned to the Neoproterozoic-Cambrian, based on the presence of echinoderm and stromatolite remains. Its thickness may reach 50 m.

H4 Unit, clastic-volcanic unit

The lithology of this unit is specific to the diapirs of the area, and is observed where the H1 unit does not crop out or is thin. There is extensive outcrop of the unit in the salt diapirs of Hamiran, Lamzan and Pozeh, with thicknesses exceeding 500 m.

The lithology of this unit is mainly red and green ripple-marked sandstones alternating with green tuffite. In some domes, it appears as fine-grained tuffite due to the abundance of volcanic rocks. This is the main reason for the volcanic appearance of the rocks alternating with the clastic sediments. It also appears as volcanic units alternating with clastic beds. Moreover, the upper part of the unit contains some anhydrite layers interbedded with black limestone. Tuffitic beds appear in the lower part of the unit amalgamated with salts and anhydrite, and in rock samples with kaolinite.

The Hormuz 'Series' is divided into two sequences of salt (1–2.5 km thick) separated by a few hundred meters of carbonates and red beds. The lithofacies boundaries of both these sequences occur above old faults in the Precambrian basement (Talbot and Alavi 1996).

The age of this 'Series' has long been disputed and different ages have been suggested. Pilgrim (1922) first regarded the Hormuz rocks to be post-Cretaceous in age, but he later (1924) assigned the salts to the Triassic and other rocks to Jurassic. Richardson (1926) suggested a post-Cretaceous age. Following the discovery of Cambrian trilobite, a Cambrian age was suggested (Lees 1927), verified by later researchers (e.g. Ahmadzadeh-Heravi 1990; Hamdi 1995). However, regional comparison with central Iran and neighboring countries confirmed the Neoproterozoic-Cambrian age of the Hormuz 'Series'.

The magmatism and tectonic events of the Neoproterozoic-Cambrian Pan-African orogeny was contemporaneous with the formation of the 'Hormuz 'Series'' (Sillitoe 1980). During this time, due to crustal extension, several intra-continental rifts were created along the northern part of the Gondwana Supercontinent, leading to alkaline magmatism involving a wide range of ultra-basic to acidic types that were eventually mixed with clastic, shallow and salt deposits (Sillitoe 1980; Berberian and King 1981; Samani 1988; Ghorbani 2007). The Neoproterozoic-Cambrian strata of well outcropped sections are summarized in Fig. 1.6.

The author is of the opinion that, in view of its stratigraphic position as well as the magmatic record of Iran, the Hormuz 'Series' must be of late Neoproterozoic age.

Figure 1.7 is a schematic illustration of the Neoproterozoic-Cambrian lithostratigraphy of the main regions of Iran. Table 1.1 also summarizes the general features of the Neoproterozoic-Cambrian rocks of Iran. The field images of selected Neoproterozoic sections are depicted in Figs. 1.8 and 1.9.

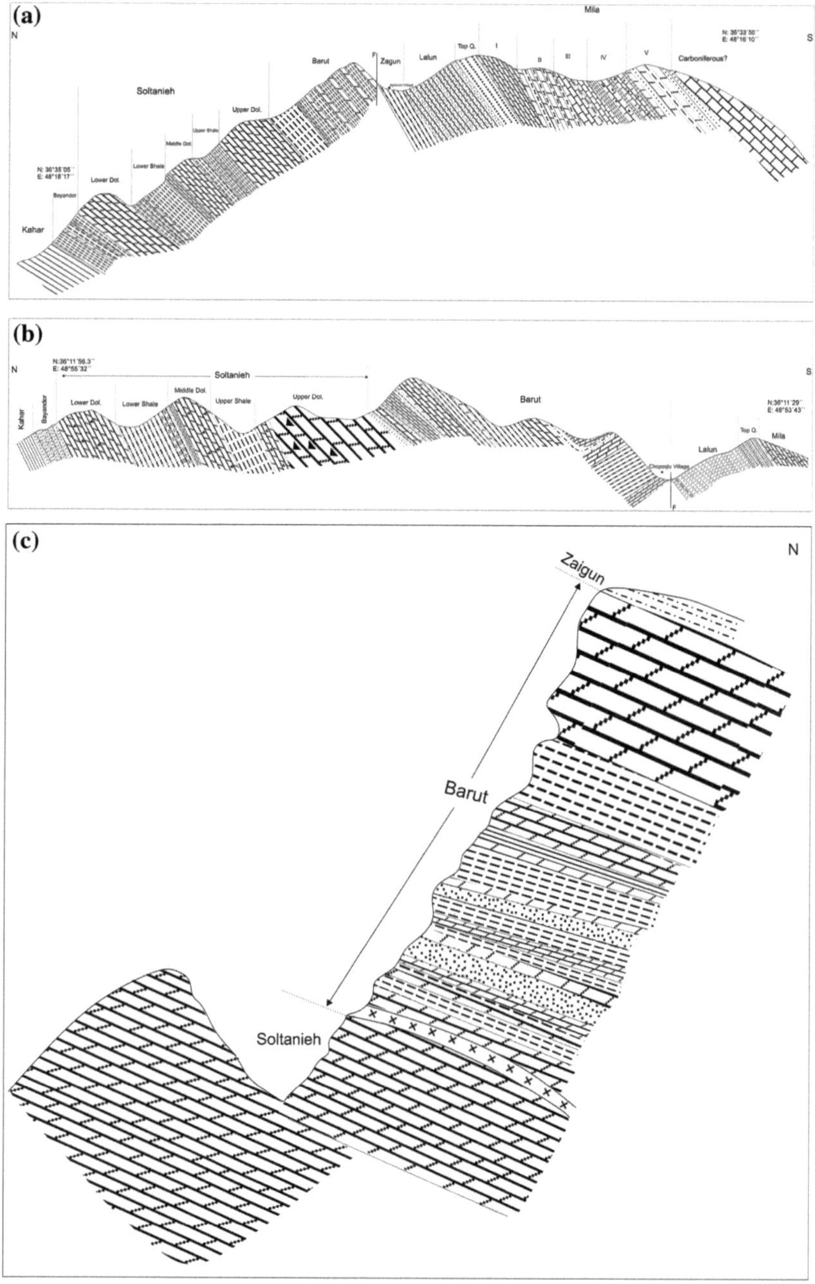

Fig. 1.6 **a** Stratigraphic section of southern Zanjan, around Barout-Aghaji Village. **b** Stratigraphic section of Neoproterozoic-lower Cambrian rocks around Chapoghlou Village, southern Abhar. **c** Barut and Soltanieh Formations of the Deh-molah section at the western Shahroud. **d** Stratigraphic section through the Soltanieh Formation at Valiabad area. **e** Paleozoic rocks in the Kushk stratigraphic section. **f** Paleozoic rocks in the Ghali-Kuh section

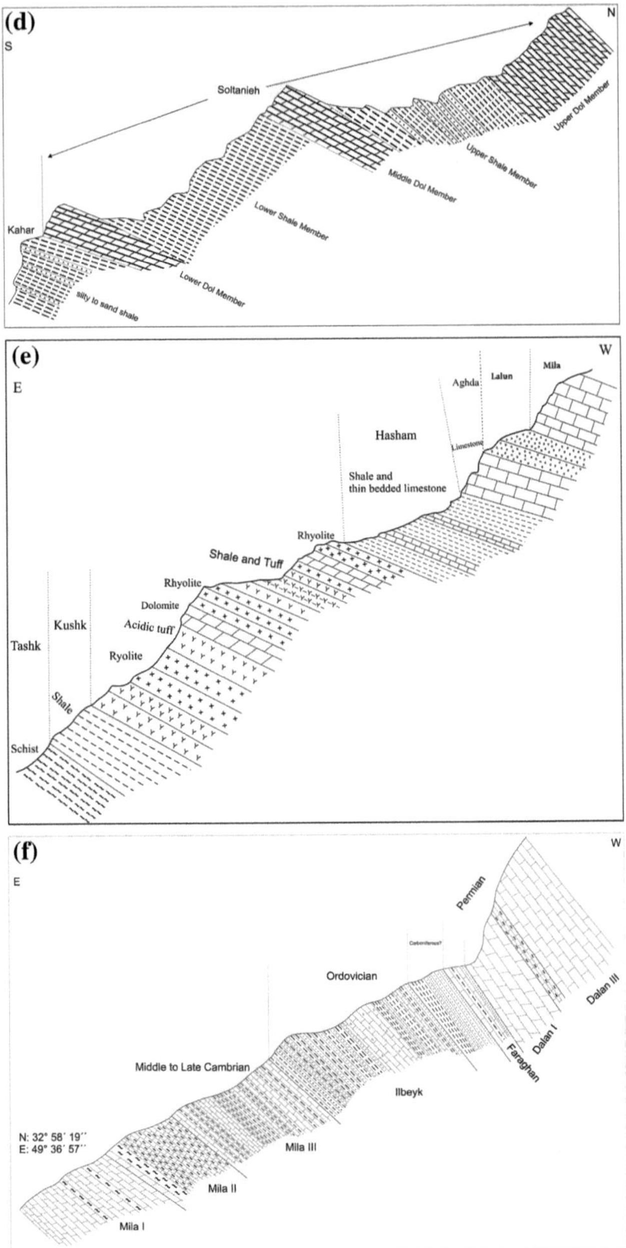

Fig. 1.6 (continued)

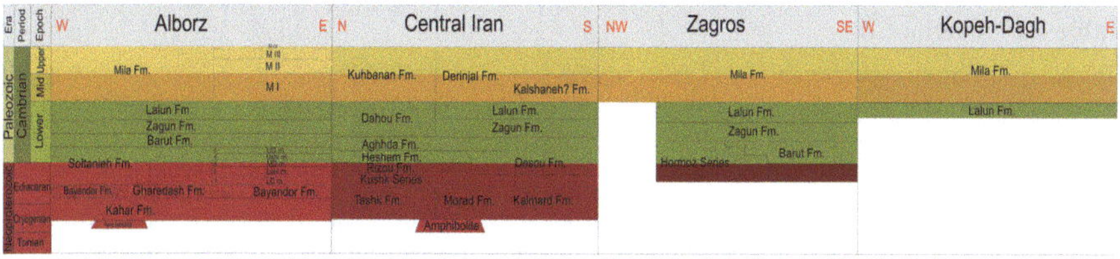

Fig. 1.7 Lithostratigraphic column of the Neoproterozoic-Lower Cambrian rocks of Iran

Table 1.1 Lithostratigraphic features of Neoproterozoic-lower Cambrian rock units of Iran

Rock unit	Chronostratigraphy	Lower/upper Boundary	Lithology and thickness at type section	Notes (references, comments)
Neoproterozoic (Alborz-Azarbaijan)				
Kahar Fm.	Neoproterozoic	Lb[a]. does not outcrop anywhere. Ub. Bayandor or Soltanieh formations and in some localities, the Ghare-Dash Rhyolite.	Silty shale, sandy shale, quartzitic sandstone, dolomite and limestone in the lower part; Sandstone and quartzitic sandstone in the upper part; Thickness: 1600 m	Hamdi (1985, p. 34), Aghanabati (1998, p. 52), Alavi-Naeini (2009, p. 50) The formation is metamorphosed (slate and phylite; greenschist facies). The Kahar Fm. was considered to be the oldest sedimentary unit of Alborz and Azarbaijan
Bayandor Fm.	Neoproterozoic	Lb. of the type section is the Doran Granite but in other areas it is underlain by the Kahar Formation. In some places such as Shahin-Dezh, the boundary is continuous Ub. is continuous and conformable with Soltanieh Fm.	Purple sandstone, mica-bearing silty shale interbedded with stromatolite bearing brown dolomite. Reported thickness: 498 m but the actual thickness is much less (ca.100 m).	Hamdi (1985, p. 21), Ghavidel–Syooki (1985, p.), Aghanabati (2004, p. 116), Aghanabati (2008, p. 144), Alavi-Naeini (2009, p. 65), Article: Ghavidel-Syooki (1995)
Gharedash Fm.	Neoproterozoic	Lb. Kahar Formation Ub. disconformable or interfingering with Bayandor Fm.	Alkali ryolithic lava, acidic tuff observable in northwestern Iran, Azarbaijan, Takab and Mahabad. Extrusive equivalent of the Doran Granite Thickness: 1140 m	Book: Aghanabati (2004), Aghanabati (2008, p. 40), Alavi-Naeini (2009, p. 64)

(continued)

Table 1.1 (continued)

Rock unit	Chronostratigraphy	Lower/upper Boundary	Lithology and thickness at type section	Notes (references, comments)
Soltanieh Fm.	Neoproterozoic-early Cambrian	Lb. Bayandor and Kahar Formations Ub. discontinuous or continuous with Barut Formation	Hamdi divided the formation into five members in ascending stratigraphic order, as follows: 1. Lower Dolomite: cherty dolomite and calcareous dolomite 2. Lower Shale: silty shale, slaty sandy shale interbedding with stratified silica-bearing limestone and calcareous shale 3. Middle Dolomite: silica-bearing limestone interbedding with shale. 4. Upper shale: lithology same as Lower Shale but tends to be coarser grained and thicker bedding. 5. Upper Dolomite: cliff-forming dolomite Thickness:1160 m	Book: Alavi-Naeini (1993, p. 50), Hamdi (1995, p. 47), Aghanabati (2008, p. 157), Alavi-Naeini (2009, p. 68) Thesis: Thiab-Ghodsi (2007), Shabestari (1994), Ashkan (1986) Article: Hamdi and Zhion (1992) Research project: Lasemi (1994) In places where the Bayandor Fm. is not exposed, the Soltanieh Fm. overlies the Kahar Fm.
Neoproterozoic (Central and Eastern Iran and Sanandaj-Sirjan)				
Morad 'Series'	Neoproterozoic	Lb. does not exposed Ub. disconformable with Rizu 'Series'	Silty shale, sandy shale to sandstone, arkosic sandstone and quartzitic sandstone Thickness: more than 500 m	Hamdi (1995, p. 24), Aghanabati (2008. P. 81), Alavi-Naeini (2009, p. 56) The Morad 'Series' mainly outcropped in the Abmorad, northwestern Kerman and northern Zarand; and considered as the equivalent of Upper Tashk Member

(continued)

Table 1.1 (continued)

Rock unit	Chronostratigraphy	Lower/upper Boundary	Lithology and thickness at type section	Notes (references, comments)
Tashk I and Tashk II Member	Neoproterozoic	Lb. is not exposed. Ub. is disconformably overlain by Rizu Fm. and in someplaces by shale beds of Kushk 'Series'	Tashk I consists of schist, phylite, slate, metagraywake, quartzitic schist accompanied with metavolcanic rocks; and Taskk II of shale, slate and slightly metamorphosed sandstone. Base is not seen. Thickness: more than 600 m	Aghanabati (2008, p. 13), Alavi-Naeini (2009, p. 48) The Tashk Fm. is a sedimentary formation accompanied by pyroclastic rocks; metamorphosed to greenschist facies during the Neoproterozoic. The formation is comparable to the Kahar Fm.
Kalmard 'Series'	Neoproterozoic	Lb. is covered Ub. is disconformably overlain by Shirgesht Fm. or sometimes Eq. to Soltanieh	Consists of tuff-bearing shale and fine-grained sandstone metamorphosed at green schist facies	Book: Hamdi (1995, p.45), Aghanabati (2008, p.47), Alavi-Naeini (2009, p. 56), (Gansser (1955), Stöcklin (1961, 1968) Gansser (1955) suggested a Cambrian to Silurian affinity, contrarily changed by Stöcklin (1961, 1968, 1071) The formation was considered as the equivalent of Kahar Fm. of Alborz and Tashk Fm. of Central Iran. Hushmandzadeh and Sahandi (1984) and Aghanabati (1975) assigned the formation to Precambrian (Neoproterozoic) affinity.

(continued)

Table 1.1 (continued)

Rock unit	Chronostratigraphy	Lower/upper Boundary	Lithology and thickness at type section	Notes (references, comments)
Rizu 'Series'	Neoproterozoic	Lb. is disconformable with Morad 'Series'; in the Kushk area continuous with Kushk 'Series' Ub. is disconformable with Heshem Fm.	Carbonate rocks in the lower part; shale, sandstone in the middle; and tuff and purple rhyolitic rock in the upper part. It is better to establish a reference section at Kushk area because of simplicity of recognizing lower and upper boundaries Thickness: 400–600 m but less at Kushk	Book: Hamdi (1995, p. 24), Aghanabati (2004, p. 118), Aghanabati (2008, p. 209), Alai-Naeini (2009, p. 58) This 'Series' is actually a volcano-sedimentary formation which is distinguished by the very facies. All of the units of the formation are of the Neoproterozoic affinity.
Desu Fm.	Neoproterozoic-Early Cambrian	Lb. is interfingering with its overlying part. Ub. is disconformable with Dahu Formation (equivalent of Lalun Formation)	A succession of limestone, dolomite, gypsum, gypsiferous shale, sandstone and locally quartzite; in the Kuhbanan area, it is mainly cherty dolomite Thickness: 160–300 m in northern Kerman	Book: Alavi-Naeini (1993, p. 87), Hamdi (1995, p. 168), Aghanabati (2008, p. 119), Alavi-Naeini (2009, p.61), Darvishi (1994) The Desu Fm. is considered as the equivalent of Barut Fm. According to the author, the upper part is comparable with the Barut Fm. the lower part is mostly of Neoproterozoic age, be considered it as a lateral facies variation of upper Rizu 'Series'.

(continued)

Table 1.1 (continued)

Rock unit	Chronostratigraphy	Lower/upper Boundary	Lithology and thickness at type section	Notes (references, comments)
Neoproterozoic (Zagros)				
Hormuz 'Series'	Neoproterozoic-Early Cambrian	Lb. is not distinguished Ub. is disturbed, resulted from doming.	Without a designated type section. Also, a complete succession is not crop out anywhere. The Hormuz 'Series' consists of multi-colored salt rocks, acidic tuff, anhydrite, gypsum and blocks of sedimentary rocks including thin-bedded limestone, cherty dolomite, red sandstone, multi-colored shale and red clay; basaltic rocks, rhyolite, keratophyre and trachyte accompanied by calcite, quartz, dolomite, hematite, epidote, pyrite, illmenite, apatite, sulfur etc. Thickness: several thickness values have been reported including Player (1969): 900–1500 m; Huber (1975): 1500 m; Verrall (1978): 2000–4000 m;	Motiei (1993, p. 8), Alavi-Naeini (1993, p. 101), Hamid (1995, p. 175), Aghanabati (2004, p. 120), Aghanabati (2008, p.235), Alavi-Naeini (2009, p. 48) The equivalent of some parts of the Rizu and Desu 'Series' and the lower Soltanieh Fm. The age of the Hormuz 'Series' has long been disputed. Some researchers regarded it as Cambrian (Kent 1970; Lees 1924; Boeckh et al. (1929), while Harrison (1930) and Motiei and Zahedi (1969) assigned it to the Cambrian. Later researchers suggested an Infracambrian age despite the presence of the Cambrian trilobite *Redlichia* (Player 1969; Kent 1970; Zharkov 1984). The Hormuz 'Series' is not exposed in the Tange-putak and Lurdegan areas where the Barut Fm. Early Cambrian age is present. Accordingly, the Hormuz 'Series' most probably belongs to the Neoproterozoic.

[a]The Lb. refers to lower boundary and Ub. to upper boundary

Fig. 1.8 A view of the tashk formation and the Rizu 'Series' along the Kushk-Behabad road

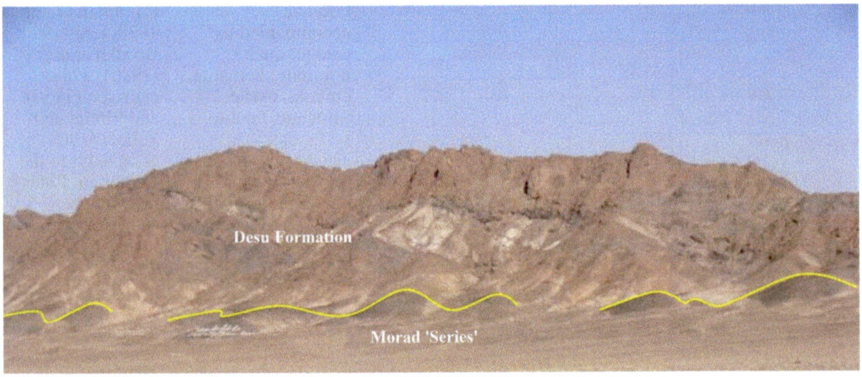

Fig. 1.9 A view of the Morad 'Series' and the Desu Fm. of the Zarand-Ravar path

References

Aghanabati, A. (1975). Etude geologique de la region de Kalmard These Doctorate thesis, No AD 11/623 Gronoble, France.

Aghanabati, A. (1998). Jurassic stratigraphy of Iran. *Geological Survey of Iran*, *2*(65), 733.

Aghanabati, A. (2004). *Geology of Iran*. Geological Survey of Iran, Report (in Persian).

Aghanabati, A. (2008). Stratigraphic Lexicon of Iran. In: Cambrian-Silurian. Geological Survey of Iran, Report (in Persian).

Ahmadzadeh-Heravi, M. (1990). New concepts of Hormoz formation stratigraphy and problem of salt diapirism in soth Iran. In *Proceeding of symposium on Diapirim with special refrence to Iran*, vol 1.

Alavi-Naeini, M. (1972). *Etude géologique de la région de Djam*, Geological Survey of Iran, Report 23, 293.

Alavi-Naeini, M. (1993). Paleozoic stratigraphy of Iran, *Treatise on the Geology of Iran*, *5*, 1–492.

Alavi-Naeini, M. (2009). *An introduction on the stratigraphy of Iran*, Geological Survey of Iran, 507.

Ashkan, F. (1986). *Depositional environment of phosphate deposits in the Soltanieh for-mation: Chashin and Lian Sections*, MSc thesis, University of Tehran, 155.

Assereto, R. (1963). The Paleozoic formations in Central Elburz (Iran). *Rivista Italiana di Paleontologia e Stratigrafia, 69*(4), 503–543.

Bagheri, S., & Stampfli, G. M. (2008). The Anarak, Jandaq and Posht-e-Badam metamorphic complexes in central Iran: new geological data, relationships and tectonic implications. *Tectonophysics, 451*(1–4), 123–155.

Berberian, M., & King, G. C. P. (1981). Towards a Paleogeography and tectonic evolution of Iran. *Canadian Journal of Earth Sciences* *18*(2), 210–265.

Boeckh H de, Lees G. M., Richardson F. D. S. (1929). Contribution to the stratigraphy and tectonics of the Iranian ranges. In: Gregory JW, editor. *The Structure of Asia.*

Blandford, W. T. (1872). Notes on the geological formations seen along the coasts of Baluchistan and Persia from Karachi to the head ofthe Persian Gulf and some Gulf Islands. Records of Geological Survey of India.

Dedual, E. (1967). Zur Geologie des mittleren und unteren Karaj-Tales Zentral-Elburz (Iran), Doctoral dissertation.

Gansser, A. (1955). New aspects of the geology in Central Iran. In *Proceedings of the 4th World Petroleum Congress, Rome, Section I/A/5*, pp. 279–300.

Ghavidel-Syooki, M. (1995). Srchahan Formation. *Earth Science Quarterly 15*(16), 74–89.

Ghorbani, M. (1999). Petrological analysis and metalogeny of the Tertiary-Quaternary magmatic rocks. Geological Survey of Iran, Report p. 263.

Ghorbani, M. (2007). *Economic geology of natural and mineral resources of Iran*, Arianzamin Publication, 492.

Ghorbani, M. (2013). *The economic geology of Iran: Mineral deposits and natural resources.* Springer Science & Business Media.

Ghorbani, M. (2018). *The magmatism and metamorphism of Iran*, in preparation.

Haghipour, A. (1974). Petrology and tectonic activity of the pre-cambrian sediments of Biabanak and Bafgh regions of Central Iran. Geological Survey of Iran, Report 34, 403.

Haghipour, A. (1978). *Etude geologique de la region de Biabanak-Bafq* (Iran Central), In *Petrologie et Tectonique du Socle Precambrian et de sa Couverture*, Geological Survey of Iran, Report 34, 403.

Haghipour, A., & Pelissier, G. (1977). Geology of the Saghand sector, Explanatory Text of the Ardekan Quadrangle Map, Geological Survey of Iran, Tehran, 10-68.

Hamdi, B. (1989). *Stratigraphy and Paleontology of the late percambrian in the Alborz mountains, North Iran, Geological Survey of Iran*, Report 59.

Hamdi, B. (1995). *Precambrian and Cambrian sedimentary rocks of Iran*. Treatise on the Geology of Iran, Ministry of Mines and Metals.

Hamdi, B., & Golshani, F. (1983). *Preliminary investigation on the lower Cambrian and Percambrian-Cambrian boundary of North Iran*, Tehran Geological Survey of Iran.
Hamdi, B., & Zhion, J. (1992). Paleozoic fossils of the Morad and Rizu 'Series' of the Central Iran. *Earth Science Quarterly, Geological Survey of Iran*, No 4.
Harrison, J. V. (1930). The geology of some salt plugs in Laristan (Sothern Persian). *Quarterly Journal of the Geological Society London ,86,* 463–522.
Huckriede, R., Kürsten, M., & Venzlaff, H. (1962). Geotektonische Kartenskizze des Gebietes zwischen Kerman und Sagand, (Iran), Bundesanstalt für Bodenforschung.
Hushmandzadeh, A, Nabavi, M. H., & Hamdi, B. (1988). *Precambrian-Lower Cambrian rocks of Iran*, In: The symposium of reservoir evaluation and ore potential of the Yazd Province, Geological Survey of Iran, Report.
Kent, P. E. (1958). Recent studies of south Persian salt plugs. *AAPG Bulletin, 42*(12), 2951–2972.
Kent, E. (1970). The salt plugs of the Persian Gulf region, United Kingdom, Transactions of the Leiceser Literary and philosophical Society, 64, 56–88.
ICS (2015). http://www.stratigraphy.org/index.php/ics-chart-timescale.
Lasemi, Y. (1994). *Petrographical analysis of the platform carbonate of Mozduran Formation*, In The 2nd symposium of geology of Eastern Iran.
Meyer, S. P. (1967). *Die Geologie des Gebietes Velian-Kechiré* (Zentral-Elburz, Iran), Thesis Mitteilungen aus dem Geologisch-Paläontologischen Institut, 79, 127.
Motiei, H. (1993). *Stratigraphy of Zagros*, Geological Book Codification Project, Geological Survey of Iran.
Motiei, H. & Zahedi, A. (1969). The Hormuz Series. Geological Survey of Iran, Report no. B112.
Müller, R., & Walter, R. (1983). *Geology of the Precambrian-Paleozoic Taknar inlier northwest of Kashmar, Khorasan Province Northeast Iran, GSI, Report 51*, 165–183, ISSN 0075–0448.
Nabavi, M. H. (1976). An introduction to geology of Iran, Geol Surv Iran, Tehran.
NIOC. (2016). *Paleontology and Biozonation of Paleozoic Sediments of Central Iran and Zagros Basins*, National Iranian Oil Company, Tehran (no 89230).
Pilgrim, G E. (1908). *The Geology of the Persian Gulf and the Adjoining Portions of Persia and Arabia, Memoirs of the Geological Survey of India, 34*, p. 177.
Pilgrim, G. E. (1922). The Sculpture deposits of southern Persia. *India Geol Surv Res, 53,* 343–358.
Player, R. A. (1969). *The Hormuz salt plugs of Southern Iran*: Ph. D thesis, Reading University, 300.
Rashid Nezhad, A. (2002). *Petrology and geochemistry of the meta-volcanosedimentary and plutonic rocks of the Muteh aera, southern Delijan, implication on the provenance of the Gold mine*, Unpublished Ph. D thesis, Tarbiat Modarres University.
Razzagh Manesh, B. (1968). *Metallogeny of copper and zinc of Taknar area* (p. 130). Thesis Rhein-Westph: Khorasan Province.
Ruttner, A., Nabavi, M. H., & Hajian, J. (1968). Geology of the Shirgesht area (Tabas area, East Iran). *Geol Surv Iran, 4,* 5–133.
Salehi-Siavashani (1980). *Precambrian-Paleozoic rocks of the northeastern Abadeh*, Geological Survey of Iran, Report.
Samani, B. A. (1988). Metallogeny of the Precambrian in Iran. *Precambrian Research, 39*(1–2), 85–106.
Samani, B., Xoeie, Ch., Taeo, G., & Kovan, T (1993). Precambrian geology of Central Iran; implication on the stratigraphy, magmatism and metamorphism, *Earth Science Quarterly, 10*, 40–63.
Seger, F. E. (1977). Zur Geologie des Nord-lamut Gebietes (Zentral-Elburz, Iran) [microform]/von Felix Subjects, Geology-Iran-Elburz Mountains.
Shabestari, Gh. M. (1994). Lithology and environment interpretation of Soltanieh-equivalent units in northern Semnan using chemical stratigraphy, Shahid Beheshti University Journal of Earth Science, 4, 85–91.
Sillitoe, R. H. (1980). Stratabound ore deposits related to Infracambrian rifting along Northern Gondwanaland. In *Proceedings of the 5th Quadrennial IAGOD Symposium I, Stuttgart*,pp. 163–171.

Stöcklin, J. (1961) *Lagoon formation and salt domes in East Iran*, Bulletin of the Iranian Petroleum Institute, 3, 29–46.
Stöcklin, J. (1968). Structural history and tectonics of Iran; a review. *American Association of Petroleum Geologists Bulletin, 52*(7), 1229–1258.
Stöcklin, J. (1971). *Stratigraphic Lexicon of Iran, Part I, Central, North and East Iran*. Geological Survey of Iran, Report 18, 384.
Stöcklin, J. (1986). The Vendian–Lower Cambrian salt basins of Iran, Oman and Pakistan, stratigraphy, correlations, paleogeography. *Science de la terre, Memorie, 47.*
Stöcklin, J., & Setudehnia, A. (1991). *Stratigraphic Lexicon of Iran. Tehran, Iran*: Geological Survey of Iran, Report no. 18.
Stocklin, J., Ruttner, A., & Nabavi, M. H. (1964). *New data on the lower Paleozoic and Precambrian of North Iran*, Geological Survey of Iran, Report 1, 13.
Talbot, C. J., & Alavi, M. (1996). The past of a future syntaxis across the Zagros. *Geological Society, London, Special Publications*, 100(1), 89–109.
Tavernier J. B., (1642). Les six voyages de Jean Baptiste Tavernier en Turquie, en Perse, et aux Indes. Pt. 1, livre 5. Paris.
Thiab-Ghodsi, A. A. (2007). *Biostratigraphy, chemiostratigraphy and depositional environments of the Upper Neoproterozoic-Lower Cambrian of the Northern Alborz and Northwestern Iran*, Unpublished Ph.D thesis, Faculty of Earth Sciences, Shahid Beheshti University.
Thiele, O., Alavi, M., Assefi, R., Hushmandzadeh, A., Seyed-Emami, K., & Zahedi, M. (1968). 1: 250000 Golpaygan quadrangle map. Geological Survey of Iran, Report.
Valeh, N., & Haghipour, A., (1970). *Geological map of Ardekan, scale 1:250,000*. Geological Survey of Iran, Tehran.
Zharkov, M. A. (1984). *Paleozoic salt bearing formations of the World*. Berlin: SpringerVerlag.

Chapter 2
Paleozoic

Being the longest Phanerozoic eras, the Paleozoic is a most geologically important era in the geology of Iran. It spans 249 Ma, and is subdivided into 6 periods the Cambrian, Ordovician, Silurian, Devonian, Carboniferous, Permian. The lower boundary of the Paleozoic coincides with the oldest beds of the Fortunian Stage and the upper boundary coincides with uppermost beds of Changhsingian Stage, overlain by Triassic rocks (Fig. 2.1).

In general terms, the Neoproterozoic-lower Cambrian deposits of Iran compare closely with those of northern Gondwana a character that continues sedimentologically into the base of the middle Paleozoic of most of Iran. At present, there is no evidence of an orogeny or even a stratigraphic hiatus at the boundary between the Neoproterozoic and the lower Cambrian with an apparently continuous transition in most of Iran (Hamdi et al. 1989). Overall, the lower Cambrian succession begins with shale, phosphate-bearing limestone, and dolomite resting conformably over Vendian dolomites. The boundary from the Soltanieh Formation into the Barut, Zaigun, and Lalun formations is transitional, so that the base of the Cambrian is very difficult to recognize in the field (Alavi-Naeini 1993). This boundary has been long-disputed by various researchers, especially in Alborz Mountains. The oldest probable record of trilobites in the form of the trace fossil, *Cruziana*, is in the red sandstone of the Lalun Formation, leading to the assignment of an Early Cambrian age for this formation and a Neoproterozoic age for the underlying Zaigun Formation. However, later studies provide an Early Cambrian age for the Zaigun Formation, based on the presence of age-diagnostic acritarchs and dinoflagellates in beds transitional from the Barut Formation to the Zaigun Formation (Figs. 2.2 and 2.3).

Stöcklin (1960) regarded the Soltanieh Formation as Neoproterozoic-early Cambrian without citing any specific biostratigraphic evidence. Seger (1977) suggested Vendian age for the Soltanieh Formation based on the achritarchs of Lower Shale Member at the northern Alamut area. Later studies of Salehi in 1960s reveal the index shelly fossils determined by Hamdi to Cambrian age. Finally, the results of comprehensive studies of Hamdi (1995) determined a Neoproterozoic-early Cambrian age for the Soltanieh Formation. This author placed the Neoproterozoic-Cambrian boundary at the base of the Middle Dolomite Member in the Central Iran. The

M. Ghorbani, *Lithostratigraphy of Iran*, Springer Geology,
https://doi.org/10.1007/978-3-030-04963-8_2

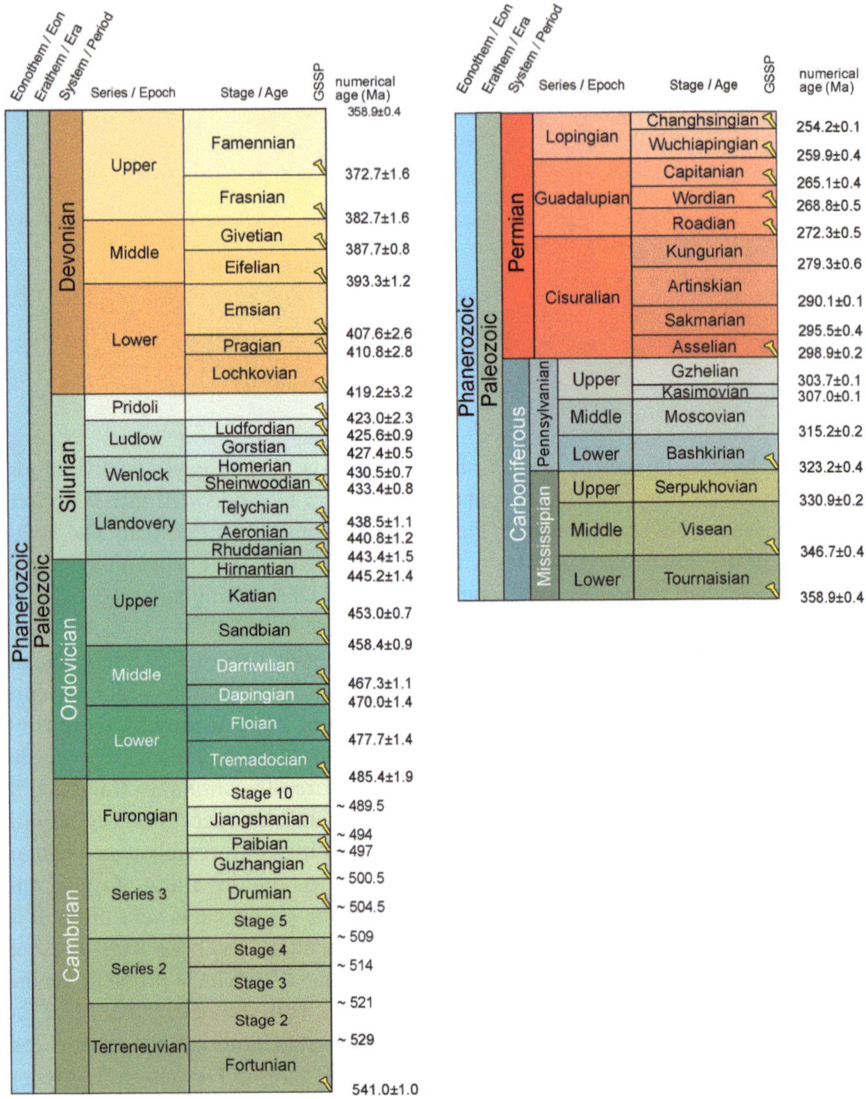

Fig. 2.1 International chronostratigraphic chart of Paleozoic (ICS 2015)

Neoproterozoic-Cambrian boundary was long studied by the researchers; of these, Huckriede et al. (1962) established the Neoproterozoic Rizu and Cambrian Desu 'Series' in Kerman area. Stöcklin (1962) subsequently introduced the Ravar Formation in the same area.

Nabavi et al. (1978) introduced the Desu, Heshem and Aghda formations during geological mapping of the Naein area. Hamedi and Alavi-Naeini (1988) concluded that among the two 'Series' introduced by Huckriede et al. (1962), only the Neoproterozoic Rizu 'Series' can be traced, and contrary to Stöcklin (1961) the Desu

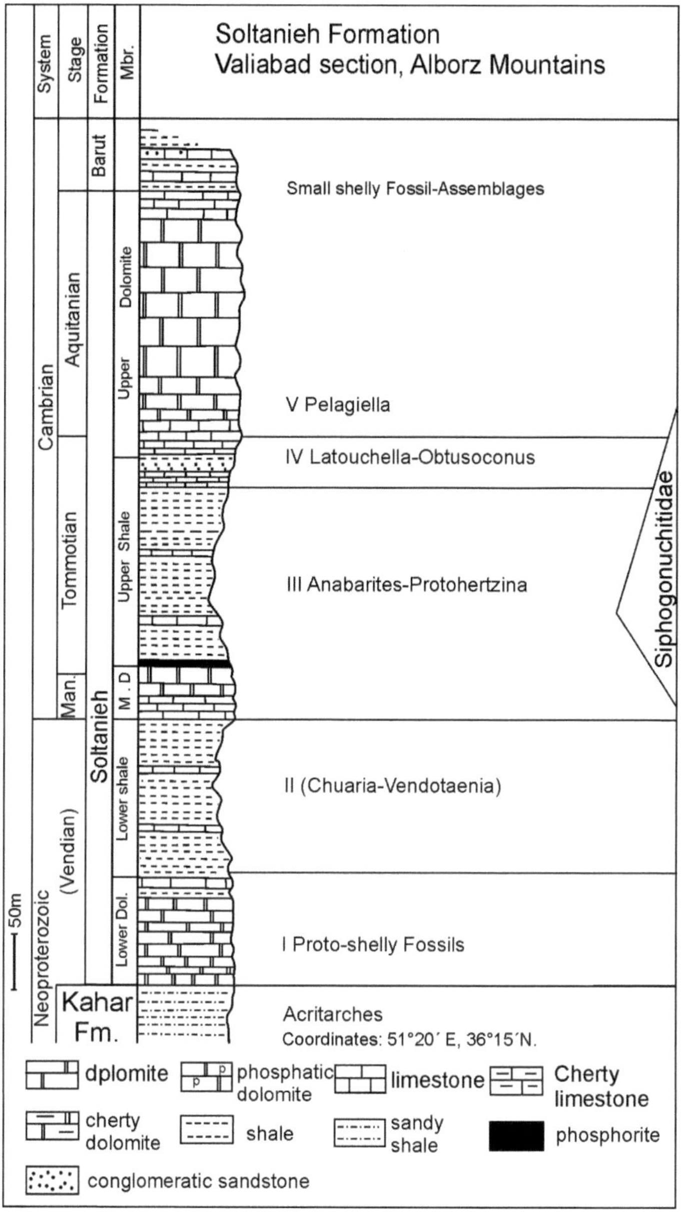

Fig. 2.2 Lithostratigraphic column of the Soltanieh Formation at the Valiabad Section (modified after Hamdi 1995)

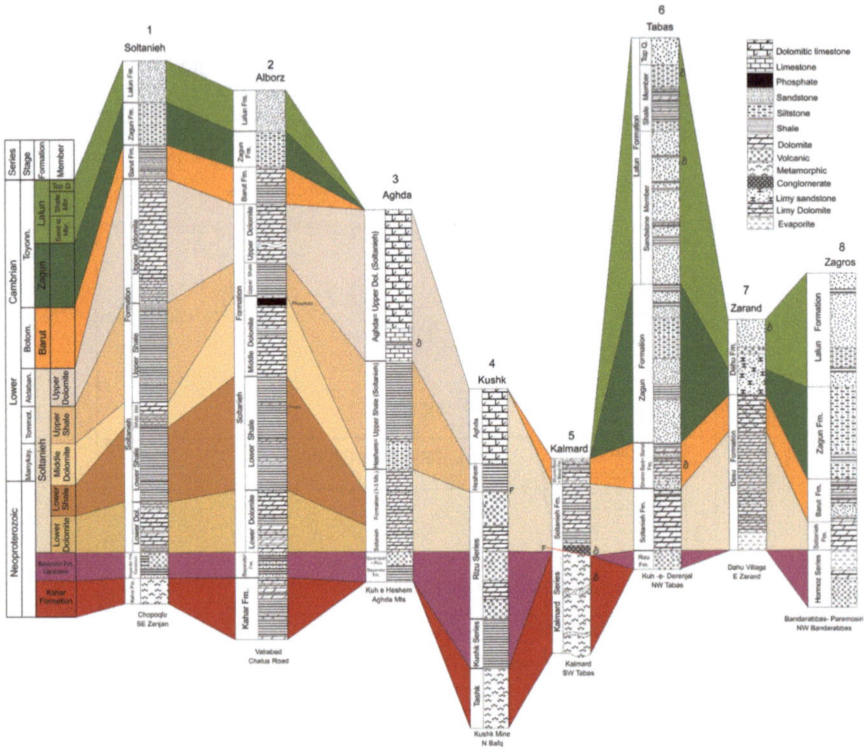

Fig. 2.3 Lithostratigraphical columns of the Neoproterozoic-lower Cambrian rocks of Iran, modified after Hamdi (1995)

'Series' is not the equivalent to the Rizu 'Series' and should be regarded as a distinct 'Series'. Overlying the Rizu 'Series', a succession of gypsum, shale, dolomite and limestone was informally introduced as the "Banestan Formation" (Hamedi and Alavi-Naeini 1984).

According to the present author, the establishment of the Banestan Formation and its correlation with the Soltanieh Formation not only resolves a geologically significant problem but also raises the ambiguities concerning the geology of Iran. The Banestan Formation of the Varaeoun (or Aghda) and the Banestan areas are actually the same as the Soltanieh Formation. The Desu and Aghda formations are overlain by either the Zaigun or Lalun Formations (or Dahu Formation) in Central Iran.

The Lalun Formation is stratigraphically and lithologically equivalent to the Purple Sandstone of the Salt Range in Pakistan; parts of the Saq Sandstone of Saudia Arabia; the Lower Quweira Sandstone of Jordan and Sudan; and the Kaplandere Formation of Turkey (Stöcklin and Setudehnia 1991). There is no evidence of any orogenic movement during early Cambrian time, as the boundaries between the

Soltanieh-Barut and Zaigun-Lalun formations are gradational and these units cannot be easily separated.

A break in sedimentation occurred in Iran at the beginning of the middle Cambrian. The subsequent marine transgression led to deposition of limestone and shale of the Mila Formation in Central Iran, with the Tabas Group, and Kuhbanan formation overlying either the Top Quartzite or older formations.

In some areas of Iran, the Upper Cambrian carbonate facies grades upward into Ordovician graptolite-bearing shale introduced as the "Lashkarak Formation" of Alborz; "Shirgesht Formation" of Central Iran; and the "Ilbeyk and Zardkuh Formations" of Zagros. It should be mentioned that Paleozoic successions of Iran include minor stratigraphic gaps, especially in Central Iran. Accordingly, the platform successions are completely exposed. The temporally-short gaps, owing to the erosional phases eroding the pre-existing shallow sediments, are traceable in the Paleozoic succession.

2.1 Cambrian

As stated above, the Neoproterozoic-Cambrian boundary in Iran coincides with the base of the Middle Dolomite Member of the Soltanieh in a continuous succession without any evidence of a tectonic event.

Outcrops of the Soltanieh Formation and its equivalents have been reported from several parts of Iran. The facies of Soltanieh Formation of the Alborz Mountain and Azarbaijan are almost similar, although being different with Central Iran ones, all of which are overlying the Neoproterozoic rocks (see below).

Equivalents of the Soltanieh Formation of Central Iran

Hamedi and Alavi-Naeini (1984) studied the equivalent Soltanieh Formation of the Kuhbanan area and regarded it as a new lithostratigraphic unit namely the "Banestan Formation".

The "Banestan" is an informal formation and is divided into four members:

1. The basal member which includes conglomerate and sandstone, of these the former consists of black rounded cherty pebbles with sandy cement. The conglomerate elements change depending on the composition of the underling rocks of the Rizu and/- or Morad 'Series'. The lateral equivalent of this part of the succession have not been reported from the Soltanieh Formation of the other areas;
2. The Gypsum-shale Member;
3. The Cherty dolomite Member;
4. Limestone member; and

As already mentioned, this formation is the same as the Desu and upper parts of the Rizu 'Series'.

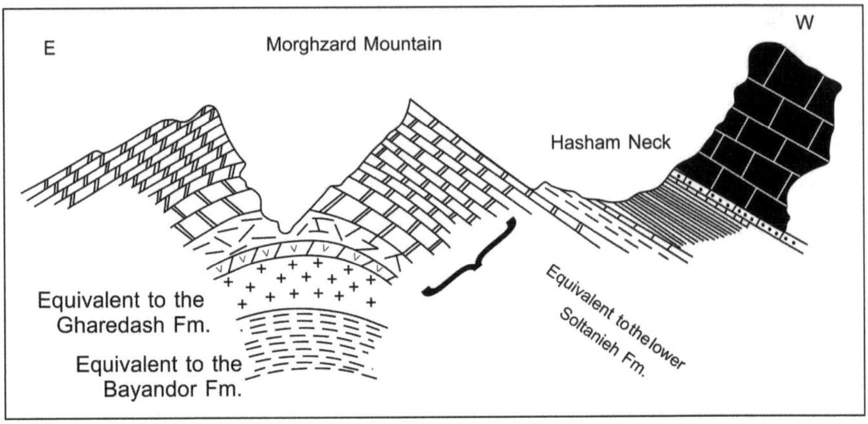

Fig. 2.4 Geological section through the Neoproterozoic-lower Cambrian rocks of southern Aghda (modified after Alavi-Naeini 1993)

Equivalent of the Soltanieh Formation of the Aghda area, northwestern Yazd
The thick succession of Soltanieh Formation cropping in the mountains of Southern Aghda can be correlated with the Soltanieh Formation of Alborz and the Soltanieh mountains. Nabavi and Amidi (in Nabavi 1976) provided a primary description of the Soltanieh Formation in the Naein area, then Nabavi (1976) described the equivalents of the Soltanieh and Barut formations in details. The equivalents of the Soltanieh Formation of the Aghda area were separately described as the Darin and Soltanieh formations.

In fact, equivalents of the Soltanieh Formation can be found in Central Iran. For example in southern Aghda, where roughly all the constituent members of the Soltanieh Formation can be traced (Fig. 2.4).

Moreover, in the Tabas area, rocks referred to as the Shorm Beds (equivalent to the Barut Formation), were reported to be underlain by the Soltanieh Formation. However, this is actually a misinterpretation of the authors who prepared the geological map of this area.

Barut Formation
The Barut Formation crops out extensively Iran including the Alborz, the Soltanieh Mountains, Central Iran, Azarbaijan and Zagros (?). It overlies the Soltanieh Formation in these areas. However, in some area it is not observed; accordingly, the Zaigun and Lalun formations immediately overlie the Soltanieh or Aghda formations.

This formation lithologically consists of argillaceous, silty and fine-sandy shale, that are strongly micaceous, of predominantly purple color with green- grey, violet and black variations. It also contains more than 50 intercalations of dolomite and limestone. These ranges in thickness from a few decimeters to a maximum of 30 m, and are characterized by thin lamination and by abundant nodules and cherty veins (Fig. 2.5).

Fig. 2.5 A view of the Barut formation at its type section

Limestone and calcareous shale predominate in the lower part which grades upward into dolomite. The dolomite beds may have the cellular and brecciated structure suggesting anhydrite dissolution (Stöcklin and Setudehnia 1991).

Equivalents of the Barut Formation in Central Iran

The Upper Shale and Upper Dolomite members show facies change in the southern Aghda, suggesting that the shale unit of the Heshem Formation is the equivalent of the Upper Shale Member of the Soltanieh Formation[1] and that thick-bedded rocks of the Aghda Formation are the equivalent of the Upper Dolomite Member (Alavi-Naeini 1993).

The Heshem and Aghda formations are equivalent to Member 4 and 5 of Soltanieh Fm., and the Barut Formation.

In Central Iran, the upper parts of Desu Formation are the equivalents of the Barut Formation as follows:

1. Upper beds of the Desu Formation;
2. The Shorm Beds of the Tabas area;
3. The Shorm Beds of Tabas area in some localities of Central Iran; and

[1] There is some evidence such as boudinage structure in the upper beds confirming that the Heshem Formation is equivalent to the Chapoghlu Shale.

Fig. 2.6 A view of the Aghda formation of the Bafgh area, around the Esfordi mine

4. Underlying rocks of the Dahu 'Series' in the Behabad area.

Heshem Formation
The Heshem Formation, which resembles the Upper Shale Member of the Soltanieh Formation, and Barut Formations of Alborz-Soltanieh, is an alternation of shale and carbonate rocks. In some areas, the composite thickness of carbonate (including dolomite and limestone) exceeds that of the shale beds whereas in other areas the shale beds constitute more of the succession. The main characteristic is the red or green color. In Heshem area, type section, its thickness was measured more than 100 m. Contrary to Barut Formation, the carbonate beds of the Heshem Formation are devoid of any cherty nodules, except of some areas representing cherty layer and nodules (Alavi-Naeini 1993).

Aghda Formation
The formation consists of black fetid limestone with lamellar structure indicating the stromatolite and algae referred by Nabavi (1976) to "algae-formed Lower Cambrian limestone". The black limestone of Aghda Formation formed most of the cliff of the southern Aghda (Such as Esfordi mine, Fig. 2.6). The Lower Cambrian Aghda Formation can be regarded as the time-equivalent of the Upper Dolomite Member and Lower beds of the Barut Formation.

Equivalents of Aghda and Heshem Formations at Bafgh area
In the Baigan Mountain of the eastern Kushk and in the Chah-Righ of Bafgh area,

a shaly-carbonate unit overlies the dolomite beds of Rizu 'Series' which can be compared with the Aghda and Heshem formations. These units show lateral facies changes. Its succession consists of shale, green, yellow and red sandstone alternating with black clay-bearing limestone and dolomite. Spheroidal algae may be major constituents of the limestone.

The equivalent succession of the Barut and the Upper parts of Soltanieh Formation of the northern Zarigan include:

1. A shale-sandstone unit regarded as the equivalent of the Heshem Formation
2. A limestone unit with 700 m thickness. This unit, the equivalent of the Aghda Formation, shows high resistance to erosion and forms most of highest parts of the area. It is lithologically divided into three members.

The lower member (200 m-thick) consists of thick-bedded black clayey limestone or limey shale rich in algal fossils specially spheroid forms, presumably the equivalent to the Heshem Formation.

The middle member (150 m-thick) consists of white, thick-bedded limestone showing an increase in clay content in its upper parts.

The upper member overlies strongly colored, thin-bedded shale, and consists of thin-bedded clayey limestone. It is regarded as the exact equivalent of the Barut Formation.

Equivalents of the Barut Formation at Zagros

Some Paleozoic successions crop out in the outer margin of High Zagros at Zardkuh and Dena especially in the margin of salt domes of the Hormuz 'Series'. In the base of the succession, underlying the Zaigun Formation, there is a unit correlated with Barut Formation of type section of the Alborz and of Soltanieh areas.

Setudehnia (1972) and Stöcklin (1974) briefly described the lithostratigraphy of this formation in the Sabzu and Bazanpir Mountains, western Esfahan Province. The Barut Formation in these areas consists of purple and red shale, and grey thin-bedded stromatolite-bearing carbonate. The thick-bedded carbonate dominates in the lower part of the formation.

The most complete outcrops of the constituent members of the Barut Formation can be observed in the Lordegan area where the lower parts are similar to Dolomite Member of the Soltanieh Formation. The massive and thick-bedded limestone belong to the topmost part of the Soltanieh Formation.

The lithological features of the Barut Formation of the Lordegan, Zagros Zone, show strong affinities to the Barut Formation of the type section. This indicates the similar paleogeographic histories of the Zagros, Alborz and Central Iran that were all covered by a monotonous shallow sea in which carbonate and muds were deposited.

Zaigun Formation

The Zaigun Formation is recognized in most of Iran including Alborz, Azarbaijan, the Soltanieh Mountains, Central Iran, Zagros and east southern Iran (Baluchistan) (Assereto 1963). In some areas of Iran such as Golpayegan, differentiation of the Zaigun Formation from Lalun Formation is difficult, so unavoidably these formations are combined as a single unit the "Zaigun-Lalun".

Lithologically this formation consists of silty to fine sandy shale of dark-red in color but also pinkish, green, dark purple and dark grey in parts. It also contains intercalations of cross-laminated, ripple-bearing, fine-grained sandstone (Assereto 1963).

No fossils have been reported from the Zaigun Formation, and its early Cambrian age was suggested based solely on its stratigraphic position.

Lalun Formation

The red arkosic sandstone of the Lalun Formation extends over all areas of Iran except for the Kalmard area where Cambrian rocks are exposed in a horst associated with the Naein and Kalmard Faults. Deposition of the formation even extended beyond the boundaries of Iran and can be traced in Pakistan (Salt Range, Kahureh Formation), Arabia (Lower parts of Saq Sandstone), Jordan (Lower Quweira Sandstone), Central Afghanistan and Turkey (Sadan, Kapalandere, Calaktepe formations).

The formation in the type section in Lalun Valley is lithologically divided into three members. In ascending stratigraphic order these are (Assereto 1963):

1. Dark-red to pink medium-grained arkosic sandstone, 497 m-thick;
2. Dark-red and variegated shale with minor sandstone, 35 m-thick;
3. White subarkose to quartzite, i.e. Top Quartzite, 50 m-thick; and

Thin cross-laminated to cross-bedding is common throughout the formation and ripple marks are frequently observed.

Sedimentological analysis of Steiger (1966) showed that:

1. The plagioclase feldspar content decreases towards the top of the formation;
2. The Lalun Formation is of continental origin;
3. Oxidation associated with weathering in an arid condition was long considered as the main cause of the red color feature of the formation. However according to Steiger (1966), this feature was caused by oxic depositional environment of the Lalun Formation. The intensity of the color significantly decreases toward the top of the formation and finally grading into the Top Quartzite suggesting less oxidizing conditions.

The overlying unit of the formation, Top Quartzite, is a fine quartzarenite observed everywhere at the top of the Lalun Formation. In contrast to the arkosic sandstone unit, it does not shows any significant thickness variations between sections (Stöcklin and Setudehnia 1991).

The formation shows little variation in lithology, confirming the hypothesis that it accumulated in monotonousness condition of sedimentation. The thickness may exceed 900 m in the Derenjal Mountains.

No fossils have been reported in the type section. However, Gansser and Hubber (1966) reported the ichnofossil *Cruziana* from the Alamkuh area, a finding verified

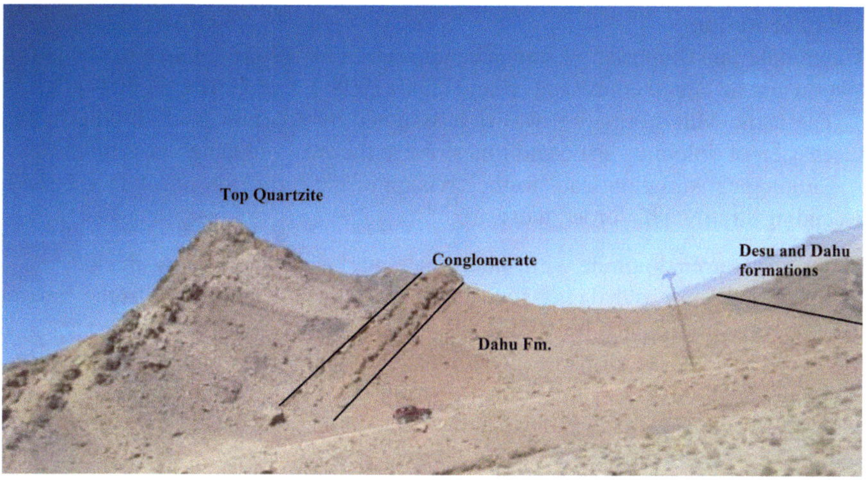

Fig. 2.7 Lower Cambrian rocks of southern Behabad

repeated by the other researchers in the Alborz Mountains (e.g. Allenbach 1966; Ruttner et al. 1968). Seilacher (in Allenbach 1966) assigned the ichnofossil to imprints of the appendages of the trilobite *Redlichia* indicating an Early-Middle Cambrian age.

An age of 510 Ma age has been determined for limestone intercalations of the upper Lalun Formation based on rubidium-strontium dating (NIOC 2016).

Dahu 'Series'

Huckriede et al. (1962) first described the Dahu 'Series' from the village of Dahu, 62 km northwest of Kerman, and subsequently correlated by Stöcklin et al (1964a, b) with the Lalun Formation. The thickness of the formation is highly variable in the area. German geologists mapped these rocks between the type section and Behabad, 170 km to the northwest. Alnog the Dahu-Behabad road, the thickness of the 'Series' significantly decreases and finally diminished near the Behabad city. Thus the Top Quartzite unit directly overlies the thick-bedded cherty dolomites of the Desu Formation[2] (Fig. 2.7).

A maximum thickness of 1000 m was measured for the formation in the Shabjareh area (600 m of which is accounted for by the shaly unit of the Zaigun Formation). In the Kuhbanan area, the total thickness is only about 100 m. The main differences between the Dahu 'Series' and Lalun Formation are:

1. The Dahu 'Series' tends to be coarser grained;
2. The Dahu 'Series' contains more conglomerate beds; and
3. Different underlying formations, so that the Dahu Formation is underlain by the Desu Formation, whereas the Lalun Formation covers the Zaigun Formation.

[2]Owing to distinct underlying and overlying strata confirmed by field observations of the present author, it is suggested here to rank the Dahu as a formation.

Mila Formation

The middle-late Cambrian of Iran is characterized by the presence of the Mila Formation and its lateral equivalents (e.g. Hamdi 1995; Bayet-Goll et al. 2014).

The name Mila Formation was first assigned by Stöcklin et al (1964a, b) to a succession of dolomite, shale and limestone to the west of Damghan, Alborz.

Lithological variation leads to the division of the formation into 5 members. In ascending stratigraphic order, these are:

Member I: Barren dolomite with subordinate marl and shale, 189 m- thick
Member II: Sparry and flaggy limestone, subordinate marl and siltstone bearing trilobite, brachiopods and hyolithids, 89 m-thick
Member III: Crystalline-sparry glauconitic limestone containing trilobite and brachiopods, 82 m-thick
Member IV: Siltstone, sandstone, sparry-glauconitic limestone, marl with trilobite and brachiopod, 89 m-thick
Member V: Barren shale, subordinate thin sandstone and limestone, basal quartzite, 129 m-thick

Stöcklin et al. (1964a, b) did not suggest any age for the Member I, due to the absence of fossils. However, because Member II is assigned to the lowest stage of the upper Cambrian, presumably, Member I must be middle Cambrian to earliest upper Cambrian in age. These authors also assigned Members III and IV to the latest Cambrian. According to these authors, Member V is barren at the Mila-Kuh. However, the corresponding shale, sandstone and limestone in the western central Alborz and in southern Azarbaijan have yielded an early-? middle Ordovician fauna (Stöcklin and Setudehnia 1991; Table 2.1). This unit also starts with a basal quartzite unit and overlies the Member IV with sharp lithological contact.

Accordingly, Member V is considered to be probably younger in age (Ordovician) than the underlying upper Cambrian Member IV.

However, an international group of geologists working on the stratigraphic sections of Cambrian rocks exposed in the central and eastern Alborz have elevated the rank of the Mila Formation to "Group" and proposed the new names "Fasham Formation" for the Top Quartzite "Deh-Sufiyan Formation" for Mbs. 1, 2 of the Mila Formation and "Deh-Molla Formation" for members 3 and 4.

The Mila Formation of Zagros

The Mila Formation crops out through most of Zagros especially in northern Zagros and the area where the Mila Formation has been brought about to surface by salt dippers such as Zardkuh, Ilbeyk Valley and Dena (Taneg-e Putak, Lordegan, Par-e Musiri and Sheikhan valley; NIOC 2016).

In these areas the Mila Formation overlies the Lalun Formation. However, in the Zardkuh, Ghalikuh and Chalishe, the Lalun Formation has not been observed, and the Mila Formation shows much facies similarity with Mila Formation of Alborz.

The main difference to the type section in some areas is the absence of the members IV and I which laterally change to the Ilbeyk and Zardkuh formations. The most complete section of the Mila Formation is located in the Tang-e Putak area where it

Table 2.1 Biostratigraphical units of Mila Fm. (Kushan 1973)

Lithostratigraphical units		Biostratigraphical units			Chronostratigraphical units
Eastern Central Iran	Alborz	Taxons (Zones)	Taxon Age	Biozone sign	
Shirgesht Fm.	Mila Fm. Mb.V	Alborzella Zone	?	g	Early Ordovician (Tremadocian)
	Mila Fm. Mb.IV	Alborzella Zone		f	Late Cambrian
Derenjal Fm.		Kaolishania Zone		e	
	Mila Fm. Mb.III	Prochuangia Zone		d	
Kalshaneh? Fm.		Drepanuca Zone		c	
	Mila Fm. Mb.II	Dorypyge Zone		b	Middle Cambrian
	Mila Fm. Mb.I	Iranoleesia Zone		a	Early Cambrian?

is overlain by the Faraghan Formation. Members a, b and c of the Mila Formation are of equal extent and thickness throughout Zagros (NIOC 2016).

The Mila Formation of Zardkuh and the Dena Mountains
The Paleozoic sequence of the Zardkuh and Dena areas, High Zagros, was studied by Setudehnia (1972) and is as follows:

Lower boundary: conformable with Top Quartzite

Upper boundary: lowermost sandstone of Faraghan Formation

The total thickness of the formation in the Dinar and Zardkuh Mountains is 525 m. It changes throughout the area and the measured thickness varies as at Tang-e Putak where its total thickness is 215 m.

The Mila Formation was divided into the following 3 members:

Member A (Early to Mid-Cambrian)
The total thickness measured 71.6 m at Tang-e-Putak and 70 m at Kuh-e Gareh. This member was not observed at the Zardkuh Mountain.

Lithologically, it consists of

- Reddish silty shale, 3 m-thick
- Grey and yellowish dolomite and limestone alternating with red-colored silty shale, 21.3 m-thick
- Dark grey medium to thick-bedded coarse-grain dolomite, 15.2 m-thick
- Thin-bedded brittle grey dolomite, 12.2 m-thick
- Brown dolomite with interbedded red-colored shale barren in fossil, 19.8 m-thick.

Member B (Mid-Cambrian)
The total thickness measured is 25.7, 136.3 and 103 m in Tang-e Putak, Zardkuh, and Kuh-e Gareh, respectively. The main lithology in the Tang-e Putak is brittle red-colored silty shale with pseudomorphs of salt nodules and greenish silty shale laminae.

In Zardkuh Mountain, this member includes red and green colored siltstone and shale with dolomite intercalations.

This member is unfossiliferous. However due to the middle Cambrian age assignment of Member C, a middle Cambrian age was suggested based on its stratigraphic position.

Member C (Mid- to Late Cambrian)
The most complete section of Member C is located at the Ilbeyk Mountain, where it is 322 m-thick. This member is lithologically composed of:

- yellow thin-bedded dolomite interbedding with shale, 6 m-thick
- grey thin-bedded limestone with some shaly intercalations, 12 m-thick
- greenish to purple shale interbedding with thin-bedded yellow and grey limestone, 27 m-thick
- thin-bedded limestone grading into limey conglomerate and limey breccia, 42 m-thickness
- trilobite-bearing shale and limestone with Hyalites and brachiopods, 68 m-thick

- fossiliferous limestone with shale and olive marl, 131 m-thick
- grey to greenish micaceous sandstone, 5 m-thick
- shell-bearing grey to greenish limestone, 1 m-thick
- olive shale and limestone with trilobite and brachiopod shell, 10 m-thick
- thick-bedded sparry limestone with shale intercalations on the upper part, 20 m thick.

Considering the exposure of three lower members of Mila formation in the Zardkuh, Ghalikuh and Chalisheh, and the fact that Ilbeyk and Zardkuh formations are equivalents of Member IV and V of Mila Formation, we can suggest that these successions are the equivalent of five members of Mila Fm. of the Alborz area (Motiei 1993).

Equivalents to the Mila Formation in Central Iran

The Mila Formation can be traced all over Iran from eastern Alborz through central Alborz, Soltanieh Mountain and the Takab area to Urmia Lake in Azarbaijan (except for the Talesh and Kopeh-Dagh where it has presumably been removed by subsequent erosion).

Normally, in these wide areas, facies and thickness changes are due to the variation of the basin conditions particularly in the eastern Central Iran, where the facies changes are so extensive that Mila Formation was ranked by some worker as the "Mila Group" (Ruttner et al. 1968).

The Mila Group of Central Iran is the exact stratigraphic equivalent of the Mila Formation of northern Iran. Because of the considerably increased thickness (about 3000 m against 585 m at Mila Kuh) and greater facies differentiation, its ranked was was raised from the formation in east Iran to group (Stöcklin and Setudehnia 1991) comprising the Kalshaneh, Derenjal and Shirgesht Formations

Kalshaneh Formation

The Kalshaneh Formation is a mixture of various carbonate rocks (limestone-dolomite), some shales and basic volcanic rocks. Three main units are recognized as follows:

1. multi-colored dolomitized or sandy shale, mainly green with subordinate yellow and grey colors;
2. black and grey finely-crystallized dolomite occasionally bearing cherty nodules and bands. This unit shows two repetitive succession of thickness, first the succession with decimeter thickness and secondly with millimeter thickness;
3. green, diabase-like hornblende-augite-dolerite and black olivine-augite-dolerite. These rocks are of younger age than the above sedimentary rocks;
4. black fetid limestone and blue-grey nodular limestone, partly marmorized, rarely with chert (Ruttner et al. 1968).

These rocks are found exclusively in the type area of the Derenjal Mountains and in a few sections of the southern Derenjal. They are underlain and overlain by the Top Quartzite and Derenjal Formation, respectively (Ruttner et al. 1968).

It seems that the true stratigraphic age of the formation may not be the middle Cambrian and it lithologically resembles the Rizu and Desu 'series' and the lower beds of the Aghda Formation of the Bafgh area. According to the present author, it is better to suppress this formation, because its lower and upper boundaries are both faulted. These faults are difficult to distinguish due to the conformable lower and upper beds of the formation. Actually, units 1–3 belong to the Rizu or Desu 'Series' and unit 4 belongs to the lower Cambrian Aghda Formation (Ruttner et al. 1968).

Derenjal Formation

The basal beds are composed of strongly colored sandstone and marl interbedded with yellow to reddish grey dolomite. Some mud cracks are observed on bedding planes of the sandstone. The middle and upper parts consist of thin-bedded limestone with black fresh fractures and yellowish to light-brown weathering surface. These parts may include marl and silt intercalation and lens-shape sandy limestone. Some of the bedding planes contain trilobite fragment (Ruttner et al. 1968).

The formation can be geographically traced in the Central Iran from the Ozbak-Kuh of north to Lakhar-Kuh of the south. Considering the lithological succession and paleontological content, the Derenjal Formation corresponds to the members 2–4 of the Mila Formation (Stöcklin and Setudehnia 1991). Due to this fact that the Kalshaneh Formation is mixture of the Rizu 'Series', Aghda and Soltanieh formations, and its upper and lower boundaries are faulted, consequently, the same formation cannot be considered as formal units and it is better to correlate it with members 1–3 of the Mila Formation, suggesting a mid- late Cambrian age.

Moving southward, the lithological variations, form the type section in the north to the reference section in the south, indicate changing depositional environments from marine to coastal settings represented by gradual increase in clastic content.

Shirgesht Formation

The Shirgesht Formation crops out in most of Central Iran, Ozbak-Kuh, Kashmar, Anarak and Kerman. In all of these areas, the formation is of more or less similar facies accompanied but with variations in the lithological thickness. The equivalents of the Shirgesht Formation can be observed in the Mahabad and Abar-Kuh of Sanandaj-Sirjan Zone. Also, the formation correlates with the upper part of Member IV, V, and Lashkarak, Gorgan Schist and Abarsaj Formation. It is believed that the Ilbeyk, Zardkuh and Seyahoo formations of the Zagros area are time equivalents of the Shirgesht Formation at its type section.

Continuous deposition across the Cambro-Ordovician Boundary has been reported from Shirgesht, Zardkuh, Ghalikuh and some parts of Alborz. It was believed that the brachiopod bearing limey layer the "*Billingsella* Bed" spanned the Cambro-Ordovician Boundary in the lowermost beds of the Shirgesht Formation at the type section (Aghanabati 2004). However, later studies disputed this age assignment and considered it to be a late Cambrian age (e.g. Ghaderi et al. 2008; Jahangir et al. 2015).

Preliminary age determinations indicated a middle Ordovician age for the middle part of the formation. The lower part is barren of fossils and an early Ordovician age was inferred based on its stratigraphic context (Stöcklin and Setudehnia 1991). However, a later study by Ghaderi et al. (2008) defined the Cambro-Ordovician boundary in the lowest 150 m of the Shirgesht Formation. Consequently, this formation is of Cambro-Ordovician age.

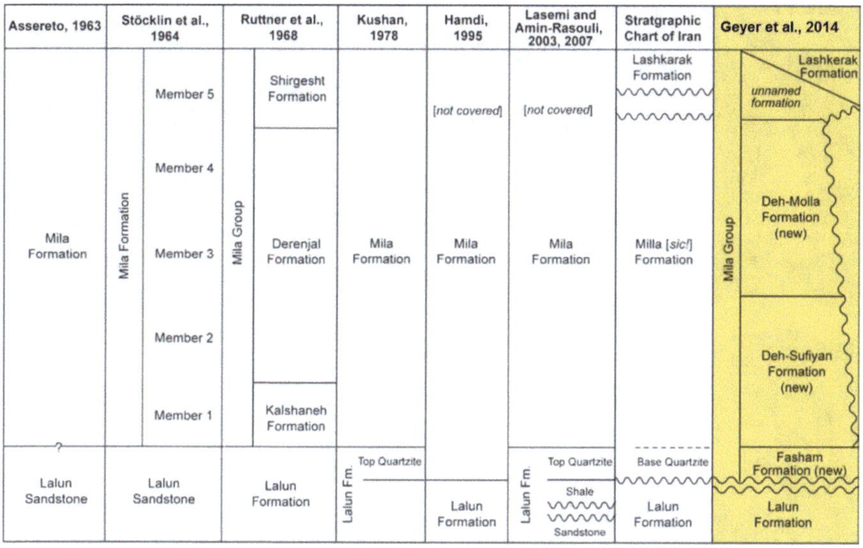

Fig. 2.8 Revised Cambrian lithostratigraphy of Iran (modified afer Geyer et al. 2014)

Following a revision carried out on the international Cambrian/Ordovician boundary, several research projects were conducted on the Cambro-Ordovician boundary interval in the Shirgesht Formation and it has been revealed that the boundary actually lies within an interval of 245–260 m the middle part of Shirgesht Formation (Jahangir et al. 2015; NIOC 2016) (Fig. 2.8).

Selected diagnostic fossils from Cambrian rocks of various zones of Iran are illustrated in Plates 2.1 and 2.2. The lithostratigraphic features of Cambrian rocks of Iran are shown in Table 2.2.

2.2 Ordovician

Ordovician rocks are exposed in most of Iran, including Alborz, Zagros, Central Iran and Sanandaj-Sirjan but these are mainly limited to the early-middle Ordovician. The absence of upper Ordovician rocks has been related to the epeirogenic phase, climate change or specially onset of the late Ordovician (Ghavidel-Syooki et al. 2011a, b). However, the marine environment of the Cambrian to middle Ordovician changes to non-marine. The former facies is represented by the succession spanning Cambro-Ordovician boundary of Alborz, east of Zagros and by the minor facies change of the Upper Cambrian and Lower Ordovician rocks. For example, the Cambro-Ordovician boundary was reported form the Mila and Ilbeyk formations of Zagros, the Tabas Group in the eastern Iran and the *Billingsella* Bed of the Shirgesht Formation

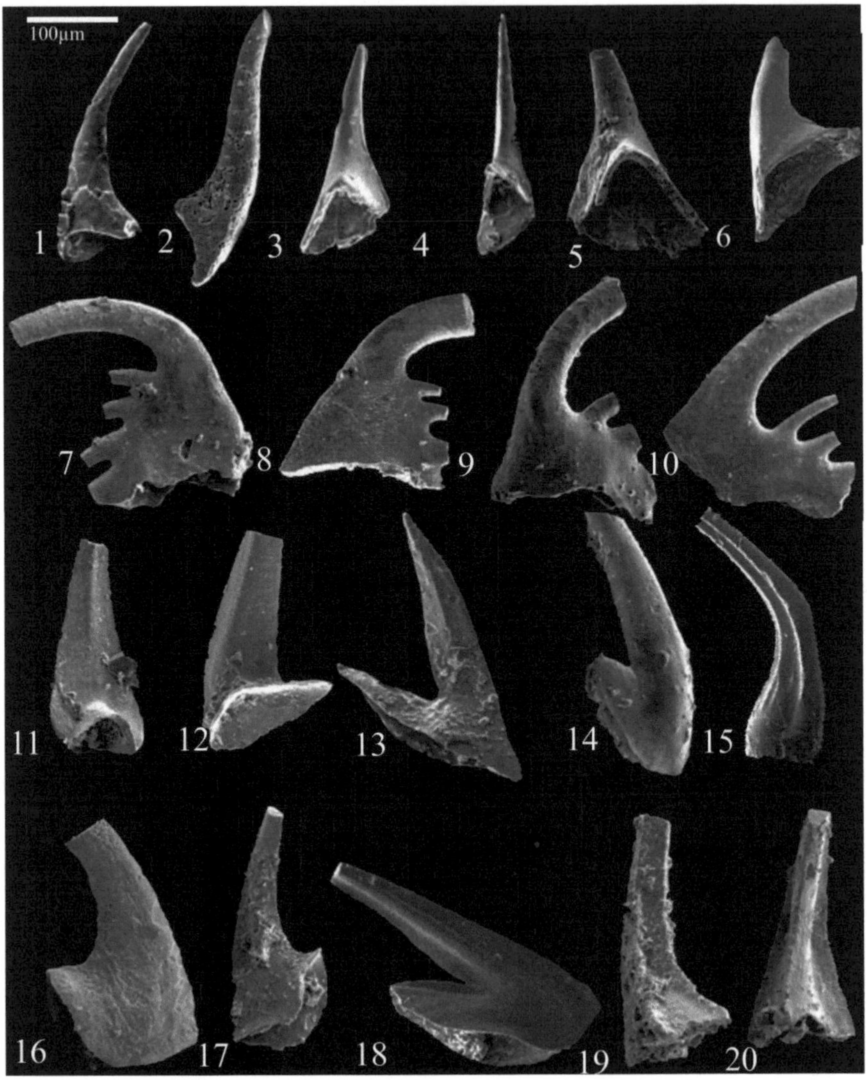

Plate 2.1 Selected diagnostic conodonts of Cambro-Ordovician of Central Iran (NIOC 2016). 1. *Proconodontus muelleri,* Shirgesht Fm., Shirgesht Section; 2, 3. *Furnishina furnishi*, Mila (V) Fm., Barut Section; 5, 6. *Furnishina asymmetrica,* Derenjal Fm., Shirgesht Section; 7, 8. *Cordylodus intermedius,* Shirgesht Fm., Shirgesht Section; 9, 10. *Cordylodus angulatus*, Shirgesht Fm., Shirgesht Section; 11, 12. *Acodus oneotensis,* Shirgesht Fm., Shirgesh Section; 13, 14. *Paroistodus proteus*, Shirgesht Fm., Shirgesht Section; 15. *Scolopodus subrex*, Shirgesht Fm., Shirgesht Section; 16, 17. *Paroistodus proteus*, Shirgesht Fm., Shirgesht Section; 18. *Paroistodus proteus,* Shirgesht Fm., Shirgesht Section; 19, 20. *Acodus deltatus*, Shirgesht Fm., Shirgesht Section

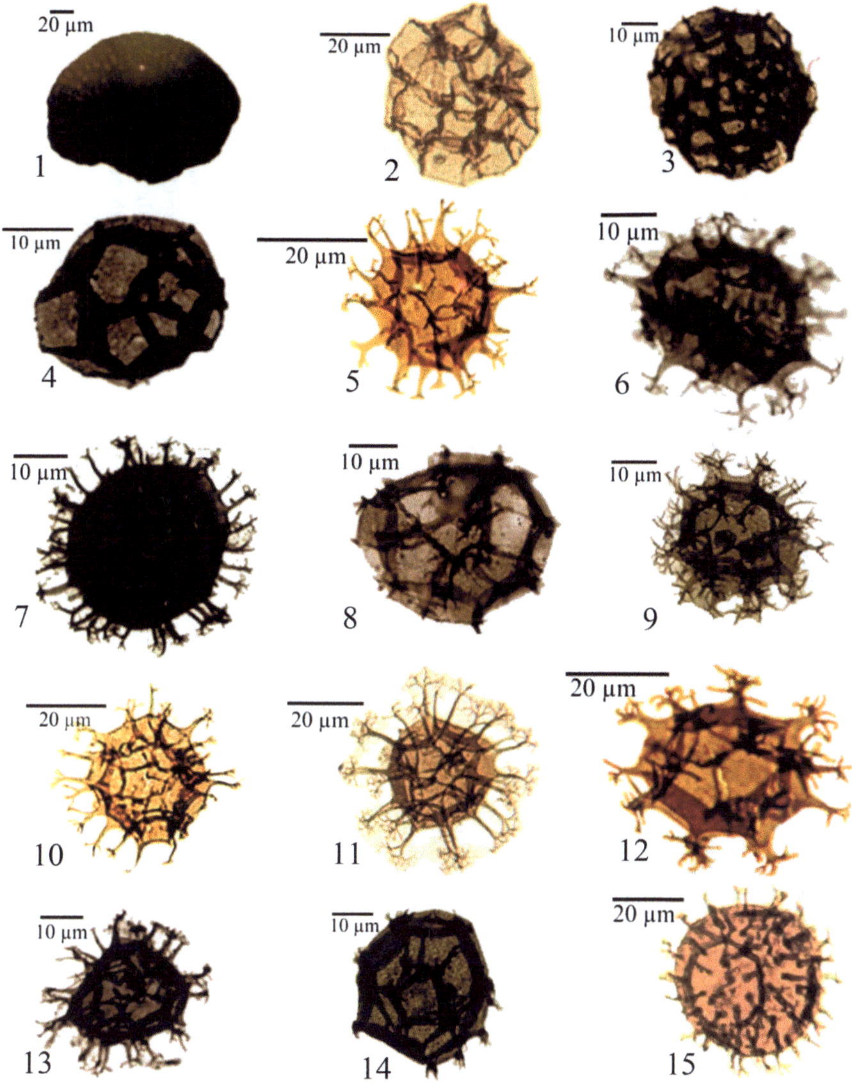

Plate 2.2 Selected diagnostic Cambrian palynomorphs of Zagros area (NIOC 2016). 1. *Soldadophycus major,* Bayandor Fm., Barut Section; 2. *Cristallinium cambriense*, Mila Fm., Zard Kuh Section; 3. *Cristallinium ovillense*, Mila Fm., Chalisheh section; 4. *Retisphaeridium howellii*, Mila Fm., Chalisheh section; 5. *Timofeevia lancarae*, Mila Fm., Darreh-Yas Section; 6. *Timofeevia pentagonalis*, Mila (III) Fm., Ghali Kuh Section; 7. *Timofeevia phosphoritica*, Ilbeyk Fm., Chalisheh section (Reworked); 8. *Vulcanisphaera nebulosa*, Ilbeyk Fm., Ghali Kuh Section; 9. *Timofeevia microretis*, Ilbeyk Fm., Ghali Kuh Section; 10. *Timofeevia lancarae*, Mila Fm., Darreh-Yas Section; 11. *Timofeevia lancarae*, Ilbyke Fm., Zard Kuh Section; 12. *Timofeevia lancarae*, Ilbeyk Fm., Chalisheh section; 13. *Timofeevia pentagonalis*, Mila Fm., Darreh-Yas Section; 14. *Timofeevia pentagonalis*, Ilbeyk Fm., Chalisheh section; 15. *Stelliferidium furcatum*, Ilbeyk Fm., Zard Kuh Section

Table 2.2 Lithostratigraphic features of Cambrian rock units of Iran

Rock unit	Chronostratigraphic affinity	Lower/Upper boundary	Lithology and thickness at type section	Note (References, comments)
Cambrian (Alborz)				
Barut Fm.	Early Cambrian	Lb. is conformable with Soltanieh Fm Ub. is gradual with Zaigun Fm The boundary was placed at the top of upper dolomite bed	The succession of the colored shale, dolomite and limestone Thickness: 714 m According to the observation of the author the more precise thickness is about 400 m	Book: Alavi-Naeini (1993, p. 65); Hamdi (1995, p. 77); Aghanabati (2004, p. 133; Aghanabati (2008, p. 309; Alavi-Naeini (2009, p. 95) Thesis: Mirshahani and Adabi (2002). The upper dolomite bed of Barut Formation was assigned as the boundary with Zaigun Formation
Zaigun Fm.	Early Cambrian	Lower boundary is the upper bed of the Barut Formation/upper boundary is Lalun Formation	6 members were established: limey shale, sandstone interbedding with siltstone, arkosic sandstone, mica-bearing sandstone, arkosic sandstone with silty sandstone intercalations Thickness: 453 m	Book: Alavi-Naeini (1993 p. 78); Hamdi (1995, p. 84); Aghanabati (2004, p. 133); Aghanbati (2008, p. 350); Alavi-Naeini (2009, p. 104) Differentiation of Zaigun Formation from Lalun Formation is relatively difficult, however the former is darker and the later tends to be coarser grained
Lalun Fm.	Early Cambrian	Lb. Zaigun Fm Ub. Mila Fm.	Red medium-grained arkosic sandstone (40–80 cm bed thickness)	Book: Alavi-Naeini (1993, p. 83); Afshar-Harb (1994, p. 19); Hamdi (1995, p. 90); Aghanabati (2004, p. 134); Aghanabati (2008, p. 375); Alavi-Naeini (2009, p. 109) In the east of Central Iran upper part consist of red and green limey siltstone and yellow nodular marl, and may include limey shale and dolomite

Table 2.2 (continued)

Rock unit	Chronostratigraphic affinity	Lower/Upper boundary	Lithology and thickness at type section	Note (References, comments)
Mila Fm.	Middle Cambrian to Ordovician	Lb. top quartzite Ub. Disconformable with Geirud Fm.	The formation has been divided into 5 members including: M1: Barren dolomite with subordinate yellow marl and shale M2: Grey to brown limestone with subordinate marl and siltstone M3: light coarse-grain glauconitic limestone containing trilobite and brachiopods M4: Siltstone, sandstone, marl and coarse-grain glauconitic limestone M5: Barren shale subordinated with thin-bedded sandstone and limestone Thickness: 585 m	Book: Alavi-Naeini (1993, p. 108); Afshar-Harb (1994, p. 20); Hamdi (1995, p. 103); Aghanabati (2004, p. 135); Aghanabati (2008, p. 454); Alavi-Naeini (2009, p. 129) In the Zagros zone the underlying formation is Zardkuh Fm. or Ilbeyk Fm. and overlying formation is Zakin Fm. or Faraghan Fm. Considering the field observation, top quartzite actually belongs to Mila Fm., accordingly, Iranian National committee of Stratigraphy recommend to regard the Top Quartzite unite as the lowermost member of Mila Formation rather than the Lalun Formation
Cambrian (central and eastern Iran, Sanandaj-Sirjan)				
Heshem Fm.	Early Cambrian	Lb. continuously with Soltanieh Fm. Ub: Aghda Fm.	A succession of green to grey mica-bearing limey shale and thin-bedded sandy shale with sandy shale nodules in the upper part especially at the contact with Aghda Fm. Thickness: 150–180 m	Book: Hamdi (1995, p. 160); Aghanabati (2004, p. 137); Aghanabati (2008, p. 410); Alavi-Naeini (2009, p. 63) This formation can be correlated with Upper Shale Member of Soltanieh Fm. The presence of black stromatolite nodules is the main character of Heshem Formation

(continued)

Table 2.2 (continued)

Rock unit	Chronostratigraphic affinity	Lower/Upper boundary	Lithology and thickness at type section	Note (References, comments)
Aghda Fm.	Early Cambrian	Lb. Conformable with Heshem Fm. Up. Disconformable with Lalun Formation	Cliff-forming black massive limestone Thickness: 35–400 m	Book: Hamdi (1995, p. 159); Aghanabati (2004, p. 137); Aghanabati (2008, p. 370)
Cambrian (central and eastern Iran, Sanandaj-Sirjan)				
Heshem Fm.	Early Cambrian	Lb. continuously with Soltanieh Fm. Ub: Aghda Fm.	A succession of green to grey mica-bearing limey shale and thin-bedded sandy shale with sandy shale nodules in the upper part especially at the contact with Aghda Fm. Thickness: 150–180 m	Book: Hamdi (1995, p. 160); Aghanabati (2004, p. 137); Aghanabati (2008, p. 410); Alavi-Naeini (2009, p. 63) This formation can be correlated with Upper Shale Member of Soltanieh Fm. The presence of black stromatolite nodules is the main character of Heshem Formation
Aghda Fm.	Early Cambrian	Lb. Conformable with Heshem Fm. Up. Disconformable with Lalun Formation	Cliff-forming black massive limestone Thickness: 35-400 m	Book: Hamdi (1995, p. 159); Aghanabati (2004, p. 137); Aghanabati (2008, p. 370)

(continued)

Table 2.2 (continued)

Rock unit	Chronostratigraphic affinity	Lower/Upper boundary	Lithology and thickness at type section	Note (References, comments)
Dahu 'Series' = Equivalent to Lalun Fm.	Early Cambrian	Lb. angular disconformity with Desu 'Series' Ub. Conformable with white colored quartzite	Purple to red sandstone subordinated with sandy claystone and conglomerate Thickness: variable in different areas, accordingly in Behabad area, it is 400–600 m and in Kuhbanan area 100 m.	Book: Alavi-Naeini (1993, p. 97; Aghanabati (2004, p. 138; Aghanabati (2008, p. 414; Alavi-Naeini (1993, p. 118). The formation can be correlated with Upper Dolomite Member of Soltanieh Formation in Alborz and Azarbaijan
Kuhbanan Fm.	Late early Cambrian-late Cambrian	Lb. Dahu 'Series' Ub. Devonian old red sandstone	Succession consist of sandstone and *Redlichia*-bearing limestone in basal beds, thick-bedded dolomite subordinated with sandstone in middle part; and shale, sandstone and red colored slate with some dolomite beds in the upper parts Thickness: about 300 m	Book: Alavi-Naeini (1993, p. 146); Hamdi (1995, p. 132); Aghanabati (2004, p. 138); Aghanabati, (2008, p. 438); Alavi-Naeini (2009, p. 157) Thesis: Ameri (2009) It seems that the lithological facies is similar with Lalun and Zaigun Formations extended to the Late Cambrian in this area
Kalshaneh Fm.	Middle Cambrian	Lb. Lalun Sandstone, the contact is disturbed by basic igneous intrusion Ub. Derenjal Formation, the contact is disturbed	A complex of sedimentary rocks (dolomite, limestone, shale, sandstone and gypsum) and volcanic ones (mostly diabase) Thickness: 1500 m	Book: Alavi-Naeini (1993, p. 137); Hamdi (1995, p. 126); Aghanabati, 2004, p. 138; Aghanabati, 2008, p. 431; Alavi-Naeini (2009, p. 150) This formation is actually consists of 3 units including Rizu, Desu and Aghda formations and the Kalshaneh Fm. is not well established

(continued)

Table 2.2 (continued)

Rock unit	Chronostratigraphic affinity	Lower/Upper boundary	Lithology and thickness at type section	Note (References, comments)
Derenjal Fm.	Middle-late Cambrian	Lb. Sharp but conformable with the black dolomite of so called Kalshaneh Fm. Ub. Shirgesht Fm.; the contact is conformable and transitional	Cream to light colored thin-bedded limestone subordinated with sparry limestone, siltstone and marl sometimes showing mud cracks and salt crystal pseudomorphs Thickness: 833 m-thick	Book: Alavi-Naeini (1993, p. 140); Hamdi (1995, p. 128); Aghanabati (2004, p. 139); Aghanabati (2008, p. 420); Alavi-Naeini (2009, p. 153) Thesis: Mehrjerdi (1995) Except of the type area, the contact is disturbed elsewhere and upper Derenjal Fm. shows the chaotic structure of the Kalshaneh Fm. actually the primary reported thickness is exaggerated
Cambrian (Zagros)				
Barut Fm.	Early Cambrian	Lb. is not observed anywhere Ub. Zaigun Fm., the contact is apparently conformable	Thin-bedded dolomite subordinated with red and purple shale Thickness: 152–300 m	Book: Motiei (1993, p. 24); Alavi-Naeini (1993, p. 105); Hamdi (1995, p. 77); Aghanabati, (2004, p. 140); Aghanabati (2008, p. 309); Alavi-Naeini (2009, p. 126) Article: Mirashahani et al. (2003). In comparison to the type section, the lithological facies show slightly changes and is very similar with Barut Fm. of Alborz. The Early Cambrian age was suggested based on the studies of Hamdi and Golshani (1983) and Hamdi (1989)

(continued)

Table 2.2 (continued)

Rock unit	Chronostratigraphic affinity	Lower/Upper boundary	Lithology and thickness at type section	Note (References, comments)
Zaigun Fm.	Early Cambrian	Lb. upper dolomite bed of Barut Fm. Ub. Zaigun Fm., the contact is apparently conformable	includes six units: calcareous shale; sandstone with siltstone intercalations, arkosic sandstone; mica bearing sandstone, sandstone with siltstone intercalations, condensed arckosic sandstone with silty sandstone intercalations Thickness: 453 m	Book: Alavi-Naeini (1993, p. 78); Hamdi (1995, p. 84); Aghanabati, (2004, p. 140); Aghanabati (2008, p. 350); Alavi-Naeini (2009, p. 104). The Lalun-Zaigun boundary is sometimes difficult to distinguish; nevertheless, the color change from the dark red to light red, and the grading change from siltstone and shale to the sandstone can be used
Lalun Fm.	Early Cambrian	Lb. Zaigun Fm., the contact is conformable Ub. Top quartzite of Mila Formation	Purple quartz sandstone subordinated with conglomerate	Book: Motiei (1993, p. 28); Alavi-Naeini (1993, p. 107); Hamdi (1995, p. 90); Aghanabati (2004, p. 140); Aghanabati (2008, p. 375); Alavi-Naeini (2009, p. 127) Thesis: Amin Rasouli (1999); Hoseini Barzi (1996) Article: Ghavidel-Syooki et al. (2008). In comparison with Lalun Fm. of Alborz no lithological changes are observed

(continued)

Table 2.2 (continued)

Rock unit	Chronostratigraphic affinity	Lower/Upper boundary	Lithology and thickness at type section	Note (References, comments)
Mila Fm.	Middle-late Cambrian	Lb. Lalun Fm. in the Lordegan. Ub. disconformable with Geirud Fm.	Setudehnia (1975) divided the Mila Fm. of Zagros into the 3 Members A, B and C: A: dolomite and subordinate shale without any characteristic fossils, 70 m thick B: red-green shale and siltstone with dolomite subordinate in the upper part, without any characteristic fossil C: a succession of thin-bedded limestone showing sandstone and dolomite beds in the upper part Thickness: 585 m	Book: Motiei (1993, p. 31); Alavi-Naeini (1993, p. 151); Hamdi (1995, p. 103); Aghanabati, 2004, p. 140; Aghanabati (2008, p. 454); Alavi-Naeini (2009, p. 162). Thesis: Amin-Rasouli (1999); Amir Sardari (1998) Article: Ghavidel-Syooki (1990a, b, 1993); Kavousi (2005); Ghavidel-Syooki et al. (2008) The Mila Fm. of Zagros does not show so much differences with the type section and all of the units of Alborz can be seen with little lithological differences

(Ruttner et al. 1968; Ghaderi et al. 2008). The Shirgesht Formation spans the Cambro-Ordovician Boundary, so that its lower part is of late Cambrian affinity.

2.2.1 Alborz

As already mentioned, the uppermost beds of Member 4 and whole Member 5 are of early Ordovician age, based on the presence of the trilobite genera *Saukia* and *Pagodio*. Presumably, lower to middle Ordovician rocks have been removed by the latter erosional phase.

Member 5 of Mila Formation at Mila-Kuh is assigned to a Tremadocian affinity (Hamdi 1995). This unit consists of:

hard, fine-grained grey quartzite bearing Tremadocian *Phycodes* sp., 47-m thick;
Green shale and green to grey sandstone with Tremadocian *Pagodia* sp., 50 m-thick;
thin-bedded grey to green shale barren in fossil, 29-m thick.

This member has been reported at various localities, for example in the Shahmirzad area. But, the thickness is always less than the type section.

Lashkarak Formation
The Lashkarak Formation was described from the succession that crops out on the eastern side of Lashkarak pass in Alamkuh area, western central Alborz (Gansser and Huber 1962). The formation consists of:

- Shale, sandstone and yellow to red thick-bedded cherty dolomite barren in fossil, 59 m-thick;
- Red nodular limestones and marls containing trilobite, cystoid and brachiopods, 25 m-thick; and
- Sandstone and shale with few limestone beds, barren of fossils, 55 m-thick.

This formation is mainly limited geographically to the small area around the type section. However, Glaus (1965) traced and mapped it from the type locality about 35 km eastward into the eastern slopes of the Chalus Valley where its thickness is much less (about 45 m).

The Lashkarak Formation can be correlated with Member 5 of the Mila Formation of Northern Iran, the Shirgesht Formation of Eastern Iran and Ilbeyk and Zardkuh of Zagros. Presumably, it is the equivalent of the Ordovician red limestone that crops out in the Talesh Area.

Gorgan Schist
The name "Gorgan Schist" was given to a complex of metamorphic rocks that crops out at the northern Alborz Mountain around Gorgan. It is mainly composed of ultramafic rock, meta-basic, chlorite schist, meta-diabase, quartzite, keratophyre and phyllite with subordinate limestone and slate.

The Neoproterozoic age of this unit was suggested by early studies (Gansser 1951, unpublished NIOC report). However, later researchers assigned the Gorgan

Schist to the Paleozoic-Triassic (Hushmandzadeh et al. 1988; Hamdi 1995, Ghorbani 1999; Rahimi 2002). However, the evidence for an Ordovician-Silurian age is more compelling than other age assignments. Ghavidel-Syooki (1995), assigned the sedimentary part of the formation to the Ordovician based on the palynological studies, an age confirmed by conodont data of Hamdi (1995).

Asalem-Shanderman and Gasht Metamorphic Complexes
Clark et al. (1975a, b) dated the metamorphic complexes of these areas. These consists of green schist and gneiss and ultra-basic rocks. Eftekharnezhad and Behroozi (1991a, b) assigned this unit to late Paleozoic, though the preliminary Neoproterozoic age assignment of Clark et al. (1975a, b). The present author suggests an early Paleozoic age, equivalent to the Gorgan Schist, based on the metallogenic and areal evidence. Furthermore, the lithological evidence suggest that the protolith of these rocks dates back to early to middle Paleozoic and that the lower parts of these complexes are comparable to the Gorgan Schist.

Ordovician rocks of the Talesh Area
The Ordovician rocks of the Talesh were studied in two areas, Masuleh to Bandaranzali and Khalkhal-Harvabad. The most complete section of the former is located at Tataroud. Here, the oldest sequence, 230–180 m-thick, is as follows:

Unit 1 Red thick-bedded-arkosic sandstone with subordinate purple shale and silty shale, barren in fossil, 100 m thick;
Unit 2 strongly deformed and altered volcanic rocks comprising green to black thick-bedded spilitic lava-flows interbedded with tuff and pyroclastic rocks, 100–140 m; and
Unit 3 fossiliferous red limestone, 30–40 m

Generally, a middle to late Ordovician age was suggested based on the fossil assemblages from Unit 3. The sandstone-shale beds overlain by the Soltan-Meydan Basalts,[3] are named "Abarsaj Formation" in the map report of the Shahroud area. The same overlying basalt beds can also be seen in the Talesh area.

2.2.2 *Kopeh-Dagh*

The only known Ordovician strata of the Kopeh-Dagh area, the Ghelli Formation, are exposed in the Bojnourd area (NIOC 2012).

This formation which attains a total thickness of 900 m is divided into 3 members as follow:

1. Lower Member (262 m-thick): 161 m agglomerate and alkaline volcanic rocks with 1 m shale and limestone intercalations, 50 m of limestone, limey shale and sandstone, 50 m green volcanic rocks and 1 m shale and fossiliferous limestone;

[3] This basalt is presumably of the Silurian affinity.

2. Middle Member (412 m-thick): 2 m siltstone, 370 m silty shale and grey to green mica-bearing sandstone and 40 m grey to green limey sandstone;
3. Upper Member (224 m-thick): silty shale, siltstone and green to grey sandstone changing upward to grey to green shale.

The fossiliferous limestone of Lower Member yields abundant brachiopods indicating an Ordovician age for the Lower unit (NIOC 2012).

This formation is comparable to the Shirgesht Formation of Central Iran.

2.2.3 *Sanandaj-Sirjan Zone*

The Ordovician of Sanandaj-Sirjan is represented by widely distributed metamorphic rocks composed of shale, sandstone with a pyroclastic protolith cropping out as erosional windows in areas such as Kuh-e sefid, Surian, Bikheir-kheng, Saman, Aligoudarz, Azna and Kabudarahang. These rocks are of Paleozoic age with their lower part belonging to the Ordovician age. Also, most parts of the Mahabad Schist, especially the unmetamorphosed parts, indicate an Ordovician age, and can be correlated with the Ilbeyk, Zardkuh, Seyahoo formations of Zagros, and the Shirgesht Formation of Central Iran.

2.2.4 *Zagros*

The Ordovician rocks of Zagros have been reported from 5 areas including Kuh-e-Faraghan (Seyahoo and upper parts of Zardkuh Formations), Kuh-e-surmeh (Seyahoo Formation), Zardkuh-e-Bakhtiari, Ilbeyk, Chalisheh, Ghali-Kuh (Seyahoo Formation) and Daregaz (Daregaz Formation), the last of which is exposed only in a small area (Ghavidel-Syooki et al. 2011a, b).

Ilbeyk Formation

Setudehnia (1972) described a succession of shale and sandstone cropping out in the Zardkuh Mountains. This formation is intruded by regional tephritic dikes in the southeast of the type section, accordingly, the thickness gradually decreases southeastward. The succession was attributed to the late Cambrian by Motiei (1993) and to the early Ordovician (Aghanabati 2004). However, it seems that the former age is more accurate. Recently, comprehensive studies carried out in the Paleozoic rocks of the Zagros area identified the *Cordylodus parvus* Zone in the limey intercalations, indicating a late Cambrian age for this formation (NIOC 2016).

In addition to the Zardkuh Mountains, this formation can be traced to the Ghali-Kuh and Chalisheh areas (NIOC 2016).

Zardkuh Formation

A succession of fine-grained shale and sandstone exclusively cropping out at the Zardkuh Mountains, was divided by Setudehnia (1972) into two informal members:

The Lower Member consists of 90 m shale and sandstone conformably overlying the Ilbeyk Formation. The main characteristic of this member is the presence of abundant trilobite.

The Upper Member includes 247 m shale and sandstone with a subordinate conglomerate in the middle part. Abundant graptolite can be found in the shaly beds of the member.

The trilobites and graptolites of the lower and upper members indicate the Ordovician age for the formation. However further studies on achritarchs, chitinozoans and scolodonts suggest late Tremadocian and Floian ages for the lower and upper members, respectively (Motiei 1993). The upper beds of the Ilbeyk Formation in the Chalisheh area must be included in the Zardkuh Formation, because of the increase in the sand contents and grading of the sandstone to siltstone.

Seyahoo Formation

This is a new name assigned to the shaley and sandy succession of Ordovician rocks of Zagros. Ghavidel-Syooki et al. (1994) studied and described the type section of Seyahoo Formation in the Faraghan Mountain, 65 km north of Bandar-abbas. It consists of 30 m of coarse-grained strongly colored sandstone with cross bedding and ripple marks in the lower part, and continues with relatively homogenous green or grey mica-bearing silty shale, fine-grained grey sandstone and brown fossiliferous limestone in the upper part. Recent studies show that lower member of the Seyahoo Formation actually correlates with lower member of Zardkuh Formation. It overlies the Zardkuh Formation and is disconformably overlain by the Zakin Formation. Abundant fossils recorded including orthceratid, trilobite, crinoid, bryozoan, palynomorphs and brachiopods. However, the age determination was mainly based on the palynomorphs indicating a late Ordovician age (Caradocian-Ashgilian) for this formation (Ghavidel-Syooki et al. 1994). Recently the clastic beds between underlying Seyahoo Formation and the overlying Sarchahan Formation were established as the "Dargaz Formation" (Ghavidel-Syooki et al. 2011a, b). However, this sequence is in fact the top part of the Seyahoo Formation which can be traced in other areas of Zagros, for example at Zardkuh and Darre-Sheikhan where it is dated as early Ordovician.

2.2.5 Central Iran

The Ordovician rocks of this area crop out in the eastern part of Central Iran such as Kalmard, Anarak and Khur, Kerman, Jam and Kashmar-Bardeskn.

Shirgesht Formation

The Shirgesht Formation of central Iran, 1160 m-thick in its type section, represents the thickest development of the Ordovician in Iran. The lower part consists of 75 m of nodular limestone with marl subordinates. This part is overlain by several tens of meters of fossiliferous red sandy limestones (Ruttner et al. 1968). The green sandstone and shale beds constitute the major parts of the Shirgesht Formation of the Central Iran.

The Shirgesht Formation, in addition to the type locality, crops out in the Derenjal Mountains and in the other areas of central Iran. It was studied and described from the Kalmard, Ozbak-kuh, Taknar as well as from Anarak and Khur in northern Zarand. The lowermost 250 m of the Shirgesht Formation is of early Cambrian age (NIOC 2016).

Shirgesht Formation of the Anarak and Kashmar Areas

The Shirgesht Formation (700 m-thick) of the Anarak area consists of fine-grained conglomerate, grey sandstone, brown limestone, siltstone and sandstone (Geological survey of Iran, Report No. 19).

In the Kashmar area, the formation represents by a 400 m-thick succession of green shale and subordinate trilobite-bearing sandstone exposed in the Bardeskan and Doruneh areas of southern Kashmar.

Ordovician rocks of the Kerman area

Huckriede et al. (1962) described a succession overlying the black *Redlichia*-bearing limestone belonging to the middle-late Cambrian (lower part) and the Ordovician (upper part). This sequence comprises of 20 m of red shale and sandstone, 2 m red sandy dolomite to dolomitic sandstone containing middle to late Ordovician brachiopod *Platystrophia* sp. and 12 m green shale. In the northern Zarand, Ordovician rocks overlie the Kuhbanan Formation. However, their facies differs from Shirgesht Formation of Anarak and Ozbak-Kuh area, in being composed of sandstones (Huckriede et al. 1962).

Ordovician rocks of the Jam area

The upper Cambrian-Ordovician rocks of the Jam area crop out in the faulted block of the Vatan Mountain. Alavi-Naeini (1972) named this succession "Vatan Shale and Limestone", consisting of 6 units; of these units 1–3 are attributed to late Cambrian and 4–6 to the Ordovician. Fossil assemblages from recrystallized black limestones and shales of Unit 4 (180 m-thick) indicate a Middle to Late Ordovician age.

Plate 2.3 illustrates selected age-diagnostic palynomorphs from Central Iran and Zagros.

Table 2.3 illustrates the main lithological features of the Ordovician rocks of Iran.

2.3 Silurian

Latest Ordovician rocks in most of Iran are of shallow marine environments because of two main processes:

1. Caledonian epeirogenic movements
2. Glaciations and marine regression

Table 2.3 Main lithological features of Ordovician rocks of Iran

Rock unit	Chronostratigraphic affinity	Lower/ Upper Boundary	Lithology and thickness at type section	Note (references, comments)
Ordovician (Alborz, Azarbaijan and Kopeh-Dagh)				
Lashkarak Formation	Early to Middle Ordovician	Lb. Lalun Formation disturbed by fault Ub. Mobarak Fm.	Consist of 3 units 1. Shale, sandstone alternating with marl and a yellow to red cherty dolomite bed in the middle part (barren in fossil) 2. nodular red limestone alternation with marl beds 3. a succession of shale, sandstone and thin-bedded limestone, barren Thickness: 175 m	Book: Alavi-Naeini (1993, p. 160); Aghanabati, (2004, p. 143); Aghanabati (2008, p. 545); Alavi-Naeini (2009, p. 167). Article: Ghavidel-Syooki et al. (2002); Ghavidel Syooki (2006); Sabouri et al. (2003)
Upper beds of Mem. IV and entire Mem. V of Mila Formation	Early Ordovician	Lb. Lalun Fm. Ub. Geirud Fm. The contact is disconformable	Quartzarenite, thin-bedded white quartzite shale, sandstone Thickness: 129 m	Book: Aghanabati (2004, p. 144); Alavi-Naeini (2009, p. 168). The lithology and age of the formation is similar with Lashkarak Formation, accordingly, National Iranian Committee of Stratigraphy recommend to use the Lashkarak Fm. instead of Mem. V of Mila Fm. of Alborz and Azarbaijan areas

(continued)

Table 2.3 (continued)

Rock unit	Chronostratigraphic affinity	Lower/ Upper Boundary	Lithology and thickness at type section	Note (references, comments)
Ghelli Fm.	Ordovician	Lb. Mila Fm. Ub. Niur Fm.	The presence of volcanic rocks lead to dividing the formation into 3 members (Afshar-Harb, 1994): 1. Lower Member (262 m): agglomerate, alkaline volcanic rocks, limestone, shale and limey shale 2. Middle Member (412 m): shale, sandstone, brown siltstone, limey siltstone and mica-bearing shale 3. Upper Member (224 m): Silty shale, sandstone, sandy silt and shale	Book: Alavi-Naeini (1993, p. 162); Afshar-Harb (1994, p. 21); Aghanabati (2004, p. 144); Aghanabati (2008, p. 585); Alavi-Naeini (2009, p. 170). Article: Ghavidel-Syooki (2006); Ghavidel-Syooki et al. (2002, 2011a, b)
Central and eastern Iran, Sanandaj-Sirjan				
Shirgesht Fm.	Late Cambrian-early Ordovician	Lb. Derenjal Fm., the contact is continuous Ub. Niur Fm.	A succession of green sandstone and shale interbedding with limestone. Thickness: 1130 m	Book: Alavi-Naeini (1993, p. 165); Aghanabati (2004, p. 145); Aghanabati, 2008, p. 522; Alavi-Naeini, 2009, p. 171. Thesis: Mehrjerdi, 1995; Bayetgoll (2009). The transition of Cambrian to Ordovician was reported at the contact of this formation with Derenjal Fm. This unit is the third formation of Mila group

(continued)

Table 2.3 (continued)

Rock unit	Chronostratigraphic affinity	Lower/ Upper Boundary	Lithology and thickness at type section	Note (references, comments)
Ordovician (Zagros)				
Ilbeyk Fm.	Late Cambrian	Lb. Mila Formation Ub. Zardkuh Fm. The both contact is conformable	Shale and Sandstone Thickness: 273 m	Book: Motiei (1993, p. 36); Alavi-Naeini, (1993, p. 179); Aghanabati (2004), p. 145; Aghanabati (2008, p. 515, Alavi-Naeini (p. 2009, p. 189) Article: Ghavidel-Syooki (1990a, b, 1993); Ghavidel-Syooki and Vecoli (2008)
Zardkuh Fm.	Ordovician	Lb. Ilbeyk Fm. the contact is continuous. Ub, Seyahoo Fm. Faraghan or younger formations with unconformable contact	Shale and fine-grained sandstone divided into two informal members as follows: Lower member consists of 90 m shale and sandstone overlying the Ilbeyk Fm. continuously characterized by trilobite Upper member with 247 m thickness consist of shale and sandstone interbedded with a conglomerate bed in the middle part	Book: Motiei (1993, p. 40); Alavi-Naeini, (1993, p. 184); Aghanabati, (2004, p. 145); Aghanabati (2008, p. 564); Alavi-Naeini, (2009, p. 176). Article: Ghavidel-Syooki (1990a, b); Ghavidel-Syooki et al. (2011a, b)
Seyahoo Fm.	Late Ordovician	Lb. Zardkuh Fm., the contact is continuous Ub. Dargaz Fm., Sarchahan Fm. Or Faraghan Fm.	Starts with 30 m colored coarse-grained sandstones containing ripple mark and cross bedding in the lower part (this sandstone is equivalent of upper part of Zardkuh Fm.) continuing with a relatively homogenous succession of olive to grey mica bearing silty shale Thickness: 700 m	Book: Aghanabati (2004, p. 146); Aghanabati (2008, p. 572) Article: Ghavidel-Syooki et al., 1994; Ghavidel et al. (2011); Tolaeie et al. (2002) In some areas such as Kuh-e-surmeh the upper contact is the Faraghan Fm.

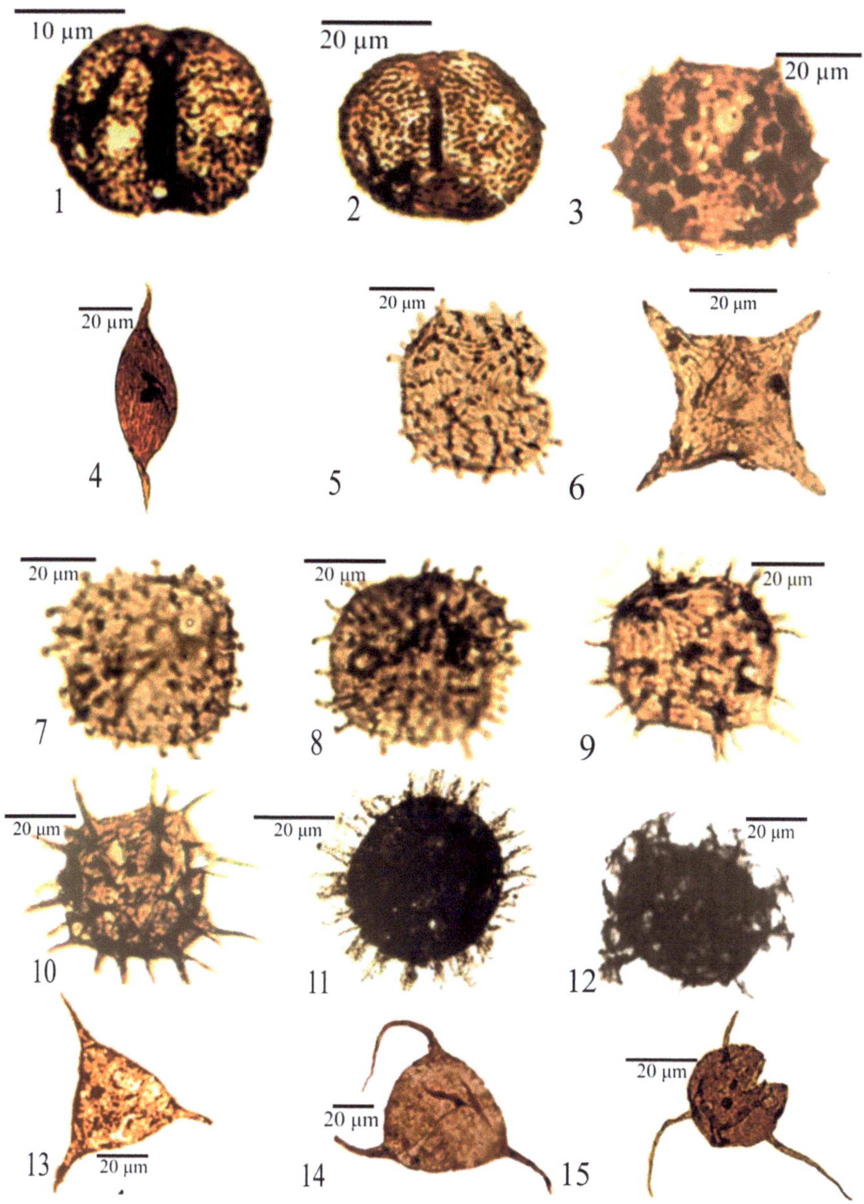
10 μm
1
20 μm
2
20 μm
3
20 μm
4
20 μm
5
20 μm
6
20 μm
7
20 μm
8
20 μm
9
20 μm
10
20 μm
11
20 μm
12
13
20 μm
20 μm
14
15
20 μm

◂**Plate 2.3** Selected Ordovician palynomorphs of the Zagros and Central Iran. 1. *Acanthodiacrodium angustum*, Zard Kuh Fm., Chalisheh section; 2. *Acanthodiacrodium angustum,* Seyahoo Fm., Darreh-Yas Section; 3. *Acanthodiacrodium* sp. cf. *A. spinum*, Zard Kuh Fm., Chalisheh section; 4. *Dactylofusa squama*, Seyahoo Fm., Darreh-Yas Section; 5. *Coryphidium bohemicum*, Zard Kuh Fm., Zard Kuh Section; 6. *Striatotheca principalis,* Zard Kuh Fm., Zard Kuh Section; 7. *Coryphidium miladae*, Zard Kuh Fm., Zard Kuh Section; 8. *Coryphidium milada*, Zard Kuh Fm., Zard Kuh Section; 9. *Coryphidium elegans*, Zard Kuh Fm., Zard Kuh Section; 10. *Coryphidium elegans*, Zard Kuh Fm., Sefidar-1 borehole; 11. *Peteinosphaeridium hystrichoreticulatum*, Zard Kuh Fm., Zard Kuh Section; 12. *Peteinosphaeridium armatum*, Zard Kuh Fm., Sefidar-1 borehole; 13. *Villosacapsula setosapellicula*, Seyahoo Fm., Faraghan Section; 14. *Veryhachium subglobosum*, Seyahoo Fm., Faraghan Section; 15. *Orthosphaeridium ternatum*, Seyahoo Fm., Zard Kuh Section

As a result, the shallow marine platforms became nonmarine with continental environments. As a consequence, Silurian strata are less widely distribution in Iran compared to those of Ordovician age.

2.3.1 Alborz

Most of the Alborz Mountains experienced continental conditions during the Silurian Period and the Ordovician Lashkarak Formation is disconformably overlain by Middle to Upper Devonian sedimentary rocks. Silurian strata in the Alborz are limited to the eastern Alborz, northern Shahroud, where they are named the Soltan-Meydan Volcanics (Jenny 1977).

Silurian of Talesh Area

As previously mentioned, the most complete lower Paleozoic succession of Talesh is exposed at the Tataroud area, 11 km west-northwest of Fuman. Here, units 1–3 are of Ordovician age, and units 4 and 5 are Silurian (Clark et al. 1975a, b).

Unit 4 of the Tataroud section of the Maouleh-Bandaranzali area consists of thick-bedded, fine-grained, hematitic limestone including white calcite veins, a large number of dextral nautiloid shells and rare specimens of gyrocone type of nautiloids.

Unit 5 of the Tataroud section, the lower part of which is Silurian in age, consists of pink to white colored, thin-bedded sandy, marly and slity limestone. Hamdi (unpublished data) reported the following conodonts in the limestone beds:

Belekella sp., *Distacodus* sp., *Pandelodus unicostatus*, *Neoprionodus* sp., *Trichonodella* sp. cf. *T. excavata.*

This association taxa suggests an early Silurian-Late Devonian age for the unit.

Derow Formation

The Silurian strata of the Khalkhal area were named the "Derow formation", as they crop out close to Derow Village. They are conformably overlain and underlain by Ordovician and Silurian rocks, respectively. Their measured thickness is 60 m. However, in some areas, this may increase to 70 m (Hamdi 1975).

This formation is composed of well-bedded red limestone with smooth to shelly fracture surface, gradually changing to crystallized red limestone. The upper part contains a 43 m-thick red limestone bearing dextral nautoiloids, brachiopods, crinoids, and conodonts suggesting a Late Silurian-Early Devonian age (Hamdi 1975).

2.3.2 *Kopeh-Dagh*

Yassini (Lapparent et al. 1973) studied the geology of Robat-e-Gharebill located at the southern margin of the Kopeh-Dagh area, along the Gorgan-Bojnourd road. These researchers described a section 1900 m thick, located 5 km north of Gharebill. The lowermost 650 m of the succession is of Silurian age and the remaining 1250 m to is assigned to the Devonian (Shemirani 2002). They described 6 horizons in this section.

There are many similarities, including thickness and lithological succession, between the type section of the Niur Formation in the Shirgesht area and the Robate-e Gharebill Formation including thickness and lithological succession. The Niur Formation is relatively well studied, but the same is not true of the Robat-e Gharebill Formation.

Soltan-Meydan Formation

During his studies on Neoproterozoic-early Paleozoic rocks of the Shahroud and Gorgan area, Jenny (1977) discovered a volcanic succession, namely the Soltan-Meydan Formation, underlain and overlain by Ordovician and Early Devonian sediments, respectively. The Soltan-Meydan Formation (700 m-thick) consists of volcanic rocks of basic tuff, basaltic and spilitic lava flow. A Silurian age was inferred based on its stratigraphic position.

2.3.3 *Zagros*

The Silurian terranes of the Zagros crop out in two area including northern Bandarabbas and southern Hajiabad as well as around the Zagros Thrust Zone (Gahkum and Faraghan Mountains). Silurian rocks have also been observed in borehole cores (such as the Darang, Kabirkuh, Asalouyeh, Aghar, Noura and Filu boreholes; NIOC 2016).

Silurian of the Gahkom Mountain

The Silurian rocks of the Gahkom Mountains start with glacially-derived conglomerate that pass upwards into black shales of the Sarchahan Formation. The Sarchahan Formation consists of 100 m of shale with subordinate conglomerate disconformably overlain by the Devonian Zakin Formation. It also crops out in the Zardkuh area.

Silurian of the Faraghan Mountains

The Silurian rocks of the Faraghan Mountains, represented by Sarchahan Formation, are similar to those that crop out in the Gahkom Mountains. The lower beds of the

Fig. 2.9 A view of Faraghan Mountain showing Sarchahan, Zakin and Faraghan formation as well as other Mesozoic rocks

Sarchahan Formation (40 m thick) consist of sandstone and conglomerate showing a gradational contact with the underlying Seyahoo Formation. Ghavidel-Syooki (2011a, b) named them as Dargaz Formation. However, there are some differences including grain size and color with Dargaz at its type section in the Gahkom. The Dargaz Formation of the Gahkom Mountains shows glacially derived characters that cannot be observed in the Faraghan Mountains. The Sarchahan Formation of the Faraghan Mountains, same as the Sarchahan Formation of Gahkom Mountains, is overlain by the Devonian Zakin Formation. This is in fact the topmost sandstones of the Seyahoo Formation.

Being the source rocks of Dalan and Kangan reservoirs of Iran, the Sarchahan and especially the Seyahoo formations are important from a hydro-carbonic point of view.

Deposition of the graptolite-bearing muds that were the precursors of the black to grey shales commenced in the Early Ordovician of Zagros and ended in the Late Silurian. There are some similarities to the Ordovician rocks of the Zardkuh Mountains but also some differences. The Seyahoo Formation of the Surmeh Mountains consists of 70 m of shale and sandstone but its lower part does not outcrop. It was only observed in the deep wells drilled in the Paleozoic rocks of Zagros, e.g. Darang, Kabirkuh.

Some geologists (e.g. Huckriede et al. 1962) reported black shale containing the Silurian graptolite *Diely mograptus* in the Asiab village area of northern Kerman. This unit can be traced with some facies variation into the Silurian of Abarkuh (Fig. 2.9).

As mentioned above, the graptolite bearing shale extends to Faraghan and the Gahkom Mountains of Northern Bandar-abas into Arabia where it latter is represented by the very thick-bedded Tabouk Formation. These rocks are also seen at Zardkuh area, although here, they are attributed to the Ordovician.

The thickness of this facies varies from a few meters in Kerman, to 100 m in Gahkom, 700 m in Faraghan Mountain and 1000 m in Saudi Arabia. Therefore, it seems that subsidence of the Zagros Basin to the south of Zagros Fault was more intense than to north. Actually depth of the basin increases southward.

2.3.4 Central Iran

Niur Formation

The most widely distributed Silurian unit exposed in Central Iran is represented by the Niur Formation. In Central Iran and also in Sanandaj-Sirjan, the Silurian sedimentary and volcano-sedimentary rocks are relatively widely distributed. In the Sanandaj-Sirjan Zone, the volcano- sedimentary rock are metamorphosed. However, in the Central Iran, especially in the eastern area, Silurian rocks are of wide extent, among of which the most important one is the "Niur Formation". The type section of the Niur Formation is situated in the mountains of northeast of Gushkamar Village in the Ozbak-kuh Range. It is composed of basal basaltic beds, thick-bedded dolomite, and brown coral limestones with minor shale intercalations conformably covered by sandstones of Padeha Formation (Ruttner et al. 1968).

The best outcrop of Niur Formation which can be regarded as a reference section is located at the Dahane-kalut, near the type section of Shirgesht Formation.

The Niur Formation in the Anarak area, Abdolhosien Mountains, consists of sandstone, dolomite and shale. In the Kashmar area, the formation is also of same lithology in the type section and the reference section but, the basaltic beds are not exposed.

The type section of the Niur Formation at Ozbak-Kuh is considered to be faulted in its lower part and is composed mainly of carbonates (Ruttner et al. 1968). Also, the lower contact of the Niur Formation, i.e. its contact with the Lower to Middle Ordovician deposits of the Shirgesht Formation, is faulted (Bruton et al. 2004; Ghobadi Pour et al. 2006). Due to these facts, the formation was not formally accepted by some researchers (Wendt et al. 2002; Hirapetian et al. 2017).

The Niur Formation is dated as lower Silurian on the basis of conodonts (NIOC 2016) and other macrofossil groups (e.g. Hirapetian et al. 2012).

The main lithostratigraphic features of the Silurian rocks of Iran are summarized below (Table 2.4; Fig. 2.10). Also, age-diagnostic fossils of this age are illustrated in Plates 2.4 and 2.5.

Table 2.4 Main lithological features of the Silurian rocks of Iran

Unit	Chronostratigra-phic affinity	Lower/ Upper boundary	Lithology and thickness at type section	Note (references, comments)
Silurian (Alborz and Azarbaijan)				
Units 4 and 5 of Tataroud section	Silurian	Lb. Units 1–3 of Tataroud Section Ub. Padeha Fm.	Overlying the basaltic bed, unit 4 starts with a trilobite-coral bearing part, continuing with some coral-brachiopod bearing beds at the middle part and terminating with sandstone beds in the upper	This book
Soltan-Meydan Fm.	Silurian	Lb. Ordovician rocks Ub. Devonian Khoshyeilagh Fm.	Marine basalts sometimes showing withdrawal evidence Thickness: 250–700 m	Book: Afshar-Hrab (1994, p. 30); Aghanabati (2004, p. 187); Aghanabati(2008, p. 617); Alavi-Naeini (2009 p. 178) Article: Ghavidel-Syooki et al. (2011a, b)
Derow Fm.	Late Silurian-Early Devonian	Lb. conformable with Ordovician rocks Ub. Devonian rocks with conformable contact	A sequence of thick-bedded red sandstone and shale bearing cephalopods and conondonts	Alavi-Naeini (1993, p. 199); Aghanabati (2004, p. 150); Alavi-Naeini (1993, 2009, p. 212)

(continued)

Table 2.4 (continued)

Unit	Chronostratigra-phic affinity	Lower/ Upper boundary	Lithology and thickness at type section	Note (references, comments)
Silurian (Central and Eastern Iran, Sanandaj-Sirjan)				
Niur Fm	Middle to late Silurian	Lb. continuous with Shirgesht Fm. Ub. Continuous with Padeha Fm.	The Niur Fm. of type section consists of dark-grey coral bearing limestone with some shale interbedding and a dolomite part. Thickness: 446 m	Book: Alavi-Naeini (1993, p. 196); Aghanabati (2004, p. 148); Aghanabati, (2008, p. 625); Alavi-Naeini (2009, p. 179) Thesis: Mehrjerdi (1995); Bakhtairi-fard (1992) Article: Ghavidel-Syooki (1994). The introduced type section of this formation is not well outcropped; accordingly, it is suggested to choose the reference section of Niur Fm. in the Shirgesht area. In Shirgesht area, the formation encompasses two basaltic layers and a succession of sandstone and dolomite beds. The basaltic parts can be traced in the all of sections, where Niur formation is exposed

(continued)

Table 2.4 (continued)

Unit	Chronostratigra-phic affinity	Lower/ Upper boundary	Lithology and thickness at type section	Note (references, comments)
Silurian (Zagros)				
Sarchahan Fm.	Early Silurian	Lb. glacially-derived conglomerate of Dargaz Fm. Ub. Unconformable with Devonian Zakin Fm.	A succession of conglomeratic sandstone, sandy limestone and limestone in the lower part; and dark grey to olive laminated shale in the upper part containing sandstone and sandy limestone intercalations	Book: Aghanabati (2004, p. 150); Aghanabati (2008, p. 606). Thesis: Baik-noori (1997) Article: Ghavidel-Syooki et al. (1994, 1995, 2004, 2011a, b); Tolaeie (2002)

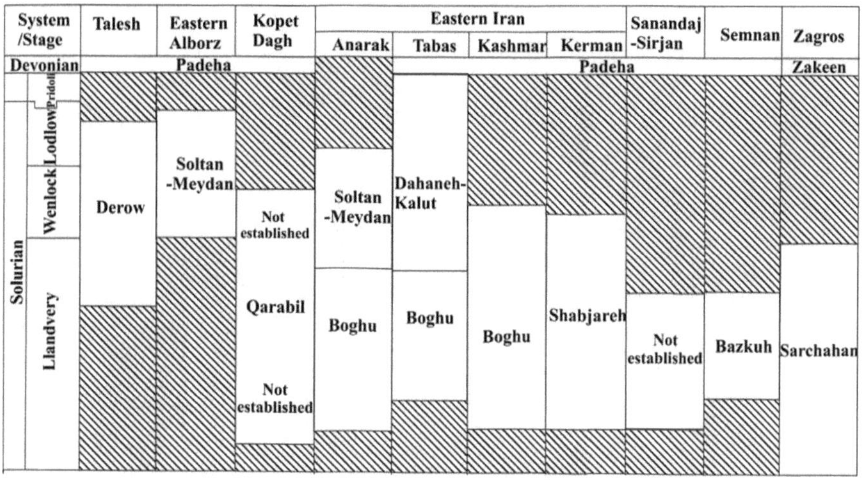

Fig. 2.10 Distribution of Silurian rocks in Iran (modified after NIOC 2016; Hirapetian et al. 2017)

2.4 Devonian

The Devonian paleogeography of Iran, in comparison to other Paleozoic systems, is very different, due to the various marine and none-marine sedimentary conditions developed. As a consequence, during the Devonian, Iran was occupied by mountains or islands surrounded by a shallow sea. This configuration was not stable during the entire Devonian, and the distribution of land and sea changed over time. The subsequent environmental changes have caused much complexity in interpretation of Devonian Stratigraphy of Iran.

This section, first introduces the Devonian lithostratigraphic units of Iran and their attributed age, and then describes the Devonian of paleogeography of Iran.

Padeha Formation

This formation was introduced in the Ozbak-kuh Mountains and was defined to include sandstone, dolomite and gypsum strata. Widely exposed in most of Central (Ruttner et al. 1968) and Eastern Iran (Alavi-Naeini 1972), Kopeh-Dagh and eastern Alborz (Bozorgnia 1973) and Kerman area (Wendt et al. 2002), the Lower Devonian Padeha Formation is composed of sandstone, dolomite and gypsum (mainly in lower and middle parts) (Ruttner unpublished).

Zakin Formation

The Devonian Zakin Formation is mainly composed of clastic sediments exposed in the Gahkom and Faraghan mountains of Zagros (Ghavidel-Syooki 1998).

Sibzar Formation

The name of this formation is proposed for the dolomite beds near the Sibzar mine in the Ozbak-kuh Mountains.

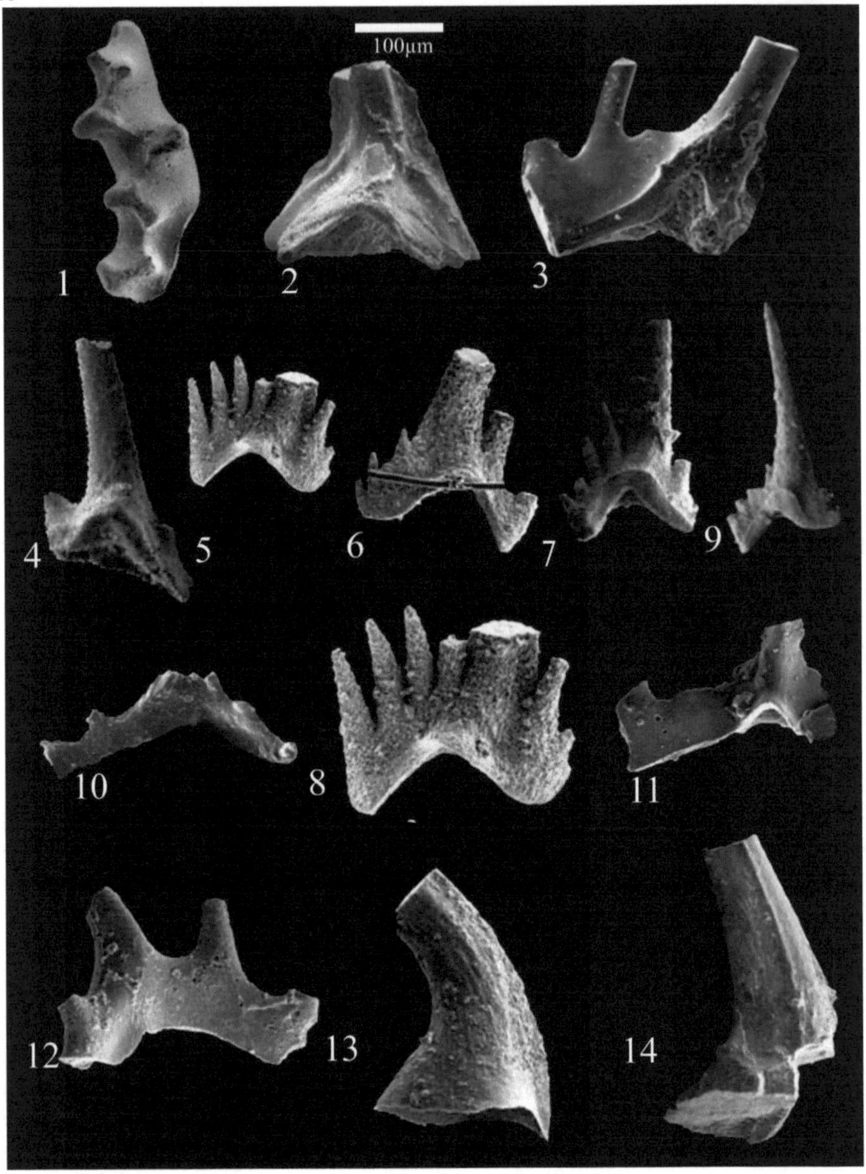

Plate 2.4 Selected diagnostic Silurian conodonts of Central Iran, Niur Formation, Shirgesht section (NIOC 2016). 1–2. *Icriodella* sp.; 3. *Lonchodina* sp.; 4. *Lonchodina* sp.; 5–6. *Lonchodina* sp.; 7–8. *Lonchodina* sp.; 9. *Lonchodina* sp.; 10. *Trichonodella* sp.; 11. *Trichonodella* sp.; 12. *Trichonodella* sp.; 13, 14. *Acanthodus* sp

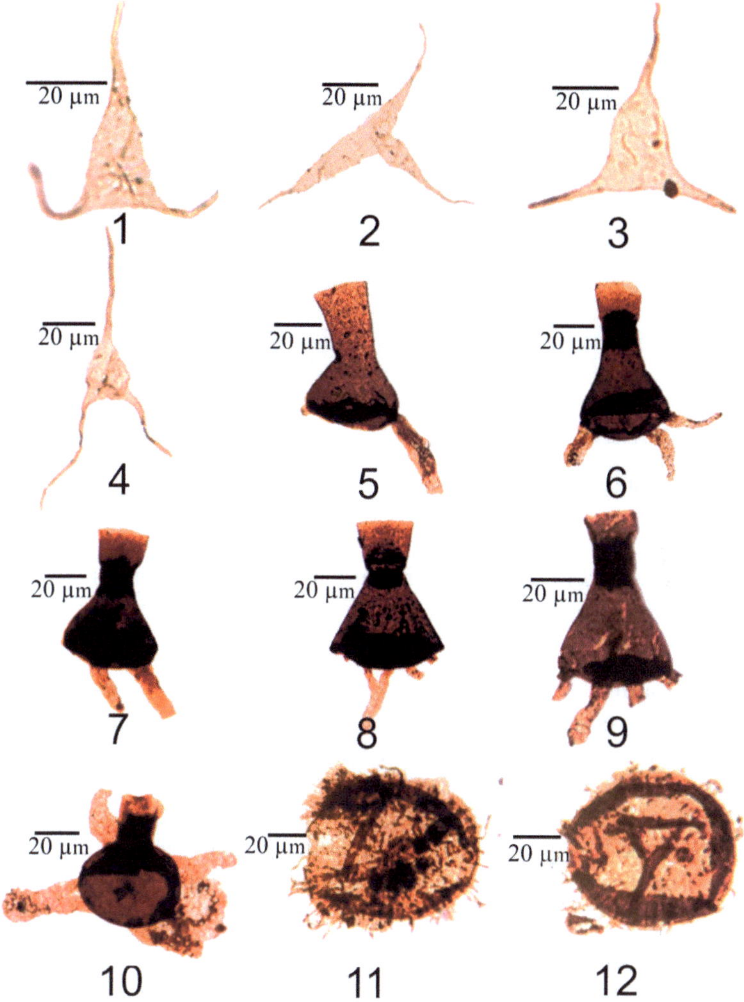

Plate 2.5 Selected diagnostic Silurian palynomorphs of Zagros area, Sarchahan Formation, Faraghan Section (NIOC 2016). 1, 2. *Domasia trispinosa*; 3. *Domasia symmetrica*; 4. *Domasia limaciforme;* 5–7. *Plectochitina paraguayensis;* 8, 9. *Plectochitina pseudoagglutinans*; 10. *Plectochitina pseudoagglutinans*; 11, 12. *Visbysphaera* sp. cf. *V. brevifurcata*

Fig. 2.11 A view of the Devonian formations of the Hoze-dorah section

The Middle to Upper Devonian Sibzar Formation (Fig. 2.11) is mainly outcropped in the east of Central Iran. However, there is no evidence of the formation in the Alborz and Zagros (Ruttner unpublished).

Bahram Formation

This formation consists of limestone and shale intercalations (Ruttner et al. 1968). Most of studies suggested Middle to Late Devonian age, based on the characteristic conodonts and brachiopods (e.g. Yazdi 2000; Bahrami et al. 2015). The formation, similar to underlying Sibzar Formation, extends mainly in the Central Iran. However, the latter is of the less distribution. Having different thickness in Central Iran, the Bahram Formation contains abundant fossil assemblage especially brachiopods and conodonts.

Shishtu Formation

Being diachronous, the Shishtu Formation was divided into two members, "Shishtu I" and "Shishtu II", of which the former was attributed to the Late Devonian and the latter to the Mississippian. It is composed of shale, marl and limestone with some sandstone intercalations (Ruttner et al. 1968). The Shishtu Formation crops out in the main areas of Central Iran such as the Shotori Mountains, Ozbak-kuh, Kashmar, Anarak, northern Kerman, Zarand and the Surian area (Bikheyr-kheng) (e.g. Wendt et al. 2002). It is generally of the same distribution as the Bahram Formation.

Khoshyeilagh Formation

The Middle to Upper Devonian Khoshyeilagh Formation is composed of a succession of basal clastic conglomerate, sandstone, shale and carbonate rocks (Bozorgnia 1973). In general, this formation consists mainly of two clastic and carbonate facies. The widest geographic extent of the formation was reported from the Alborz and Kopeh-Dagh. It can be observed in some areas of Central Iran such as Behabad, Soh in Kashan, Abadeh and Darabad, but with some lithological variations.

Geirud Formation

Lithologically resembling the Khoshyeilagh Formation, the Geirud Formation is a siliciclastics-carbonate succession characterized by the presence of some basaltic layers and a phosphatic horizon.

The Geirud Formation has been subdivided into four members A, B, C and D, of which member A, overlying the Mila Formation, belongs to the Late Devonian and the remaining 3 members were dated as Mississippian. The widest geographic extend of the formation was reported from Alborz (Assereto 1963). Member A-C of this formation are included in the Mobarak Formation and the Geirud Formation is restricted to the Unit A (Stöcklin and Setudehnia 1991).

2.4.1 *Early Devonian of East-North Central Iran*

The succession of the Padeha Formation consisting mainly of sandstone, dolomite and sometimes gypsum beds extends from north to south Central Iran including the Binaloud, Ozbak-kuh and Shirgesht areas. Moving southward, the formation tends to diminish, until in the Kerman appears once again. The Early Devonian shallow sea covered the northern Kerman in the southern area of Abar-Kuh and Anarak, Abdol-hosien Mountain. The same shallow marine conditions dominated in the northern area. Seemingly, in the Early Devonian some areas including Shotori, Naibandan, Lakarkuh Kalmard, Saghand, Poshtbadam, and Ardakan-khorangh experienced continental conditions or were occasionally submerged in the shallow sea. The Lower Devonian sea of these areas was characterized by shallow water condition and mixed carbonate-clastic sedimentation (NIOC 2016).

In all of the mentioned areas, the Padeha Formation ends with a red shale horizon, a succession of red sandstone and shale or a succession of, gypsum, shale and red sandstone bearing ripple mark and mud crack structures obviously indicating a rapid marine regression in the latest Early Devonian.

The Kereshk Group of western Kerman includes three units. Of these the upper beds of Unit 2 and lower beds of Unit 3 belong to the Early Devonian and consist of a succession of yellow and red sandstone, limestone, dolomitic limestone, grey siltstone and fossiliferous limestone.

In the western Kopeh-Dagh and eastern Alborz areas, the Devonian sea shows a transgression indicated by the presence of the Padeha Formation extending from the east of Gonbad-kavous, Robate-Gharebill to east and north of Semnan, Jam and Toroud areas (Alavi-Naeini 1972; Bozorgnia 1973). Having the same lithology as in Central Iran, the Padeha Formation of these areas proves the presence of the Devonian shallow sea of Central Iran (e.g. Wendt et al. 2002).

The exposures of the Padeha Formation in the Yazd, Kashan and northeastern Esfahan appear as a succession of dolomite and sandstone. They indicate the extension of Early Devonian sea into the Central Iran.

2.4.2 Early Devonian of Alborz

Except for some areas in the eastern Alborz mentioned earlier, the Early Devonian sedimentary rocks in central and western Alborz have not yet been reported. Shemirani (2002) assigned a Devonian age to the strata cropping out in the Almkuh and Talesh areas. However, there is no evidence to confirm this age assignment. Nevertheless, the Lower Devonian beds are missing in Alborz or are of low thickness, apparently implying dominance of continental condition in most of the Alborz Mountains.

2.4.3 Early Devonian of Northwestern Iran and Northern Azarbaijan

Muli Formation

During the Early Devonian, northern Azarbaijan was occupied by a shallow sea represented by the thick-bedded Muli Formation in the Maku area (Alavi-Naeini and Bolourchi 1973). The lower unit of Muli Formation, dated as the Early Devonian, starts with quartzite gradually passing upward into sandstone and dolomite. These are followed by dolomite with quartzite intercalations and dolomitic sandstone. The succession shows many similarities to the Padeha Formation in Central Iran, implying dominance of same sedimentary conditions. The upper unit of the Muli Formation consists completely of thick-bedded dolostone comparable with the Sibzar Formation of Central Iran (Alavi-Naeini and Bolourchi 1973).

2.4.4 Middle Devonian of Central Iran

The Early Devonian continental conditions of Central Iran continued to the beginning of the Middle Devonian (Eifelian Hiatus). However, all of these areas, due to the regional subsidence were submerged again towards the beginning of Givetian. This transgression forms the Sibzar Formation, showing different thickness, for example 10–30 m in Kashmar and Taknar; 30 m in Binaloud and Shirgesht; 100 m in Ozbak-kuh; 130 m in Jam; 100 m in Toroud; and 100 m in Anarak. This thickness variation verifies the diachroneity of the Middle Devonian transgression in Central Iran, following the Eifelian hiatus.

The Sibzar Formation of the Shotori Mountains is either missing or of a low thickness consisting of dolomitic limestone rather than dolomite. Accordingly, the Bahram Formation immediately overlies the Padeha Formation. This implies that the Middle Devonian marine transgression of the Shotori Mountains has laterally occurred in comparison to those of Central Iran (Fig. 2.11).

Fig. 2.12 A view of equivalent dolomite to the Bahram Formation as well as the Padeha, Kuhbanan and Dahu formations and Desu 'Series' at Kuhbanan Section

The Middle Devonian of Central Iran is recognized by the presence of thick-bedded fossiliferous, mainly brachiopods, limestone of the Bahram Formation cropping out in most of Central Iran, such as Ozbak-kuh, Shirgesht, Shotori Mountains, Anarak and Kashmar.

The Middle Devonian of northern Kerman is represented by the presence of the sandstone and dolomite beds, of which the latter is presumably comparable with the Sibzar Formation.

The overturned anticline of Northern Kuhbanan encompasses 250 m-thick sequence of dolomite and dolomitic limestone which is comparable with the Sibzar and Bahram formations (Fig. 2.12).

Khoshyeilagh Formation

The Khoshyeilagh Formation starts with a basal conglomerate and disconformably overlies the Padeha Formation. It is composed of the following units in the eastern Alborz (Wendt et al. 2002).

Unit 1 from base to top consist of conglomerate, dolomite accompanied by shale and sandstone intercalations, siltstone, sandstone, tuffaceous limestone, diabase, grey siltstone and thin-bedded fossiliferous limestone.

Unit 2 is composed of a succession of grey limey siltstone, fossiliferous limey sandstone, thick-bedded dolomite, dolomitic limestone and grey fossiliferous marly limestone.

The latter unit belongs to the Givetian and the former units to the Late Devonian.

The Middle Devonian rocks are not observed in western and central Alborz. However, recently some areas were reported to be Middle Devonian in age.

The sedimentary environments of the Early Devonian of northern Azarbaijan do not indicate much change in the Middle Devonian, except for the lower beds of Muli Formation, attributed to Middle Devonian. The Lower Devonian evaporate beds tends to change to carbonates, equivalent to the Sibzar Formation of Central Iran, in the Middle Devonian. Overlying the equivalent beds to the Sibzar Formation, the middle part of lower beds of Muli Formation contains dolomitic limestone and fossiliferous limestone probably comparable with the Bahram I Member in Central Iran. As a consequence, the middle Devonian depositional environments of Maku area are considered to be the same as those in Central Iran.

2.4.5 *Late Devonian of Iran*

Towards the onset of Late Devonian time, a general sea level transgression occurred in most of areas already experiencing continental condition or submerged in shallow water, developing wide mixed carbonate-clastic platform in most of the areas (e.g. Sardar Abadi et al. 2017). This transgression was accompanied by deposition of transgressive conglomerate, sandstone or basal quartzite disconformably overlying the older rocks. The evidence of widespread transgression extensively recorded in the major parts of Iran, including Alborz, Azarbaijan, Zagros and Makran. As a consequence, the upper Devonian basin was of wider extent in comparison to that of Early and Middle Devonian. However, there is no evidence to confirm the occurrence of such an event in the Zagros Zone.

2.4.6 *Central Iran*

The sedimentation of the Late Devonian sea is traceable through most of Central Iran. The Late Devonian of Central Iran is characterized by the presence of the Bahram dated to the Middle to Late Devonian (Bahrami et al. 2015; NIOC 2016). Due to the local sedimentary environment, in some localities such as Ardakan-yazd and Soh in the Kashan area, the carbonate sedimentation changed to carbonate-clastics indicated by a succession of limestone-sandstone or dolomite-sandstone. Late Devonian sedimentation in Central Iran terminated with deposition of the Shishtu 1 Member consisting of shale, marl, limestone and sandstone. However, its composition and age vary from place to place. It is now believed that this facies and age variation could be a result of horst and graben structure of the passive margin of Central Iran during late Devonian time (Bahrami et al. 2017). The Shishtu 1 Member is underlain and overlain by Bahram Formation and Mush Horizon (Late Devonian-Mississippian), respectively.

The Upper Devonian formations of Central Iran in Shotori Mountains (Hoze-dorah section), Ozbak-kuh (Nasrolah Mountain), Anarak (Abdolhosien Mountain), Behabad and Kashmar show many resemblances and record continuous Devonian-Carboniferous transition in some sections, especially in Hoze-dorah (Bahrami et al. 2011).

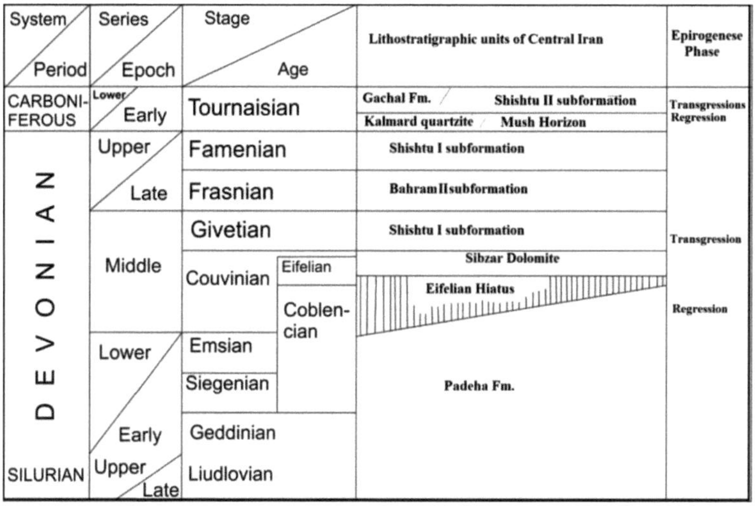

Fig. 2.13 Devonian lithostratigraphic units of Central Iran

The Devonian rocks were reported from northern Kerman in Badamu Mountain, Abmorad and Kuh-e Tizi to northern Zarand beginning with a succession of limestones and dolomites, and terminating with sandstones and shales. This succession implies the presence of same shallow water condition of eastern central Iran in the northern Kerman.

In the southern and northwestern Kerman, Rafsanjan Area, the Upper Devonian rocks are defined by Formation 5 of Kereshk Group. Unit 4 is an intraformational breccia characterizing Eifelian Hiatus of this area. Unit 5 is composed of grey thick-bedded bioclastic limestone containing Late Devonian fossils in the lower part and Late Carboniferous ones in the upper part. However, the D-C boundary was not defined due to rare fossil occurrence.

The Late Devonian rocks of the northern parts of Central Iran are characterized by thick-bedded fossiliferous limestones of the Bahram II Member. The Shishtu I Member has considerable thickness in the Binaloud, Ozbak-Kuh and Shirgesht areas of northeastern Central Iran, gradually decreases easternward, such that in the southern Damghan and Damghan-Semnan areas, it cannot be traced, and the Late Devonian rocks are only represented by the Bahram II Member which is traceable form east to west of Central Iran, Jajarm, Toroud and Jam (NIOC 2016) (Fig. 2.13).

In fact, the Devonian of Central Iran was dominated by a shallow sea, and rare continental condition only temporarily prevailed.

2.4.7 *Alborz*

The Late Devonian marine transgression affected most of the Alborz including the areas previously submerged and those that experienced continental condition. However, presumably being a horst structure, the extensive region of Azarbaijan, Zanjan, Takab and Shahin-Dezh continental condition continued the Late Devonian.

The Eastern Alborz contains the Khoshyeilagh Formation which is dated back to Middle to Late Devonian.

According to Shemirani (2002) the "Khoshyeilagh Formation" extends from Khoshyeilagh (type section) to Ghelli, Jajarm, Kalariz (northern Damghan), and the northern Attari Fault of the Jam area, i.e. Peyghambaran Mountain. Contrary to Early Devonian time, the sea in Central Iran dominated in the Eastern Alborz and western Kopet Dagh. In Late Devonian time, the Alborz basin extended into the northern part of Central Iran and Eastern Kopet Dagh. This implies that, not only were the boundaries between two or more neighboring basins unstable, but they also interfered with each other to change the disposition of the basin's coasts. The Late Devonian transgression deposited Member A of the Geirud Formation that disconformably overlies the Mila Formation or older units, suggesting the dominance of continental conditions in the area from Ordovician to Middle Devonian times.

The Middle to Late Devonian rocks of northern Azarbaijan are represented by the Muli and Ilanghare formations (Alavi-Naeini and Bolourchi 1973) which are the equivalents of the Bahram and Shishtu I Formations in Central Iran, respectively.

Three Members have been defined in the Ilanghare Formation as follows:

Member A consisting of 3 units, sandy dolomite in the lower part, a succession of shale and limestone in the middle part and thick-bedded limestone in the upper part.

Member B starts with thin-bedded, fossiliferous limestone followed by grey laminated limestone with oolitic and bioclastic limestone intercalations.

Member C includes a succession of grey shale, sandy limestone and limestone.

The Late Devonian sedimentary evidence from Alborz indicates that the sedimentary environment of Azarbaijan have many resemblances to that of Central Iran. The sedimentary gap of Early to Middle Devonian is also observed in the Central Alborz. However, this is possibly related to a subsequent erosional phase. It is worth mentioning that some parts of Central Iran and Alborz record continuous Devonian-Carboniferous deposition (e.g. Gholamalian et al. 2002; Habibi et al. 2007).

Table 2.5 illustrates the main lithological features of the Devonian rocks of Iran. Also, selected age-markers of this system in Central Iran and Zagros are shown in Plates 2.6 and 2.7.

Table 2.5 Main lithostratigraphic features of Devonian rocks of Iran

Rock unit	Chronostratigraphic affinity	Lower/Upper boundary	Lithology and thickness at type section	Note (references, comments)
Devonian (Alborz and Azarbaijan)				
Muli Fm.	Early-Middle Devonian	Lb. Metamorphic beds assigned to Neoproterozoic, however, recently Ordovician age is defined Ub. Informal Ilanghare fm. (equivalent to Shishtu I)	The formation contains two distinct units, among which the lower unit consists of red quartzite, succession of white quartzite, dolomitic-quartzitic sandstone, iron-bearing quartzite accompanied with a succession of yellow quartzite, light dolomite, and dolomitic and quartzitic sandstone Upper unit consist of dolomite with sandstone, quartzite, limestone and shale intercalations. Thickness: 1250 m	Book: Alavi-Naeini (1993, p. 246); Aghanabati (2004, p. 157); Aghanabati (2008, p. 673); Alavi-Naeini (2009, p. 204). Thesis: Asadi (1997); Jafarizadeh (1997)

(continued)

Table 2.5 (continued)

Rock unit	Chronostratigraphic affinity	Lower/Upper boundary	Lithology and thickness at type section	Note (references, comments)
Devonian (Central and eastern Iran, Sanandaj-Sirjan)				
Khoshyeilagh Fm.	Middle to Late Devonian	Lb. covered in the type section; however, Iranian geologists believe that the Lb. is Padeha Fm. with conformable and discontinuous contact, nevertheless in the type area Padeha Fm. does not outcropped and the Lb. is possibly Soltan-Meydan basalts. Ub. apparently is gradual with Mobarak Fm.	The formation is divided into three rock units: Lower Carbonate Unit: Limestone, marly, silty and clayey sandstone, bioclastic limestone, and dolomitic limestone Middle clastic unit: red to brown sandstone Upper Carbonate unit: succession of limestone, bioclastic limestone and limey shale Thickness: 1354 m	Book: Alavi-Naeini (1993, p. 220); Afshar-Harb (1994, p. 33); Aghanabati (2004, p. 155); Aghanabati (2008, p. 735); Alavi-Naeini (2009, p. 190) Thesis: Siasie-torbati (2003); Zaman (2001); Ahari-pour (1996; Amiri-Bahman, 2009). Article: Ashouri (2006); Ghavidel-Syooki (2007); Birounro et al. (2003)
Padeha Fm.	Early Devonian	Lb. sandy limestone and limey sandstone of Niur Fm. Ub. Sibzar Fm. with disconformity	It is divided into 4 units Unit 1: gypsum, marl and shale Unit 2: dolomite and sandstone Unit 3: Sandstone Unit 4: Marl and gypsum Thickness: 492 m in type section and less than 492 m in reference section	Book: Alavi-Naeini (1993, p. 251); Afshar-Harb (1994, p. 30); Aghanabati (2004, p. 152); Aghanabati (2008, p. 679); Alavi-Naeini (2009, p. 209). Thesis: Rozegar (2008); Shafei (2008); Aharipour (2010) Article: Saidei (2010); Ghavidel-Syooki (1992, 1996); Ghavidel-Syooki et al. (2007)

(continued)

Table 2.5 (continued)

Rock unit	Chronostratigraphic affinity	Lower/Upper boundary	Lithology and thickness at type section	Note (references, comments)
Sibzar Fm.	Middle Devonian	Lb. Padeha Fm. with sharp and discontinuous contact Ub. Bahram Fm.	A succession of grey to black dolomite Thickness: 100 m in type section, but changes in different areas such as Shirgesht measured 30 m	Book: Alavi-Naeini (1993, p. 262); Aghanabati (2004, p. 153); Aghanabati (2008, p. 703); Alavi-Naeini (2009, p. 214)
Bahram Fm.	Middle to late Devonian	Lb. Sibzar Fm. with continuous contact in Shirgesht area Ub. In Derenjal and Kalout the Ub. is eroded and in the Ozbak-kuh and Shotori Mountains is conformable with Shishtu Fm.	The Bahram Fm. includes limestone, shale with sandstone and shale intercalations alternating with sandy dolomite Thickness: less than 200 m in type section, 150 m in Anarak, 150 m in Kashmar, 150 m in type section of located at the Ozbak-kuh mountains	Book: Alavi-Naeini (1993, p. 265); Aghanabati (2004, p. 153); Aghanabati (2008, p. 712); Alavi-Naeini (2009, p. 217). Thesis: Rozgar (2008), Shafiei (2008) Article:Ghavidel-Syooki (1996).

(continued)

Table 2.5 (continued)

Rock unit	Chronostratigraphic affinity	Lower/Upper boundary	Lithology and thickness at type section	Note (references, comments)
Rahdar Fm.	Late Devonian	Lb. Shirgesht Fm. the contact is disconformable Ub. Gachal Fm. with disconformable contact In Bakhshi Section, the Lb. is nonconformable with pre-Ordovician granodiorits and in Kalmard area with angular conformity is Neoproterozoic rocks	The formation consists in type section of quartzite, quartz sandstone with sandy shale intercalations, sandy dolomite, dolomite, bioclastic limestone, dolomitic limestone. The lithology of reference section, Bakhshi Section, is represented by quartzite, dolomitic sandstone, limestone, dolomite with marly limestone and dolomitic sandstone Thickness: 58 m in type section and 37 m in reference section	Book: Aghanabati (2008, p. 795) Being presumably the equivalent of Padeha Fm., the formation needs to be revised
Devonian (Zagros)				
Zakin Fm.	Devonian	Lb. the Zakin Fm. overlies the Sarchahan Fm. with disconformable contact indicating Middle-Late Silurian sedimentary gap Ub. Faraghan Fm. with disconformable contact (paleosoil)	The Zakin Fm. consists of basal microconglomerate, a succession of coal bearing shale and sandstone. The main content of the formation is sandstone Thickness: 285 m	Book: Aghanabati (2004, p. 159); Aghanabati (2008, p. 661) Thesis: Najfi-naisiyani (2004) Article: Ghavidel-Syooki (1998) Zakin Fm. can be correlated with Padeha Fm. of Central Iran

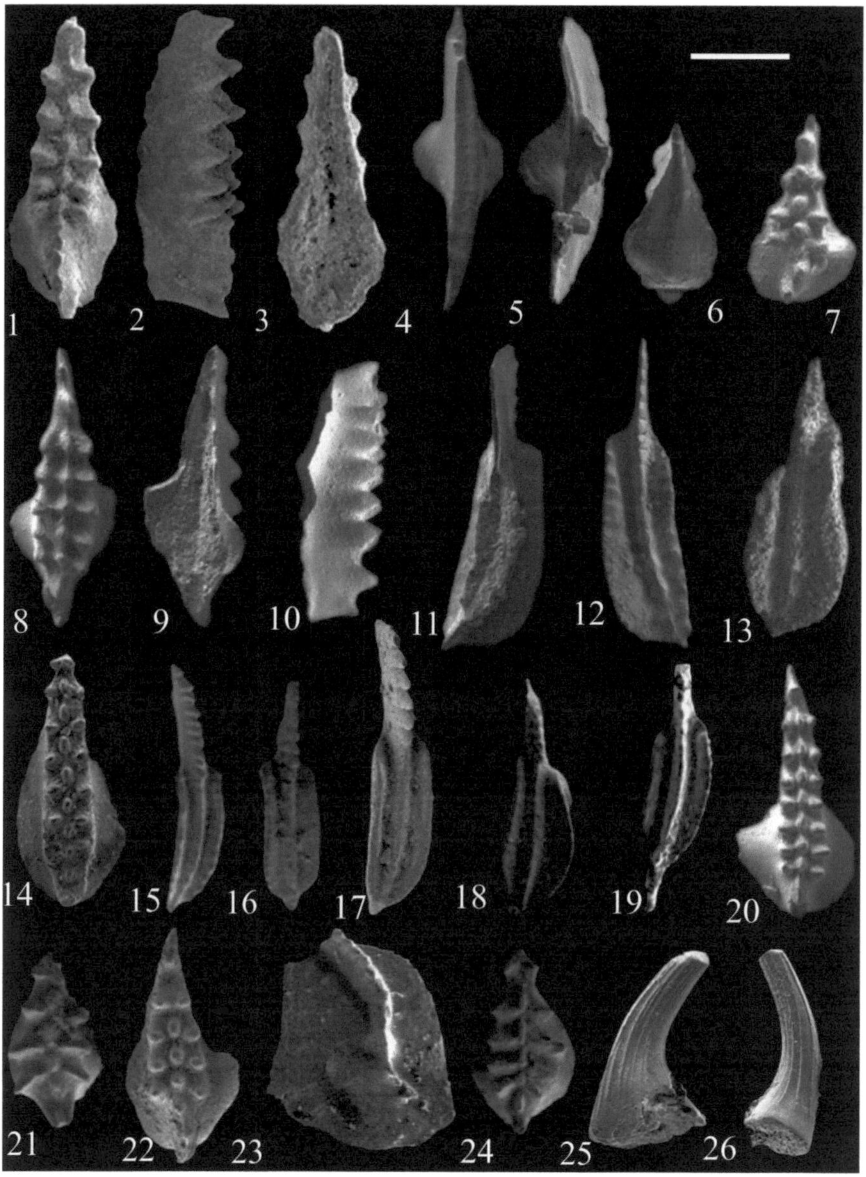
1
2
3
4
5
6
7
8
9
10
11
12
13
14
15
16
17
18
19
20
21
22
23
24
25
26

◄**Plate 2.6** Selected diagnostic Devonian conodonts of Central Iran (NIOC 2016). 1–3. *Icriodus brevis*, Bahram Fm., Behabad Section; 4, 5. *Bipennatus bipennatus*, Bahram Fm., Behabad Section; 6, 7. *Icriodus obliquimarginatus*, Bahram Fm., Soh Section; 8–10. *Icriodus expansus,* Bahram Fm., Soh Section; 11, 12. *Polygnathus linguiformis linguiformis,* Bahram Fm., Soh Section; 13. *Polygnathus xylus,* Shishtu (I) Fm., Anarak Section; 14. *Icriodus alternatus alternatus*, Shishtu Fm., Poldasht-Ilanlou section; 15. *Polygnathus praepolitus,* Shishtu (I) Fm., Anarak Section; 16. *Polygnathus aspelundi*, Shishtu (I) Fm., Anarak Section; 17. *Polygnathus politus,* Bahram Fm., Poldasht-Ilanlou Section; 18. *Polygnathus alatus,* Shishtu (I) Fm., Anarak Section; 19. *Polygnathus brevilaminus,* Shishtu Fm., Poldasht-Ilanlou section; 20. *Icriodus expansus*, Bahram Fm., Soh Section; 21. *Icriodus cornutus,* Shishtu Fm., Poldasht-Ilanlou section, 22. *Palmatolepis* sp., Shishtu Fm., Surian Section; 23. *Icriodus iowaensis*, Shishtu (I) Fm., Anarak Section; 24, 25. *Pelekysgnathus inclinatus*, Shishtu Fm., Poldasht-Ilanlou section

Plate 2.7 Selected Devonian palynomorphs of the Zagros, Zakin Formation, Faraghan Section (NIOC 2016). 1. *Rhabdosporites langii*; 2. *Chelinospora vermiculata*; 3. *Emphanisporites angulatus*; 4–5. *Emphanisporites rotates*; 6. *Retusotriletes distinctus*; 7. *Emphanisporites* sp; 8. *Grandispora riegelii*

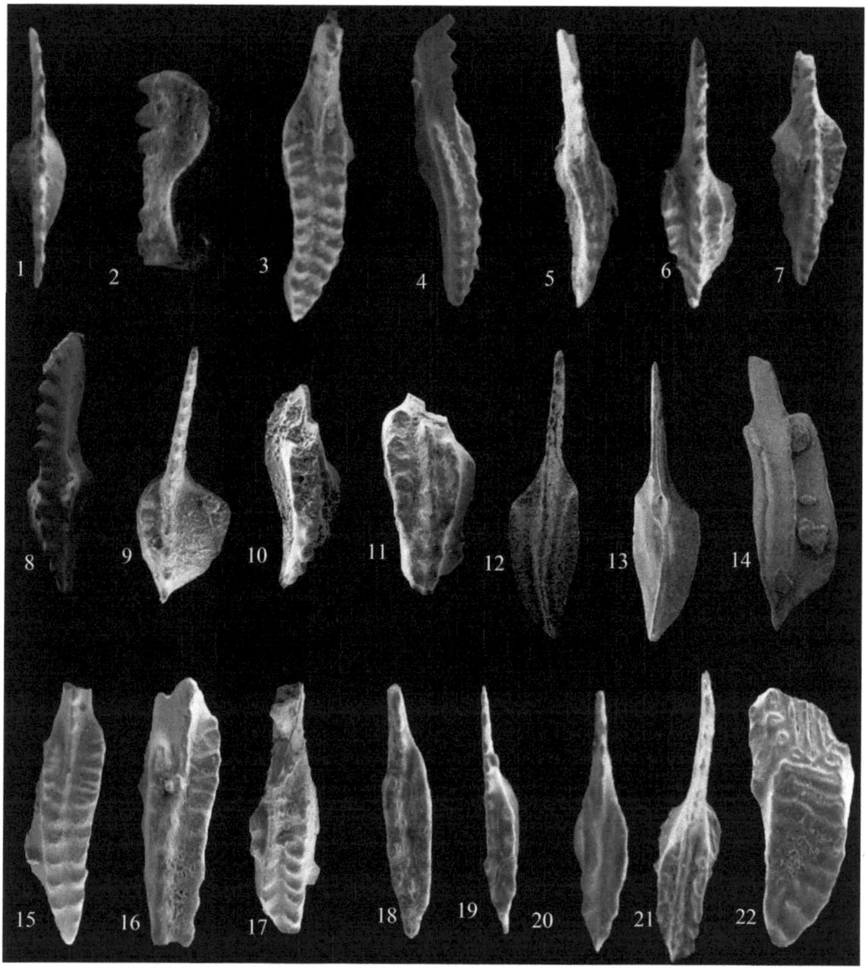

Plate 2.8 Selected diagnostic Carboniferous conodonts of the Central Iran (NIOC 2016). 1, 2. *Bispathodus stabilis*, Shishtu (II) Fm., Anarak Section; 3. *Declinognathodus noduliferus noduliferus*, Sardar Fm., Anarak Section; 4, 5. *Gnathodus girtyi simplex,* Shishtu (II) Fm., Anarak Section; 6. *Gnathodus girtyi girtyi*, Shishtu (II) Fm., Anarak Section; 7. *Gnathodus pseudosemiglaber*, Shishtu (II) Fm., Anarak Section; 8. *Gnathodus typicus*, Shishtu (II) Fm., Anarak Section; 9. *Lochriea commutate*, Shishtu (II) Fm., Anarak Section; 11, 12. *Gnathodus texsanus pseudosemiglaber*, Shishtu (II) Fm., Shotori section 15–17. *Declinognathodus noduliferus noduliferus*, Shishtu Fm., Ozbakkuh (Zaladu) Section; 18, 19. *Rhachistognathus minutus minutus*, Sardar Fm., Ozbakkuh (Zaladu) Section; 20. *Idiognathoides sulcatus*, Sardar Fm., Ozbakkuh (Zaladu) Section; 21. *Gnathodus girtyi girtyi* Sardar Fm., Ozbakkuh (Zaladu) Section; 22. *Idiognathodus* sp. cf. *I. delicatus*, Sardar Fm., Ozbakkuh (Zaladu) Section

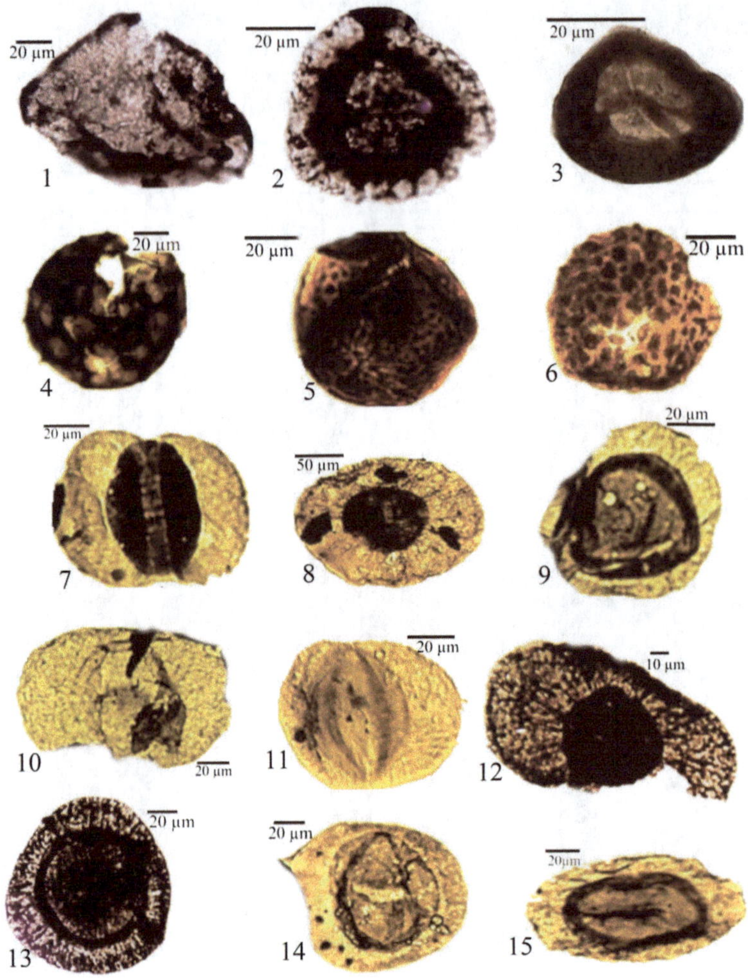

Plate 2.9 Selected diagnostic Carboniferous palynomorphs of Zagros and Central Iran (NIOC 2016). 1. *Spelaeotriletes triangulus*, equivalent of Berwath Fm., Naura-1 Borehole; 2. *Cingulizonates bialatus*, equivalent of Berwath Fm., Naura-1 Borehole; 3. *Densosporites anulatus*, Faraghan-Zakin Fm., Zirreh-1 Borehole; 3. *Cordylosporites magnidictyus*, Sardar Fm., Howz-e Dorah Section; 4. *Cordylosporites magnidictyus*, Sardar Fm., Howz-e Dorah Section; 11. *Verrucosisporites depressus*, Sardar Fm., Howz-e Dorah Section; 5. *Verrucosisporites gobbetti*, Sardar Fm., Howz-e Dorah Section; 7. *Protohaploxypinus amplus*, Sardar Fm., Howz-e Dorah Section; 8. *Florinites occultus*, Sardar Fm., Howz-e Dorah Section; 9. *Plicatipollenites trigonalis*, Sardar Fm., Howz-e Dorah Section; 10. *Potonieisporites jayantiensis*, Sardar Fm., Howz-e Dorah Section; 11. *Potonieisporites magnus*, Sardar Fm., Howz-e Dorah Section; 12. *Caheniasaccites flavatus*, Sardar Fm., Kerman Section; 13. *Plicatipollenites malabarensis*, Sardar Fm., Kerman Section; 14. *Potonieisporites novicus*, Sardar Fm., Howz-e Dorah Section; 15. *Potonieisporites* sp., Sardar Fm., Howz-e Dorah Section

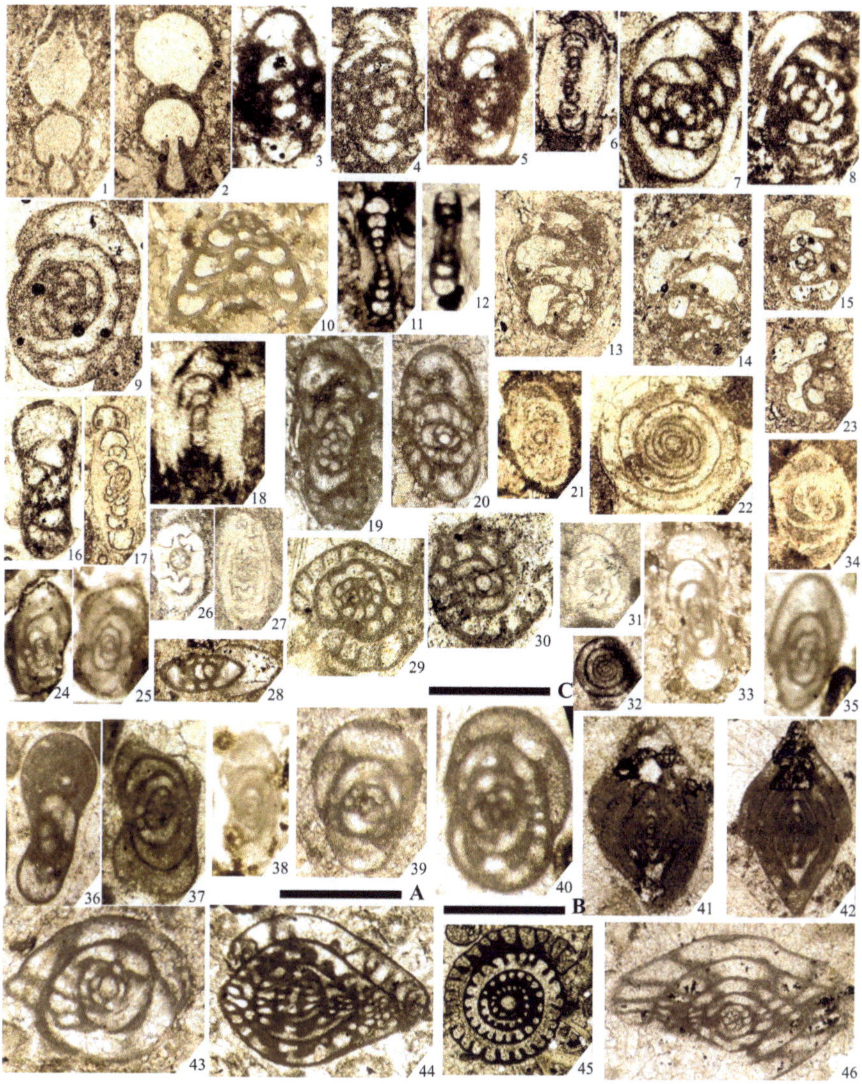

2.5 Carboniferous

The Late Devonian marine transgression in Iran continued into the Mississippian as indicated by the diachroneity of Upper Devonian rocks in Alborz and Central Iran, the Geirud Formation in Alborz and the Shishtu Formation in Central Iran. In the Latest Visean, a marine regression occurred that resulted in the emergence of the main parts of Iran in the Late Carboniferous.

◂**Plate 2.10** Selected diagnostic Carboniferous foraminifers of Alboz and Central Iran (foraminifers of Alborz are adapted from Zandkarimi et al. (2017a, b, and Central Iran from NIOC (2016)). Scale bar: A = 1mm for figs. 1"12, B = 500μm for figs. 13, 14, 15, 18 and C = 200 μm for figs. 16, 17, 19"21. 1, 2. *Darjella monilis,* Mobarak Fm., Alborz; 3, 4. *Eoparastaffella simplex* cf. *lata,* Mobarak Fm., Alborz; 6, 18. *Uralodiscus rotundus*, Mobarak Fm., Alborz; 7, 8. *Dainella* ex gr *chomatica*, Mobarak Fm., Alborz; 9. *Lysella conferta*, Mobarak Fm., Alborz; 10. *Tetrataxis* sp., Sardar Fm., Central Iran; 11. *Ammarchaediscus bozorgniae*, Sardar Fm., Central Iran; 12. *Ammarchaediscus* sp., Sardar Fm., Central Iran; 13, 14. *Eotextularia diversa*, Mobarak Fm., Alborz; 15. *Septabrunsiina kingirica*, Mobarak Fm., Alborz; 16. *Septabrunsiina krainica,* Mobarak Fm., Alborz; 17. *Paraarchaediscus koktjubensis*, Mobarak Fm., Alborz; 19, 20. *Pojarkovella* cf. *pennarroyensis*, Mobarak Fm., Alborz; 21. *Neoarchaediscus* sp., Dozdehband Fm., Alborz; 22. *Archaediscus moelleri*, Mobarak Fm., Alborz; 23. *Septabrunsiina* sp., Mobarak Fm., Alborz; 24. *Plectostaffella seslavica*, Sardar Fm., Central Iran; 25. *Eostaffella pseudostruvei angusta*, Sardar Fm., Central Iran; 26. *Brenckleina* sp., Dozdehband Fm., Alborz; 27. *Brenckleina rugosa*, Dozdehband Fm., Alborz; 28. *Millerella* ex gr. *marblensis*, Sardar Fm., Central Iran; 29, 30. *Eostaffellina protvae*, Dozdehband Fm.; Alborz; 31. *Asteroarchaediscus baschkiricus*, Dozdehband Fm., Alborz; 32. *Pseudoammodiscus* sp., Saddar Fm., Central Iran; 33. *Millerella* cf. *excavate*, Sardar Fm., Central Iran; 34. *Neoarchaediscus akchimensis,* Dozdehband Fm., Alborz; 35. *Eostaffella ovoidea*, Sardar Fm., Central Iran; 36. *Millerella* sp., Sardar Fm., Central Iran; 37. *Plectostaffella* (*Varistaffella*) *varsanofevae*, Sardar Fm., Central Iran; 38. *Eostaffella libera*?, Sardar Fm., Central Iran; 39. *Pseudostaffella* sp. cf. *P. antiqua*, Sardar Fm., Central Iran; 40. *Pseudostaffella* cf. *paracompressa*, Sardar Fm., Central Iran; 41, 42. *Ozawainella mosquensis*, Sardar Fm., Central Iran; 43. *Semistaffella* sp., Sardar Fm., Central Iran; 44. *Aljutovella* sp. aff. *A. artificialis*, Sardar Fm., Central Iran; 45. *Fusiella* sp. aff. *F. paradoxa*, Sardar Fm., Central Iran; 46. *Aljutovella subaljutovica*, Sardar Fm., Central Iran

Fig. 2.14 A view of Permo-Triassic transition of Shahreza Section

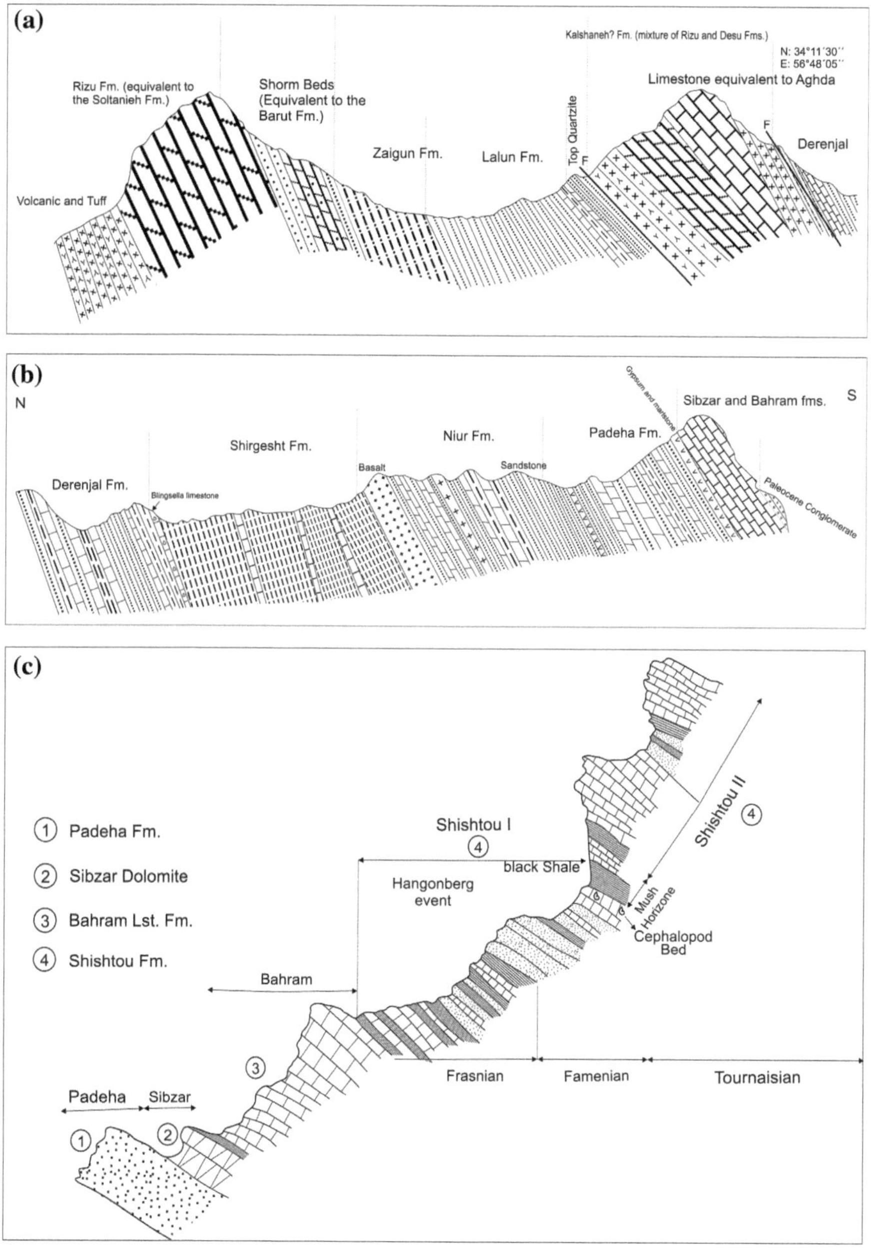

Fig. 2.15 **a** Shirgesht stratigraphic section. **b** Shirgesht stratigraphic section (Dahenh-kalut area). **c** Paleozoic formations of Hoze-doerah section, Central Iran. **d** Paleozoic formations of Hoze-dorah section, Central Iran (continued) **e** Paleozoic rocks of Sarbagh section northern Zarand **f** Paleozoic rocks of Kuhbanan section, Central Iran. **g** Paleozoic formations of Kuh-e-Tizi section, Central Iran, northern Kerman. **h** Paleozoic-Lower Mesozoic formations of Behabad section, Central Iran. **i** Lower Cambrian formations of Lalun Section, northeastern Tehran. **j** Paleozoic formations of the Geirud section, northern Tehran. **k** Paleozoic formations of Aberoun section, Rafsanjan. **l** Paleozoic formations of the Shirgesht section, Tabas Block

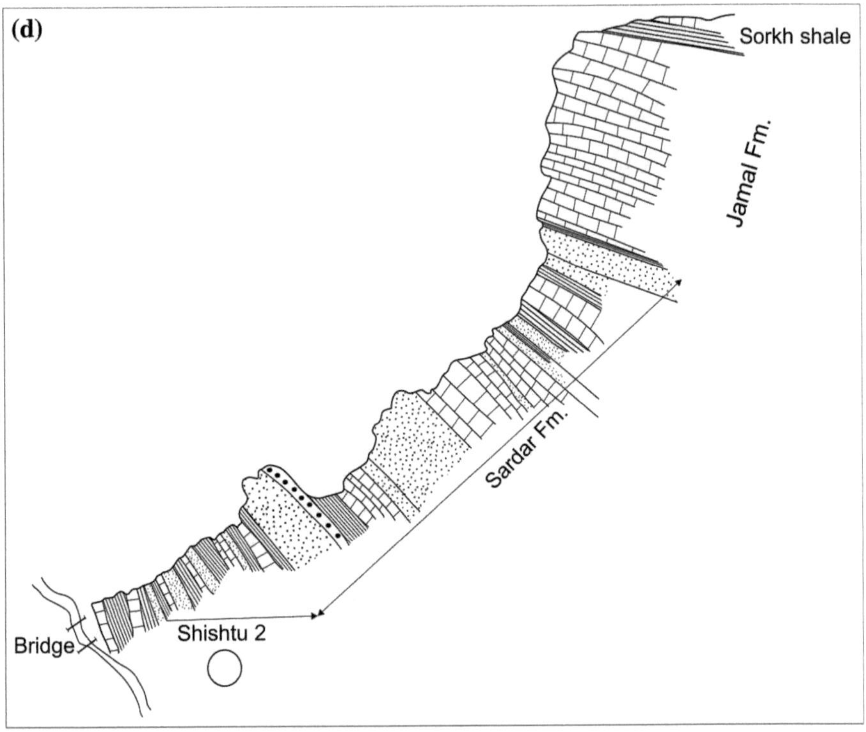

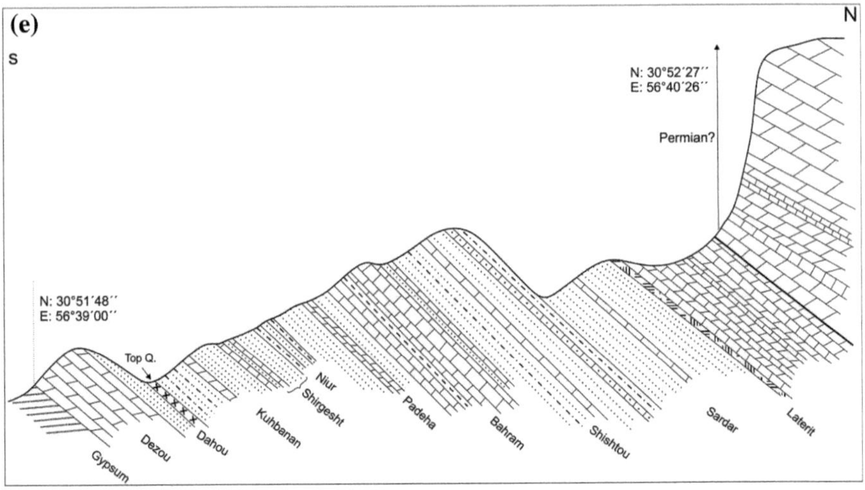

Fig. 2.15 (continued)

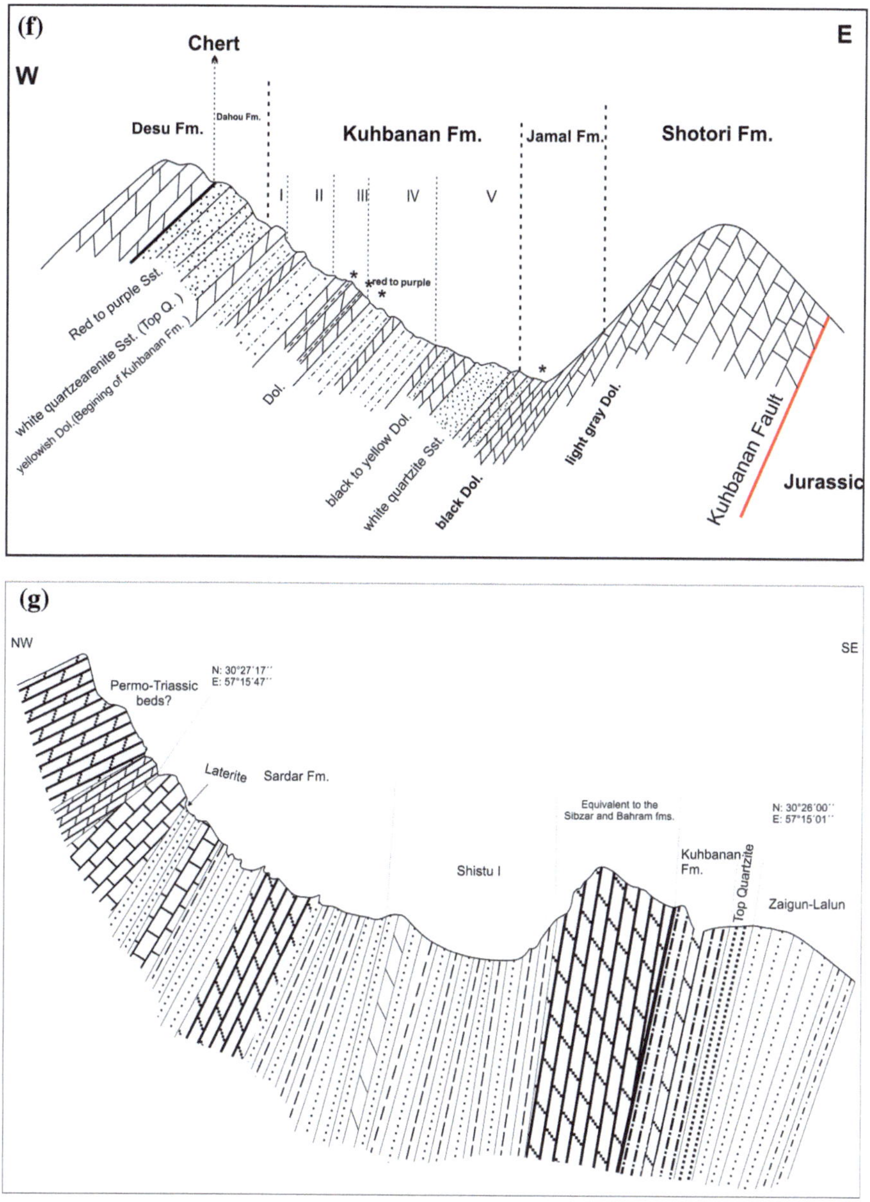

Fig. 2.15 (continued)

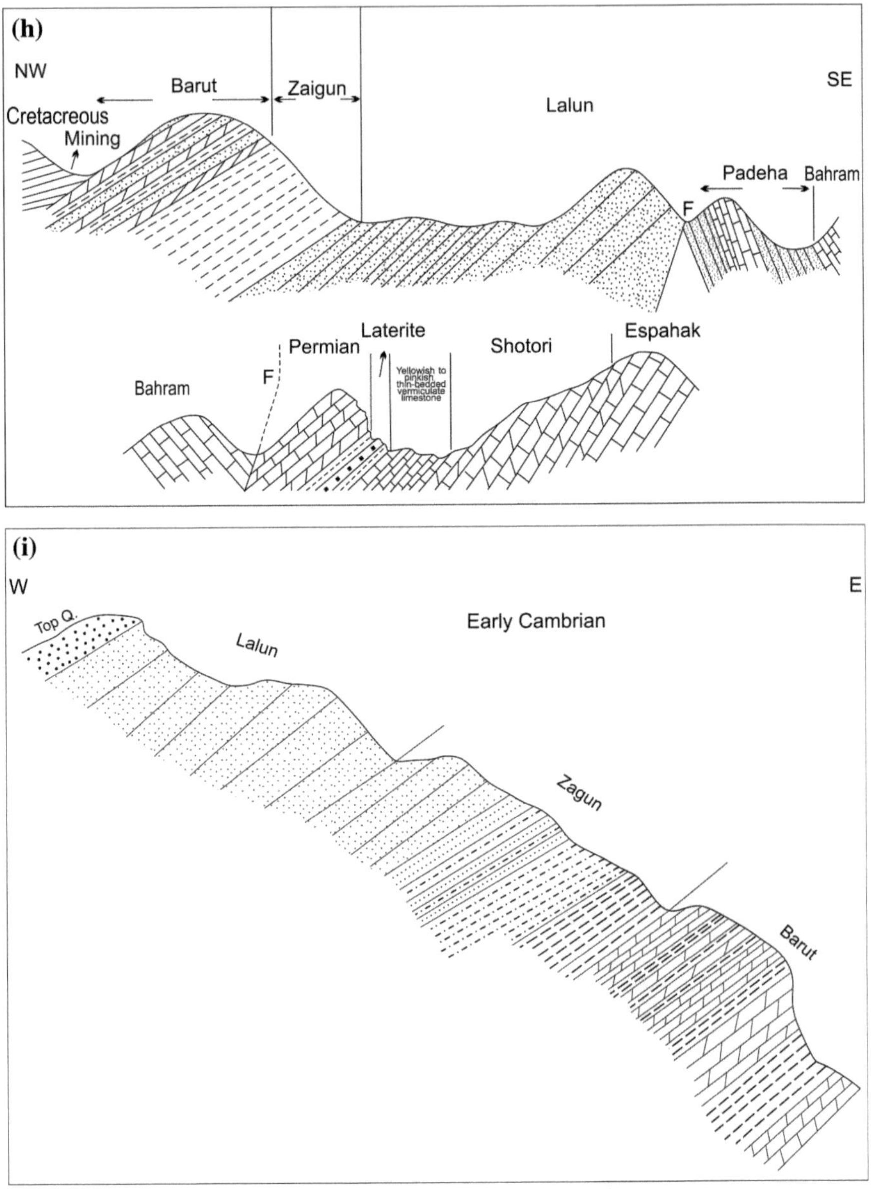

Fig. 2.15 (continued)

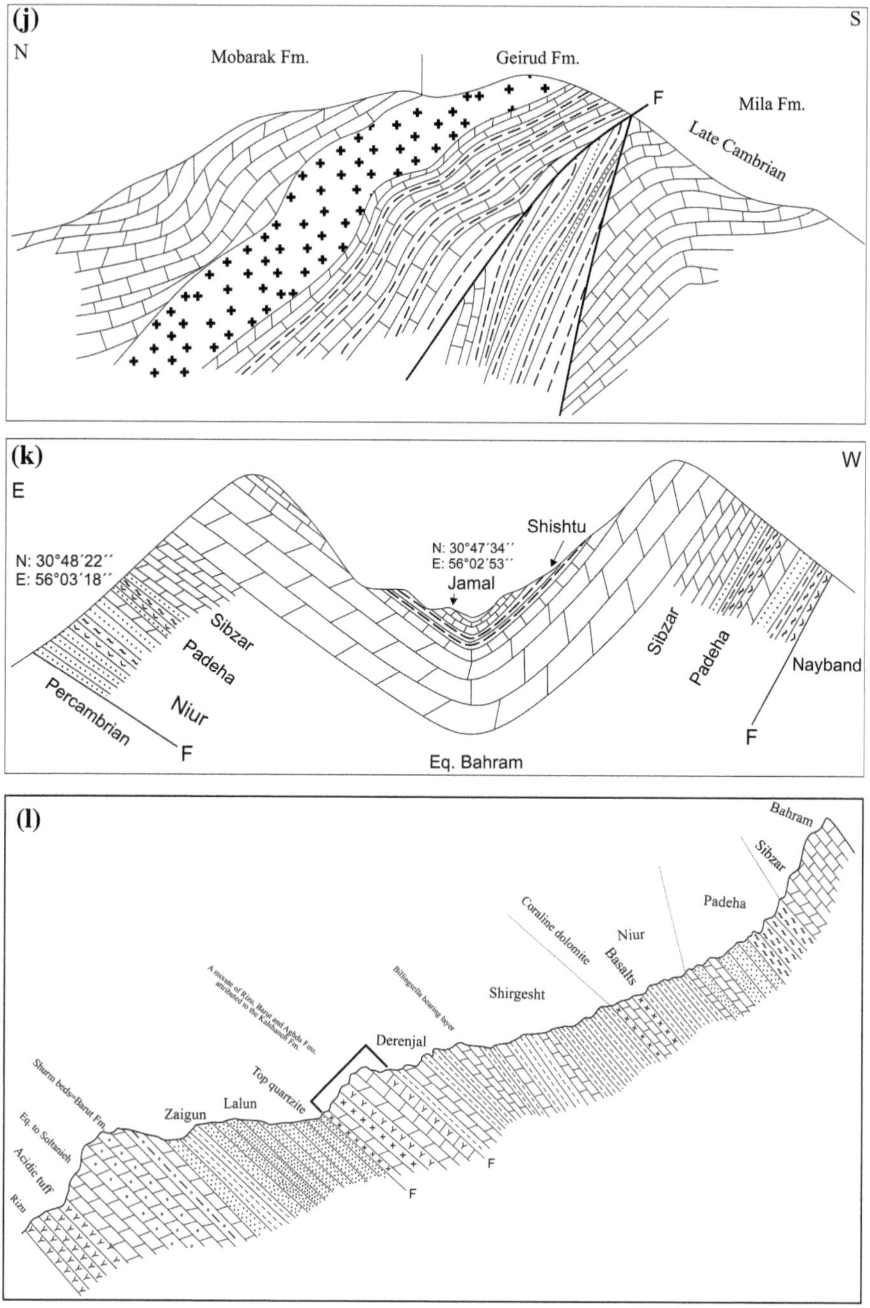

Fig. 2.15 (continued)

The following chapter describes the Carboniferous sedimentary units of Carboniferous of Iran.

2.5.1 *Alborz*

Geirud Formation
As already mentioned, this formation is divided into members A, B, C, D, of which the Member A belongs to the Late Devonian and the others to Mississippian (Assereto 1966).

Geographically, the Geirud Formation extends in the Central Alborz.

Mobarak Formation
The Mississippian Mobarak Formation is exposed in the Alborz and Central Iran, Shahreza-Asadabad (Esteghlal Mine). The formation was subdivided into three members lithologically very similar to the Members B, C and D of the Geirud Formation. The name "Mubarak Limestone" was given by Assereto (1963) to the dark fossiliferous limestone with shale intercalations in the Emamzadeh Hashem area, northern Mobarakabad. The latter studies used "Mobarak Formation" for the same unit. The spelling "Mobarak Formation" conforms to the official transliteration and was accepted by the National Iranian Commission on Stratigraphy (Bozorgnia 1973).

The Mobarak Formation of Central Iran may contain cherty layers in its lower part. The Dozdehband Formation of northern Alborz, and the Bagherabad and Ghezelghaleh formations of Eastern Alborz were considered as the equivalent of Mobarak Formation. This formation have been dated by means of various fossils as Mississippian (Bozorgnia 1973; Habibi et al. 2008; Zandkarimi et al. 2014a; Zandkarimi et al. 2017a, b, ; Plate 2.13).

2.5.2 *Central Iran*

Shishtu Formation
This formation is subdivided into two members "Shishtu I" and "Shishtu II", which the former is assigned to the Late Devonian age (Yazdi 1999, 2001; Gholamalian and Kebriaei 2008; Gholamalian et al. 2009) and the latter to the early Mississippian (Bahrami et al. 2011, 2014).

The Shishtu II Member consists of a succession of dark-grey limestone and shale and is the age equivalent of the Mobarak Formation, but with different sedimentary facies (e.g. Wendt et al. 2002).

Sardar Formation
Located in the Shotori Range, Eastern Iran, the Sardar Formation is mainly composed of clastic sediments (shale and sandstone) with carbonate intercalations in its upper

part (Ruttner et al. 1968). This formation attains a significant thickness in the east of Central Iran, where a thickness of 900 m has been recorded in the area located between Seshanghoshti and the Bagh-e vang Mountains. It can be correlated with Carboniferous strata of west Central Iran, for example in the Surian area. Of late Mississippian to early Pennsylvanian age, the Sardar Formation varies greatly in its thicknesses and lithology from place to place, with the upper part dominated by limestone in the Ozbak-kuh, Anarak and Kerman (NIOC 2016).

Leven and Gorgij (2005) established the Ghaleah Formation in the Ozbakkuh Mountains as corresponding to the Sardar 1 Subformation in the former nomenclature of Ruttner and Stöcklin (1966) and the Absheni Formation corresponding to Sardar 2. Consequently, the Ghaleah Formation consists of grainstone with shale interlayers and basal quartz and Absehni Formation of shales and sandstones with thin interlayers of diverse limestones and dolomites.

Gachal Formation

The Mississippian depositional environments of the Kalmard area, eastern Iran, significantly changed with deposition the Gachal Formation represented by evaporitic and calcareous and evaporate facies creating the condition, in which the Gachal Formation has been deposited (Aghanabati 1975). It is composed of carbonate sediments, shallow water limestone and dolomite, and gypsum with considerable thickness. Gypsum layers imply the onset of the continental condition. In comparison to other areas in Iran, especially the Tabas area, Shotori Mountain and Shirgesht areas, the presence of the lithologically different formations of the Kalmard Region is presumably related to the sedimentation on the Kalmard Horst which was created by the movement of the Kalmard Fault in the Neoproterozoic time. This formation has been recently dated to be Tournaisian in age (Vachard and Arefifard 2015).

As already mentioned, in the Late Carboniferous, according to many authors (e.g. Berberian and King 1981), a Pennsylvanian epeirogenic phase affected the Alborz, Central Iran, Sanandaj-Sirjan and other parts of Iran. However, the nature of this event has not been adequately explained and, contrary to the views of some authors, the absence of Pennsylvanian rocks may be simply explained by the absence of comprehensive sedimentological and paleontological studies. Furthermore, the early to late Carboniferous Sardar Formation is traceable throughout most of Central Iran, and is accompanied by major facies variations in its upper part.

Following the deposition of the Mush Horizon, carbonate facies of the Shishtu II and clastic-carbonate facies of the Sardar I Member were deposited in the Pennsylvanian transgressive sea.

In the Ozbakkuh Mountains, the Shishtu II Member consists of fossiliferous limestone and of pebble bearing limestone accompanied by marl intercalations. The Carboniferous-Devonian strata are separated by a cephalopod-bearing calcareous horizon (Stöcklin et al. 1965), namely Goniatitic Horizon II. This member has been dated as the Tournaisian to early Viséan based on the conodonts and palynomorphs (Bahrami et al. 2011, 2014; Arianasab et al. 2015).

In the Shirgesht area, the Shishtu II Member and the Mush Horizon are overlain by crinoid-bearing limestone and grey to light oolitic limestone, with the latter containing bryozoans, productids brachiopods and corals. The conodont and foraminiferal assemblages suggest a Tournaisian-Early Visean age for the Shishtu II Member (NIOC 2016; Plates 2.7, 2.8 and 2.9). This is restricted to the southern part of Shirgesht area, and is not exposed in the Derenjal Mountains. The Sardar Formation is composed of a succession of dark-green shale and brown sandstone. In some localities, a quartzite unit is observed which may be accompanied by grey to brownish sandy limestone containing bryozoans, brachiopods and cephalopods indicating a Late Visean-Early Serpukhovian age. This formation can only be observed on the western side of the Shirgesht area, between Sheshanghoshti and the Bagh-e vang Mountains. The considerable thickness and great facies variation suggest that deposition of this formation may have ended in the Late Namurian. However, in the Lakarkuh area of southern Naibandan, the Mississippian is represented by the Sardar Formation with the absence of the Shishtu II Member presumably confirming the presence of a Mississippian to middle Visean sedimentary gap in the area (NIOC 2016).

Deposition in the Mississippian Sea of the southern and southwestern Kerman yields different facies. Deposition of Kereshk Group of the Rafsanjan area still continued to the Mississippian. The Unit 5 of the Kereshk Group informally named "upper carbonate unit" consisting of thick-bedded bioclastic limestone containing brachiopods, bryozoans, ostracods, algae and crinoids (NIOC 2016).

In central and west Central Iran, the Mississippian rocks crop out in the northeastern Esfahan and Northeastern Shahreza, Asadabad area (Boncheva et al. 2007).

In the east of Chah-reiseh Village, Kaftar Mountain, a relatively thick sequence of Late Paleozoic rocks consists of 4 members, of which members 1–2 and the lower part of member 3 belong to the Frasnian-Famennian, Late Devonian, whereas the upper part of Member 3 and the whole of Member 4 are of Mississippian age (Shemirani 2002).

The upper part of Member 3 is composed of 100–110 m of sandstone accompanied by fossiliferous marly limestone, limey sandstone and fossiliferous limestone. All fossil assemblages indicate a Mississippian age (NIOC 2016).

Member 4 of this sequence consists of yellow, cliff-forming, massive limestone and then black-grey dolomite with some sandy marl intercalations in which a few specimens of brachiopods and gastropods have been found.

In the Carboniferous part, the basal sandstone, indicates a shallow marine transgression in the area. This depositional environment, persisted unchanged till the end of Mississippian. It was then followed by a regressive phase occurred in the middle Visean, exposing the area through the Early Permian (Late Carboniferous Hiatus).

In the west of Central Iran, at Surmaq-Dehbid-Abadeh, Mississippian rocks are exposed in several localities. These rocks can be divided into two lower and upper parts.

Overlying the upper Devonian quartzarenite, the lower unit is composed of hard recrystallized black limestone containing corals and crinoids. The fossil assemblages indicate a Mississippian age. It is similar to limestone of Mobarak Formation of Central Alborz.

The upper unit (132 m-thick) consists of grey quartzitic and limey sandstone with fossiliferous dolomite limestone, coquina beds, shale and fossiliferous limestone beds. The sandstone beds contain abundant cross bedding and indeterminate gastropods. This unit attributed to the Mississippian, is comparable to the Sardar I Member of eastern Central Iran. It is disconformably overlain by Permian limestones.

The Mississippian rocks of the Kalariz area, northern Damghan, are represented by a succession of carbonate and clastic beds lithologically resembling those the Khoshyeilagh area but thicker. In this area, the Visean marine regression of Alborz began in the Early Visean, while in the Central Alborz it occurred middle Visean in age (Zandkarimi et al. 2014a). As a consequence, the youngest rocks of the Mobarak Formation ends with the upper Tournaisian beds (Brenckle et al. 2009).

Located to the north of the Anjilou Fault, in the Torud Area, the Peyghambaran Mountain contains a Carboniferous succession overlying the Upper Devonian rocks. This consists of a sequence of limestone, laminated marly limestone, black calcareous-clayey shale and quartzitic sandstone. Foraminifers indicate middle to late Tournaisian age. The Mobarak Formation in the area is disconformably overlain by lower Permian sediments. As similar succession crops out in the Jam area, north of Attari Fault.

The Mobarak Formation of Shahmirzad Section, Semnan, is of Tournaisian age and begins with some quartzite and regressive sandstone beds overlying Famennian Limestone, Member A of Geirud Formation. It indicates a marine transgression of the base of the Tournaisian. The succession starts with sandstone and limey sandstone, and fossiliferous limestone in the middle part terminating with alternation of shale and limestone. Finally, the late Visean regression leads to sedimentary cessation and exposure of the area.

Mississippian sediments are widely distributed in the Alborz Range, especially, in the eastern Alborz, Khoshyeilagh area. They are represented by the Member B and C of Geirud Formation, equivalent of Mobarak Formation, and are composed of limestone beds, marly limestones and marls with marly limestone, overlying the Middle to Upper Devonian Khoshyeilagh Formation. In the Middle Visean, the area was exposed by a regional marine regression phase whose consequences were already mentioned in the description of Central Iran. This phase continued through the Early Permian (e.g. Bozorgnia 1973).

The Mississippian outcrops of Alborz were first observed in the Mobarakabad section, eastern Tehran, the type section of Mobarak Formation (Assereto 1963). These are composed of black fossiliferous limestone with black basal marl intercalations and are disconformably underlain and overlain by Member A of Geirud Formation and Permian units, respectively (Assereto 1963).

2.5.3 *Western Alborz*

Mobarak Formation
Members B and C of the Geirud Formation crop out in the Gaduk, Nesen and Dozdehband areas in northern Alborz; and the Mobarakabad, Abnak, Ruteh and Geirud areas of southern Alborz (Bozorgnia 1973).

In the western Alborz, at Alamkuh and Ghazvin to Rasht, the early Carboniferous is represented by thick limestone of the Mobarak Formation having the same lithological and paleontological characteristics as in central Alborz.

2.5.4 *Azarbaijan*

The Upper Devonian to Mississippian succession of Northern Azarbaijan is represented by thick units of the informally-defined Ilanghare Formation of the Maku area (Poldasht, Ilanqarreh and Ilanlou villages). It consists of 4 members including Member A (Middle Frasnian), Member B (Late Frasnian), Member C (Famennian) and Member D (early Carboniferous). The last named member is 113 m in thick and consists of following units (Alavi-Naeini and Bolourchi 1973):

Lower unit beginning with well-bedded grey limestone with shale and sandy limestone intercalations; and terminating with brittle grey shale, sandstone and brachiopod-bearing limestone.

Middle unit consists of 3 m-thick white basal quartzite and 2 m phosphate-bearing fossiliferous sandy limestone.

Upper unit composed of micaceous silty shale overlain by 2 m fossiliferous sandy limestone. All of the paleontological evidence suggests an Early Carboniferous age.

The lower unit of the Formation can be correlated with the Shishtu II Member and the Middle and Upper units to the Sardar I Member.

2.5.5 *Zagros*

There are no published descriptions of the Carboniferous in, though, Setudehnia (1972) reported the presence of some Carboniferous rocks in the Lorestan area. However, later studies by NIOC revealed that these rocks are actually of Permian age. According to the present author, due to extensive exposures of Carboniferous rocks in other parts of Iran including Central Iran, Alborz, Azarbaijan and Sanandaj-Sirjan, and also in Arabia, Carboniferous rocks almost certainly occur in Zagros. This is confirmed by new discoveries of early Carboniferous palynomorphs in boreholes including West Aghar-1, Naura-1 and Zeireh-1, in the Zagros area (Sabouri et al. 2014).

Late Carboniferous
In the Late Carboniferous, a positive epeirogenic phase affected some parts of Iran leading to emergence and the establishment of continental conditions. However, a late Carboniferous sea still occupied small parts of eastern Central Iran including the Shehshanghoshti-Baghvang, Binaloud, Ozbak-kuh, Anarak areas (Leven and Gorgij 2011a); some parts of Alborz, Shirinabad (Gaetani et al. 2009); and Asadabad (Bahrami et al. 2017) and Surian in Sanandaj-Sirjan (Boncheva et al. 2007), represented by Member 2 of the Sardar Formation (equivalent to the Absehni Formation) and the Zaladou Formation (Leven and Gorgij 2006a, 2011a).

According to present author, the late Carboniferous and even early Carboniferous stratigraphic gaps in many parts of Iran occurred due to the epeirogenic phase and subsequent exposure of the region that resulted in the erosion of older rocks.

Selected, stratigraphically important Carboniferous microfossils are illustrated in Plates 2.8, 2.9 and 2.10.

Table 2.6 summarizes the main lithostratigraphic features of Carboniferous rocks in Iran.

2.6 Permian

This chapter summarizes the Permian stratigraphy of Iran which is characterized throughout the country, by an initial, diachronous marine transgression.

2.6.1 Central Iran

The Permian ocean deposited the Jamal Formation in Central Iran. The Permian of east of Central Iran, Shotori Mountain, in the Kaleh-sardar and Shirgesht areas, begins with Baghe-vang Formation which overlies the Sardar Formation. The lithostratigraphic units of eastern Central Iran, from the north to south are as follows:

The Jamal Formation of Ozbak-kuh and Shirgesht has been divided into two members. The lower member is composed of grey to black limestone with a few pink to red intercalations bearing cherty bands and pebbles, as well as fusulinids, corals, bryozoan and brachiopods. This member was referred by some geologists to the Baghe-vang Formation. The Upper Member includes black cherty limestone, changing upward to grey limestone and dolomitic limestone.

In the Shotori Mountains and Boshraveyeh area, the Jamal Formation, which rests disconformably on the Sardar Formation, begins with brown shale and limey sandstone beds, while lower Permian beds similar to those in the Baghe-vang area crop out at Kal-e-Sardar.

In southwestern and southern Kerman, on the Sirjan-Bandarabas road and Khajou Mountain, Lower Permian rocks include turbidite deposits consisting of microconglomerate and sandstone with graded bedding structure interbedded with grey to

Table 2.6 Main lithostratigraphic features of Carboniferous rocks of Iran

Rock unit	Chronostratigraphic affinity	Lower/Upper boundary	Lithology and thickness at type section	Note (references, comments)
Carboniferous (Alborz and Azarbaijan)				
Geirud Fm.	Late Devonian-Mississippian	Lb. Mila Fm. with parallel unconformity contact Ub. Dorud Fm. with disconformable contact.	In the type section, the formation is composed of 4 members, viz: Member A, Member B, Member C and Member D. in Subsequent revision on the Paleozoic succession of Alborz, Iranian Commission of Stratigraphy abandoned the name Geirud as imprecise and preferred the name Mobarak for the Mississippian beds including Members B, C and D Thickness: 230 m	Book: Aghanabati (2004, p. 156); Aghanabati (2008, p. 771) Article: Ghavidel-Syooki (1994); Kebriaei-zadeh et al. (2003); Bagheri et al. (2003) Research project: Lesemi (1993) The measured thickness of formation has been frequently changed in the context of studies. Accordingly, primary study measured 760 m in thickness (NICS 1964), but changed after subsequent revision limited the Geirud Fm. to Member A (NICS 1966) to 335 m

(continued)

Table 2.6 (continued)

Rock unit	Chronostratigraphic affinity	Lower/Upper boundary	Lithology and thickness at type section	Note (references, comments)
Chaikasan Fm.	Carboniferous	Lb. Devonian rocks Ub. Lower Permian Rocks	The formation is divided into three units Lower unit: igneous rocks, conglomerate, sandstone, siltstone and pyroclastic rocks Middle unit: Siltstone, pyroclastic conglomerate Upper unit: Igneous rocks, pyroclastic conglomerate Thickness: 425 m	Book: Aghanabati (2008), p. 812
Mobarak Fm.	Mississippian	Lb. Mila Fm. concordant contact but with major sedimentary gap Ub. Marl beds of Dorud Fm.	The formation is divided into 4 different lithostratigraphic units as follows: Unit 1: marly limestone and black marl Unit 2: thick-bedded limestone Unit 3: hard and massive limestone Unit 4: Limestone and sandstone Thickness: 450 m	Book: Alavi-Naeini (1993, p. 292); Afshar-Harb, (1994, p. 40); Aghanabati (2004, p. 163); Aghanabati (2008, p. 850); Alavi-Naeini (2009, p. 226) Article: Ghavidel-Syooki et al. (2011a, b); Sahabi et al. (2003); Lasemi et al. 2005; Zandkarimi (2014a, 2017a, b). Thesis: Amiri bahman-beiglu (2009); Kazeruni (2009), Mossadegh (2000); Najafi (2001); Mohammadi (1993); Fahim (2009); (2006)

(continued)

Table 2.6 (continued)

Rock unit	Chronostratigraphic affinity	Lower/Upper boundary	Lithology and thickness at type section	Note (references, comments)
Bagherabad Fm.	Late Carboniferous	Lb. Mobarak Fm. Ub. In the type section is covered, however, in some area Ub. is Member 2 of Qezelghaleah Fm.	The formation is divided into 4 Members as follows: Member 1: micritic, oolitic and sparry limestone Member 2: sandstone, limestone, sandy limestone and quartz limey sandstone Member 3: micritic and oolitic limestone Member 4: alternation of sandy limestone and siltstone, micritic limestone, micritic limestone. quartz limestone Thickness: 260 m	Book: Alavi-Naeini (1993, p. 304); Aghanabati (2004, p. 164); Aghanabati (2008, p. 878); Alavi-Naeini (2009, p. 230)
Khashachal Fm.	Late Carboniferous	Lb. Mobarak Fm. with paraconformity Ub. Dorud Fm. with Paraconformity thickness: 426 m	a succession of conglomerate, dolomite, quartzite, shale and sandy limestone	Book: Aghanabati (2008, p. 947)

(continued)

Table 2.6 (continued)

Rock unit	Chronostratigraphic affinity	Lower/Upper boundary	Lithology and thickness at type section	Note (references, comments)
Qezel-ghaleah Fm.	Late Carboniferous	Lb. Mobarak Fm. Ub. Dorud Fm.	A succession of limey shale (lower part), sandy limestone (middle part) and shale and sandstone (upper part) Thickness: 224 m	Book: Alavi-Naeini (1993, p. 306); Aghanabati (2004, p. 164); Aghanabati (2008, p. 952); Alavi-Naeini (2009, p. 231)
Carboniferous (Central and Eastern Iran, Sanandaj-Sirjan)				
Kaseh Fm.	Mississippian	Lb. equivalent of Geirud Fm. with continuous and gradual contact Ub. A quartzitic bed whose overlying formation was assigned to Permian or Carboniferous Both contact are conformable	The described features of Kaseh Fm. of the Kaseh Mountain are marl and shale, alternation of sandstone and dolomite, alternation of dolomite and limestone, alternation of sandstone and marl; and rhyolite and quartz sandstone. Thickness: 529 m	Book: Aghanabati (2008, p. 817)

(continued)

Table 2.6 (continued)

Rock unit	Chronostratigraphic affinity	Lower/Upper boundary	Lithology and thickness at type section	Note (references, comments)
Sink Fm.	Mississippian	Lb. Mila Fm. Ub. covered but probably is Permian limestone and sandstone	The formation is divided into two distinct units: Lower unit: quartzite and quartz sandstone Upper unit: dolomite sometimes accompanied with limestone intercalations Thickness: 212 m	Book: Aghanabati (2008, p. 822)
Shishtu Fm.	Late Devonian-Mississippian	Lb. Bahram Fm. Ub. Conglomerates of Sardar Fm.	The formation encompasses the following rocks: Sandy shale, quartzite, silty shale, limestone, sandy dolomite, sandstone, dolomite, shale, sandy limestone, alternations of shale and limestone, quartzite sandstone, quartzite; alternation of limey-sandy shale, limey shale, sandy shale; alternation of limestone and shale, limestone, chert and limestone; and alternation of quartzitic sandstone and sandstone. Thickness: the Shishtu Fm. measures 543 m in thickness, from which the lower 326 m belongs to Shishtu I Member and upper 217 to Mississippian Shishtu II Member	Book: Alavi-Naeini (1993, p. 282, 308); Aghanabati (2004, p. 154, 165); Aghanabati (2008, p. 828); Alavi-Naeini (2009, p. 231) Thesis: Rastgar (1995); Yamini (1995) It is considered as the third formation of Ozbak-kuh group

(continued)

Table 2.6 (continued)

Rock unit	Chronostratigraphic affinity	Lower/Upper boundary	Lithology and thickness at type section	Note (references, comments)
Sardar Fm.	Carboniferous	Lb. is not outcropped due to locating in the core of an anticline. The previous report of author suggests Shishtu II. at the reference section, the Lb. is a conglomerate bed Ub. Baghe-Vang Fm. in the type section; in the reference section and other sections the Ub. is Jamal Formation separated by a Quartzitic key bed	The formation is composed of silty, sandy and clayey shale with intercalations of sandstone, siltstone, sandy limestone, limestone and quartzite Thickness: 660 m	Book: Alavi-Naeini (1993, p. 318); Aghanabati, (2004, p. 166); Aghanabati (2008, p. 888); Alavi-Naeini (2009, p. 238) Thesis: Akbarian-reshvanlou (1992) The formation is considered as the fourth formation of Ozbak-kuh Group
Gachal Fm.	Mississippian	Lb. Rahdar Fm. Ub. Khan Fm. with disconformable contact	The formation is divided into 5 members: Member A: quartzite, dolomitic sandstone, dolomite, limestone Member B: Dolomite Member C: Gypsiferous marl, brittle dolomite, shale Member D: Oolitic limestone Member E; Sandstone, sandy limestone, marly deposits Thickness: 538	Book: Alavi-Naeini (1993, p. 329); Aghanabati (2004, p. 167); Aghanabati (2008, p. 928); Alavi-Naeini (2009, p. 240) Article: Mahdavi Feiza-Badi et al. (2001, 2003)

brown fossiliferous sandstone, limestone, and conglomerate. These also include diabase sills.

Khan Formation

Permian rocks in the Kalmard area are of different facies, called the "Khan Formation". Three major successive transgressive and regressive phases can be traced in the area. In the Kalmard area of the Lakharbakhshi Mountain, the Permian Khan Formation is divided into three members (Aghanabati 1975) as following:

Member A consists of clastic rocks, resting on an eroded surface of the Gachal Formation, and fusulinid-bearing limestone indicating the deepening of the basin. The upper beds contain marly rocks indicating the regression phase. This member has been dated to Early Permian.

Member B begins with transgressive clastic beds, quartzitic sandstone changing upward into limestone with brachiopod-bearing marly intercalations (deepening of basin). The upper part is composed of shale and shaly sandstone bearing plant fossil remnants, indicating the second regression phase.

Member C is separated from the underlying members by an erosional surface. The plant fossils of the Member B presumably indicate a brief exposure of area. This member starts with a red clastic bed implying the onset of the third marine transgression. A laterite layer covers the old beds, suggesting the continental erosion of the Guadalupian-Lopingian boundary deposits. The succession continues with dolomite beds followed by dark grey limestone bearing Upper Wuchiapingian fossils and terminates with an erosional surface overlain by laterite lenses and Triassic rocks.

The Jamal Formation outcrops can be observed in many localities of central and east Central Iran. Permian rocks are exposed in the central parts of Central Iran, Bafgh-Biabanak, Saghand-Poshtbadam as well as the Kharanegh-Ardakan and Shir-Kuh of Yazd, and the Soh area of Kashan.

In some areas of Central Iran including Dehbid, Abadeh, Surmaq and Shahreza, Permian the rocks is characterized by very thick and most complete succession (Taraz 1973).

In the Biabanak-Bafgh and Saghand- Poshtbadam, transgression of Permian Sea after the Mila Formation, produced conglomerate and then shale. (Haghipour 1971) compared these basal Permian beds with Sardar Formation of northeastern part of this areas.

Seven units (equivalent to the Surmaq, Abadeh and Hambast formations) can be distinguished in the Hambast section, near Abarkuh Town:

1. White massive limestone (450 m-thick)
2. Black to grey limestone bearing chert layers in the lower part and cherty pebbles in its upper (260 m-thick)
3. Dark grey limestone (110 m-thick)
4. Alternation of black limestone, marl and purple to grey shale (278 m-thick)
5. Cliff-forming black limestone (56 m-thick)
6. Grey limestone (17.5 m-thick)
7. Thin-bedded red ammonite-bearing limestone (19 m-thick)

These units contain abundant fossils permitting well-defined age determination. Units 1–3, belong to the Surmaq Formation; units 4, 5 to the Abadeh Formation; and units 6–7 to the Hambast Formation. The Lower Permian sediments of Khan Formation are not exposed at this area.

Taraz (1974) introduced a new Tethyan stage, namely the "Abadeian" including units 4, 5 of the above succession (Fig. 2.14).

The Permian of Shahreza (Darreh-Cheleh village) and Asadabad (near Rameshehe village) includes interbedded sandstone, shale, and limestone beds referred by Baghbani (1993) to the "Vazhnan Formation". Based on the small numbers of fusulinids, he attributed the formation to the Asselian-Sakmarian interval (Leven and Gorgij 2011b).

The Permian-Triassic boundary has been reported from two areas of Central Iran, viz. "Abadeh" (Taraz et al. 1981; Heydari et al. 2003) and "Shahreza" (Stepanov et al. 1969; Tichert et al. 1973; Ghaderi et al. 2013).

The Abadeh section can be correlated first with both the Permian succession of Zal area, Azarbaijan, and with that of the Kalmard area.

Unit 1 of Abadeh Section consisting of 12 m-thick basal Artinskian limestones is comparable with Member B of the Khan Formation in the Kalmard area. Therefore, the basal part of Kalmard section appears more complete than the Abadeh section. Units 2 and 3 of Abadeh Section are also comparable with the lower part of unit C of Khan Formation of Kalmard area assigned to the Guadeloupian age. However, the equivalent of the Abadeian Stage, units 4, 5, has not been reported (Aghanabati 2004). Permian-Triassic sedimentation was continuous in the Julfa (Azarbaijan) and Abadeh areas (Central Iran), while the other parts of Central Iran were experiencing a phase of continental erosion (Taraz et al. 1981) (Fig. 2.15).

Permian rocks comprising sandstone and red conglomerate are also observed in the Behabad area, where they are disconformably underlain and overlain by the Dahu and Shotori formations, respectively (NIOC 2016).

2.6.2 Sanandaj-Sirjan

The succession associated with the Permian marine transgression in the Sanandaj-Sirjan Zone, is characterized by volcano-sedimentary rocks accompanied by volcanics. These rocks were studied by Thiele et al. (1968) and named the "Ab-barik Facies" in the southeastern Golpayegan, southern Azna and Boyeen-Zahra areas.

The northern Divandareh contains questionable rocks assigned by some workers to the Permian. However, in opinion of the present author, these rocks are older and presumably belong to the Ordovician. Hoseini (2011) assigned similar rocks exposed along the Babaali-Glali Road (Hamedan), and at Zafarabad (Divandareh), Aligoudarz and Shahrekord, a late Paleozoic-Early Triassic age based on radiometric evidence.

2.6.3 *Alborz*

Dorud Formation

Resulting from the early Permian transgression of Alborz, the Dorud Formation was divided by Assereto (1963) into following 4 units:

Unit 1. A Succession of clayey marls and marly limestones as well as red thin-bedded quartzite;

Unit 2. A Succession of conglomerate and shale with siltstone and sandstone intercalations;

Unit 3. Limestone and limey conglomerate;

Unit 4. Siltstone, shale and sandstone.

The basal unit of the latter appeared to be Carboniferous rather than Permian (Bozorgnia 1973). The present accepts that the Dorud Formation is tripartite including units 2–4 of Assereto (1963).

The Ruteh Formation is composed of massive to medium-bedded bioclastic limestones (Assereto 1963). The top of the formation in type section is eroded implying the subaerial exposure of the area (Aghanabati 2004).

Following deposition of the Ruteh Formation, most of Alborz was exposed by the Cimmerian orogeny (e.g. Zanchi et al. 2009. However, marine conditions persisted in the northern Alborz indicated by the presence of the Upper Permian Nesen Formation (Glaus 1964; Gaetani et al. 2009; Zandkarimi et al. 2014b, 2017c).

Nesen Formation

The Upper Permian Nesen Formation represents another complete sedimentary cycle of Alborz, the third sedimentary cycle of the Permian. It starts with basal limey-shaly strata that mark the onset of the transgression and continues with black to grey limestone followed by alternations of shale and limestone (Glaus 1964) in its part. The ripple mark structures of upper parts imply a decrease in water depth of the basin.

The three mentioned formations extend through most of the northern Alborz, while the southern part, as already indicated, encompasses the Dorud and Ruteh formations seen at Torud, Jam, Shahmirzad of Semnan, all of the central Alborz and western Alborz (e.g. Soltanieh of Zanjan), Azarbaijan and Talesh.

A complete succession of Permian-Early Triassic rocks can be observed in the northern Azarbaijan,[4] Julfa area, showing that in the Aras-northern Azarbaijan zone, the latest Permian sedimentation continued into the earliest Triassic without any regression (Tichert et al. 1973). In this area, the Permian marine transgression begins with sedimentation of Dorud Formation forming complete sedimentary cycle (NIOC 2016). Overlying Dorud Formation, some fossiliferous limestone beds, Gnishik Beds of Caucasian terminology, are exposed which are lithologically and paleontologically comparable to the Ruteh Formation of Alborz. These beds (300 m-thick) were assigned to Guadeloupian age (Ghaderi et al. 2014).

[4]According to the author, the northern part of the Soltanieh-Tabriz Fault is regarded as northern Azarbaijan and southern part of the same fault as Southern Azarbaijan.

The highly fossiliferous thick-bedded grey cherty limestone of the Middle Permian (Midian) Khachik Formation overlies the equivalent of the Gnishik Formation (Partoazar 1995). It is followed by Julfa Formation that consists mainly of a succession of clayey to sandy loose shale in the lower part and alternations of loose marl, firm marl and limestone in the upper part (Partoazar 1995). These parts are of equal thickness, and of continuous common boundary. Accordingly, Julfa Formation is divided into two subunits, viz. "Lower Julfa Bed" and "Upper Julfa Bed" (Partoazar 1995).

A succession of marine sediments, referred as the 'Transitional Beds', overlies the Julfa Beds and shows the marine condition of Permian-Triassic of Julfa area. It includes dark-purple shale, interbedded and mudstone (17 m-thick) bearing Permian-Triassic fossils.

2.6.4 Zagros

The Permian marine transgression of the Zagros Basin is represented by the Faraghan and Dalan formations.

Faraghan Formation

Setudehnia (1975) was the first to describe a clastic sequence subsequently named by Szabo and Kheradpir (1978) from the Faraghan Mountain of northern Bandarabbas. Overlying the Silurian graptolite shale of the Gahkom Formation, this formation is composed of coarse quartz conglomerate with sub-rounded pebbles in a finer sand matrix. These grade upwards into medium- to coarse-grained cross-bedded quartzarenite. At the other localities of Zagros Basin, underlying formation are Ordovician or Cambrian in age. The sequence ends with variegated shale interbedded with medium grained quartz sandstone.

There is disagreement regarding the age determination of the Faraghan Formation. Earlier researchers considered it to be Devonian-Carboniferous in age but this interpretation was rejected by Ghavidel-Syooki who assigned it to the Permian. Following the palynological research of Ghavidel-Syooki (1986), NICS[5] proposed a Permian age for this formation.

The shale facies is in the form of intercalations whose thickness increases towards the top.

An Early Permian age for the sandstone member was considered most likely and, in consequence, the lower member was attributed to the early Permian and the upper member to the middle Permian (NIOC 2016; Plate 2.12).

Dalan Formation

The Dalan Formation consists of a lower carbonate member, a dominantly anhydrite middle member, and an upper carbonate member with anhydrite intercalations, suggesting a fluctuating inner neritic to coastal sabkha (?) depositional

[5]National Iranian Committee on stratigraphy.

environment (Szabo and Kheradpir 1978). The lower carbonates contain abundant macrofossils, such as corals, crinoids, brachiopods, and gastropods, and abundant foraminifera, such as *Langella conica*, *Condonella* sp., *Pachyphloia iranica*, *P. cukurkoyi*, *Hemogordius* sp., *Gymnocodium* sp., *Mizzia* sp., and *Ammodiscus* sp. (Kolodka et al. 2011). In some areas it may also include oolitic, onkolitic, and pelletal and in part dolomitized limestone? The middle anhydrite member (Nar Member), if present, is barren of fossils. The carbonates above the anhydrite are oolitic-pelletal to micrite. The faunal assemblage includes foraminifera indicative of a middle neritic depositional environment including the above mentioned foraminifera and also *Endothyra* sp., *Frondina* sp., *Ichtyolaria primitiva*, *Climacammina* sp., *Paraglobivalvulina* mira, and Dagmarita chanakchiensis. The presence of Paraglobivalvulina mira, and Dagmarita chanakchiensis is indicative of a Djulfian (Late Permian) age for the uppermost part of the Dalan Formation (Kolodka et al. 2011). These foraminifera have been reported from the Djulfian section of the Alborz Mountains in northern Iran (Bozorgnia 1973) and from Transcaucasia (Reitlinger 1965). This contradicts the Guadalupian age previously assigned to the entire Dalan Formation by Khalili (in Szabo 1976).

Table 2.7 illustrates the main lithological features of the Permian rocks of Iran. Stratigraphically-significant fossils are illustrated in Plates 2.11, 2.12 and 2.13.

2.7 Selected Paleozoic Sections

In the following, in order to illustrate the distribution of lithostratigraphic units of Alborz, Central Iran and Zagros and their relationships, several selected sections, resulted from field observations of the author, are presented.

Table 2.7 Main lithological features of the Permian rocks of Iran

Rock unit	Chronostratigraphic affinity	Lower/Upper Boundary	Lithology and thickness at type section	Note (references, comments)
Permian (Alborz and Azarbaijan)				
Dorud Fm.	Early Permian	Lb. in the type section is Mississippian M. C of Geirud Fm. Ub. is doubtful and needs revision, however, according to Assereto (1963) is Ruteh Fm., accordingly at the top of Dorud Fm. is a bed of grey quartz sandstone, conformably overlain by Ruteh Fm.	Includes 4 units: Unit 1: alternation of clayey marl and marly limestone and red thin-bedded quartzite Unit 2: succession of quartz conglomerate and shale with siltstone and sandstone intercalations Units 3: limestone and limey conglomerate Unit 4: Siltstone, shale and sandstone	Book: Alavi-Naeini (1993, p. 343); Afshar-Harb, (1994, p. 46); Aghanabati (2004, p. 172); Aghanabati (2008, p. 1008); Alavi-Naeini (2009, p. 244) Thesis: Arefifard (1997); Hasani (1999); Okhravi (1991); Mehrban (1983) Article: Gaetani et al. (2009) Bozorgnia (1973) reported a major gap between unit 1 and unit 2 and assigned the unit 1 to Visean to Serpukhovian, while the unit 2 belongs to Permian. These resulted later proved by Assereto et al. (1978) who removed the unit 1 form initial Dorud Fm. Therefore, the present accepted Dorud Formation is tripartite including units 2–4 of Assereto (1963)

(continued)

Table 2.7 (continued)

Rock unit	Chronostratigraphic affinity	Lower/Upper Boundary	Lithology and thickness at type section	Note (references, comments)
Ruteh Fm.	Middle Permian	Lb. unit 4 of Dorud Fm. with conformable contact Ub. Elika Fm. of type section with parallel disconformity	Widely distributed white cliff-forming limestone	Book: Alavi-Naeini (1993, p. 347); Afshar-Harb (1994, p. 47); Aghanabati (2004, p. 174); Aghanabati (2008, p. 1063); Alavi-Naeini (2009, p. 247) Thesis: Kohansal (2009); Arefifard (1997), Hasnai (1999) Article: Okhravi et al. (2002) In the northern Alborz, the formation is overlain by some basaltic beds attributed to Midian age
Nesen Fm.	Late Permian	Lb. the melaphir and basalts overlying the Ruteh Fm. Ub. Elika Fm. with conformable boundary	Succession of black marly shale and nodular limestone Thickness: 230 m	Book: Alavi-Naeini (1993, p. 355); Afshar-Harb, (1994, p. 47); Aghanabati (2004, p. 175); Aghanabati (2008, p. 1257); Alavi-Naeini (2009, p. 251) Thesis: Shokarvi (1997)
Alibashi Fm.	Late Permian	Lb. Shamar member of Julfa Fm. with continuous and conformable boundary Ub. Unit E of Elika Fm. with continuous and conformable boundary	According to Partoazar (1995) the formation is divided into two members: shaly limestone with shale intercalations; and succession of marly-silty limestone and shale Thickness: 55 m	Book: Aghanabati (2008, p. 1164)

(continued)

Table 2.7 (continued)

Rock unit	Chronostratigraphic affinity	Lower/Upper Boundary	Lithology and thickness at type section	Note (references, comments)
Julfa Fm.	Late Permian	Lb. grey limestone beds with Murgabian age. Ub. Transitional beds	Partoazar (1995) divided the formation into two members namely "Silgerd" and "Shamar" respectively. The former consist of marl, marly limestone, calcareous shale and dolomitic limestone; and the later of limestone with chert nodules. He considered the Shamar Member as a new rock unit "Alibashi Fm." Thickness: 272 m	Book: Aghanabati (2008, p. 1175) According to Stepanov et al. (1969), the Julfa Formation are equivalent of Ruteh Fm., while Partoazar (1995) compared it with the Surmaq Fm. of Abadeh area
Qeshlaq Fm.	Late Permian	Lb. Ruteh Fm. (upper part of Ghosnavi Group). Ub. Elika Fm.	The Gheshlagh Fm. is of clastic facies and encompasses succession of shale, siltstone and sandstone, whose upper part contains coal and laterite beds Thickness: 180 m	Book: Aghanabati (2008, p. 1250) Article: Gaetani et al. (2009)

(continued)

Table 2.7 (continued)

Rock unit	Chronostratigraphic affinity	Lower/Upper Boundary	Lithology and thickness at type section	Note (references, comments)
Permian (Central and Sanandaj-Sirjan)				
Khan Fm.	Permian	Lb. Member D of Gachal Fm. with parallel disconformity Ub. Sorkh-Shale Fm. with disconformable contact and an iron-bearing quartzite horizon	The formation includes two members, each of them are further subdivided into 3 rock units M 1: massive quartz sandstone, marly limestone with some brachiopod-bearing marl intercalations M2: Quartzite, quartzitic sandstone, limestone and dolomite Thickness: Halavan Section (type section of M1): 300 m Lakhar-bakhshi (type section of M2): 223 m	Aghanabati (2008, p. 1250) Article: Leven and Gorgij (2011a) Thesis: Arefifard (2006)
Jamal Fm.	Permian	Lb. Sardar Fm. Ub. Sorkh Shale Fm.	The revised formation contain 3 informal members, viz. "lower quartz sandstone", "middle sandstone" and "upper dolomite" Thickness: Jamal Fm. of Stöcklin et al. (1965) measured 473 m in thickness, and in the revision, 547 m	Book: Alavi-Naeini (1993, p. 411); Afshar-Harb (1994, p. 177); Aghanabati (2004, p. 174); Aghanabati (2008, p. 1181); Alavi-Naeini (2009, p. 264). Lowermost formation of Tabas Group

(continued)

Table 2.7 (continued)

Rock unit	Chronostratigraphic affinity	Lower/Upper Boundary	Lithology and thickness at type section	Note (references, comments)
Mahalat beds (facies?)	Permian	Lb. Mila Fm. with disconformable contact. Lb. in other localities maybe Lalun, Soltanieh and Kahaformations or even metamorphic rocks assigned to Neoproterozoic age Ub. Lower Triassic-Middle Jurassic rocks of Shemshak Fm.	Most of the Permian succession of Mahallat-Muteh is dolomite rocks. These rocks from base to top are: 1. Fine-grained conglomerate, sandstone 2. light to grey dolomite 3. recrystallized limestone bearing some fusulinds In the report of Golpayegan map the facies measured 1000–2000 m in thickness	Alavi-Naeini (1993, p. 442); Aghanabati (2008, p. 966)
Harat Complex	Permian	This complex is located in the Sanandaj-Sirjan zone and due to the deformation and metamorphism the boundaries are not defined	Consist of two metamorphosed clastic and carbonatic members (200–300 m-thick). The former is composed of phyllite, albite, cholorite, epidote and quartz schist; and the later of limey and dolomitic marmarite	Book: Aghanabati (2008, p. 993) The complex overlies the Tutak complex attributed to the Ordovician-Devonian age. According to Hosseini (2011) the Tutak complex is overlain by the Surian complex whose upper part is Harat complex

(continued)

Table 2.7 (continued)

Rock unit	Chronostratigraphic affinity	Lower/Upper Boundary	Lithology and thickness at type section	Note (references, comments)
Ab-barik facies	Permian	Lb. are not well defined, although thought to be Neoproterozoic rock with faulted and disconformable contact Ub. Shale and sandstone assigned to Late Triassic based on the *Heterster* echinoid species	The unit is mainly composed of sandy shale interbedding with limey sandstone, limestone, shale and slate with intercalations of sandstone and basic to andesitic igneous rocks Thickness: 400–600 m	Book: Aghanabati (2008, p. 998)
Basam Fm.	Early Permian	Lb. shallow marine/coastal rocks attributed to Carboniferous (Moscovian stage) with disconformable contact Ub. Surmaq Fm.	Basal microconglomerate, sandstone, purple and red shale, quartzite with sandy limestone intercalations Thickness: 108 m	Book: Aghanabati (2008, p. 1004);Baghbani (2008) considered continuous and conformable lower and upper boundary with Vazhnan and Surmaq formations, respectively

(continued)

Table 2.7 (continued)

Rock unit	Chronostratigraphic affinity	Lower/Upper Boundary	Lithology and thickness at type section	Note (references, comments)
Vazhnan Fm.	Early Permian	Lb. Carboniferous rocks with erosional disconformity Ub. cliff-forming limestone of unit 1 of Surmaq Fm. with conformable contact	Consists of 7 rock units: sandstone with calcareous cement; limestone with basal sandstone; siltstone, sandstone with limestone intercalations; sandstone; quartzitic sandstone with thin-bedded limestone intercalations in the middle part; limestone changing upward to dolomite; sandstone with shale intercalations. Thickness: 142	Book: Alavi-Naeini (1993, p. 431); Aghanabati (2008, p. 1053) The formation is the first unit of the Shorjestan group. The formation is informally introduced whose thickness is highly varied from place to place and is often less than 70 m
Surmaq Fm.	Late Early Permian-early late Permian	Lb. disturbed by fault. Ub. Abadeh Fm. with continuous contact	The formation encompasses three lithological units as follows: Unit 1: this unit is further subdivided into three subunit Unit 2: include 4 subunit, viz. limestone with few chert nodules, limestone bearing cherty lenses and nodules, alternation of chert and limestone, alternation of chert and limestone (the ration of chert exceeds limestone) Unit 3: limestone Thickness: 650 m	Book: Alavi-Naeini (1993, p. 432); Aghanabati (2008, p. 1101). The second formation of Shorjestan group, Report: Taraz (1972)

(continued)

Table 2.7 (continued)

Rock unit	Chronostratigraphic affinity	Lower/Upper Boundary	Lithology and thickness at type section	Note (references, comments)
Abadeh Fm.	Late middle-early late Permian	Lb. Surmaq Fm. Ub. Hambast Fm. with gradual contact	According to Iranian-Japanese Research Group (1981), the formation encompasses units 4, 5 as well as upper part of unit of Taraz (1972) 3 consisting of a succession of shale and limestone, shale and limey shale with limestone intercalations. Baghbani (1996a, b) considered the formation to include unit 3 (limestone) and 4 (alternation of shale and limestone) Thickness: 475 m	Aghanabati, 2008, p. 1146 Third formation of Shorjestan group. Technical Report: Taraz (1972)

(continued)

Table 2.7 (continued)

Rock unit	Chronostratigraphic affinity	Lower/Upper Boundary	Lithology and thickness at type section	Note (references, comments)
Hambast Fm.	Late Permian	Lb. Abadeh Fm. with conformable and gradual contact Ub. Lower Triassic beds with conformable and continuous contact	According to Iranian-Japanese Research Group (1981), the formation encompasses the unit 6 (a succession of shale and limestone) and unit 7 (limestone) of Taraz (1972), while Baghbani (1990) considered the formation to include units 5 (limestone with chert intercalations in the lower part and shale intercalations in the upper part), unit 6 (alternation of shale and limestone) and unit 7 (limestone with thin-bedded shale intercalations) in the Hambast Fm. Thickness: 34–50 m according to the Iranian-Japanese research group (1981) and 62 m according to Baghbani (1990)	Book: Aghanabati (2004, p. 178); Aghanabati (2008, p. 1284). Thesis: Mosavi-abnavi (2009) Report: Taraz (1972) The fourth formation of Shorjestan group

(continued)

Table 2.7 (continued)

Rock unit	Chronostratigraphic affinity	Lower/Upper Boundary	Lithology and thickness at type section	Note (references, comments)
Permian (Zagros)				
Faraghan Fm.	Early Permian	Lb. Cambrian to Devonian rocks with angular unconformity Ub. The upper boundary is disturbed in most of Zagros localities, the Ub. is Dalan Fm.	Basal quartz conglomerate with semi-rounded pebbles, and sandstone, colored shale and quartz arenite with limestone intercalations The Faraghan and overlying Zakin formations are of the such similarity that cannot be simply distinguished, consequently leading to overestimation in measured thickness (300–500 m thick) The quartz arenite part measures 30–100 m in thickness.	Book: Motiei (1993, p. 45); Alavi-Naeini (1993, p. 449); Aghanabati, (2004, p. 181); Aghanabati (2008, p. 1040); Alavi-Naeini (2009, p. 276) Article: Ghavidel-Syooki (1985, 1988, 1990a, b, 1993, 1994) Ghavidel-Syooki revealed a major gap between Upper Frasnian-Famennian Zakin Fm. to Permian Faraghan Fm. First formation of the Dahrom Group. Some researcher assigned the formation to Permian and others to the Permo-Carboniferous (Setudehnia 1972)

(continued)

Table 2.7 (continued)

Rock unit	Chronostratigraphic affinity	Lower/Upper Boundary	Lithology and thickness at type section	Note (references, comments)
Dalan Fm.	Middle to late Permian	Lb. Faraghan Fm. conformable and gradual contact Ub. Kangan Fm.	The formation is mainly composed of two carbonate and evaporite part and is divided into 3 Members: M1: Lower Carbonate (informal), limestone and dolomite M2: Nar Member, anhydrite and dolomite M3: Upper Carbonate (informal): oolitic and micritic limestone and dolomite accompanied by anhydrite and dolomitic anhydrite. The lithofacies of the formation varies from place to place including coastal carbonate facies, coastal organic carbonate, open-marine carbonate and near shore clastic-carbonate facies	Book: Motiei (1993, p. 52); Alavi-Naeini (1993, p. 451); Aghanabati, (2004, p. 182); Aghanabati (2008, p. 1224); Alavi-Naeini (2009, p. 277) Article: Ahmadi (2011); Lotfpour et al. (2002); Kheradmand (2000); Davydov and Arefifard (2013) Thesis: Balazadeh (1983); Lotfpour, (2003); Mosavi (2003); Mahboubipour (2004); Abbasi (2009); Tabatabei (2003); Braham (2006)

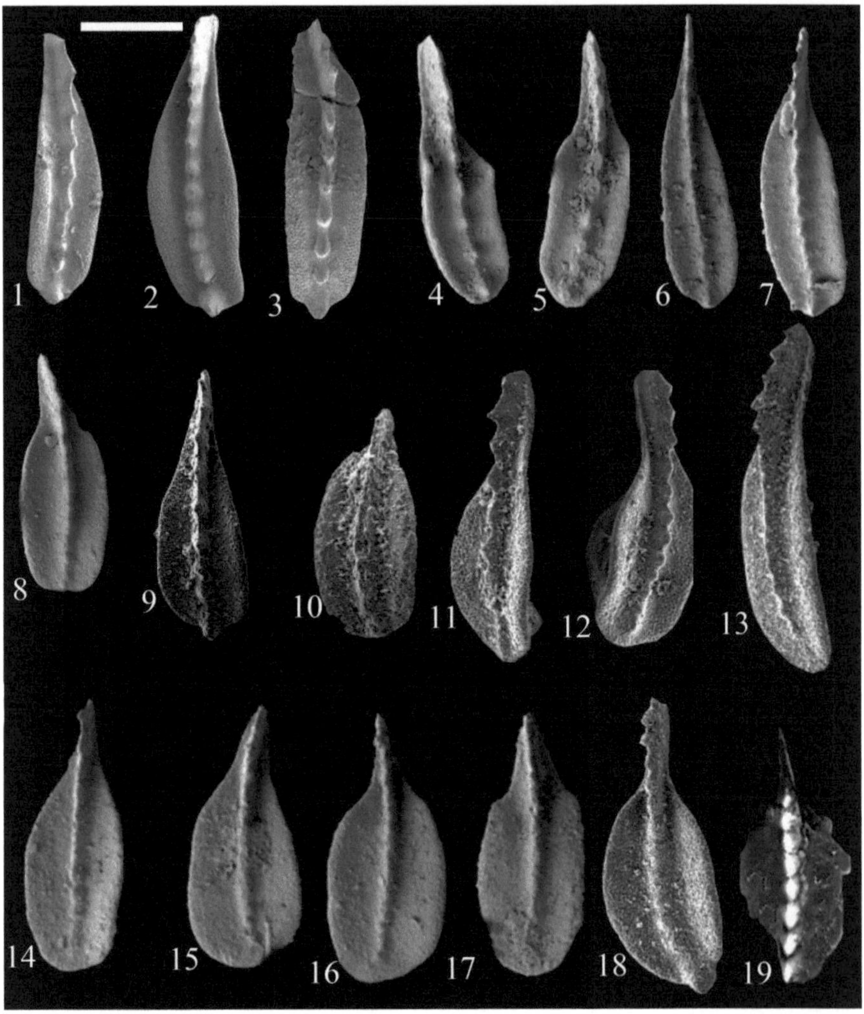

Plate 2.11 Selected diagnostic Permian conodonts of Central Iran and Azarbaijan (NIOC 2016). 1, 2. *Clarkina postbitteri*, Jamal Fm., Baghvang Section; 3. *inogondolella* sp., Jamal Fm., Baghvang Section; 4, 5. *larkina leveni*, Hambast Fm., Abadeh Section; 6, 7. *Clarkina liangshanensis*, Hambast Fm., Abadeh Section; 8. *Clarkina transcaucasica*, Hambast Fm., Abadeh Section; 9. *Clarkina subcarinata*, Hambast Fm., Jolfa section; 10. *Clarkina* sp. cf. *C. orientalis*, Hambast Fm., Jolfa section; 11, 12. *Clarkina changxingensis* Hambast Fm., Jolfa section; 13. *Clarkina changxingensis*, Hambast Fm., Jolfa section; 14, 15. *Clarkina orientalis*, Hambast Fm., Abadeh Section; 16, 17. *Clarkina abadehensis*, Hambast Fm., Abadeh Section; 18. *Clarkina yini*, Hambast Fm., Jolfa section; 19. *Hindeodus preaparvus* Elika Fm., Jolfa section

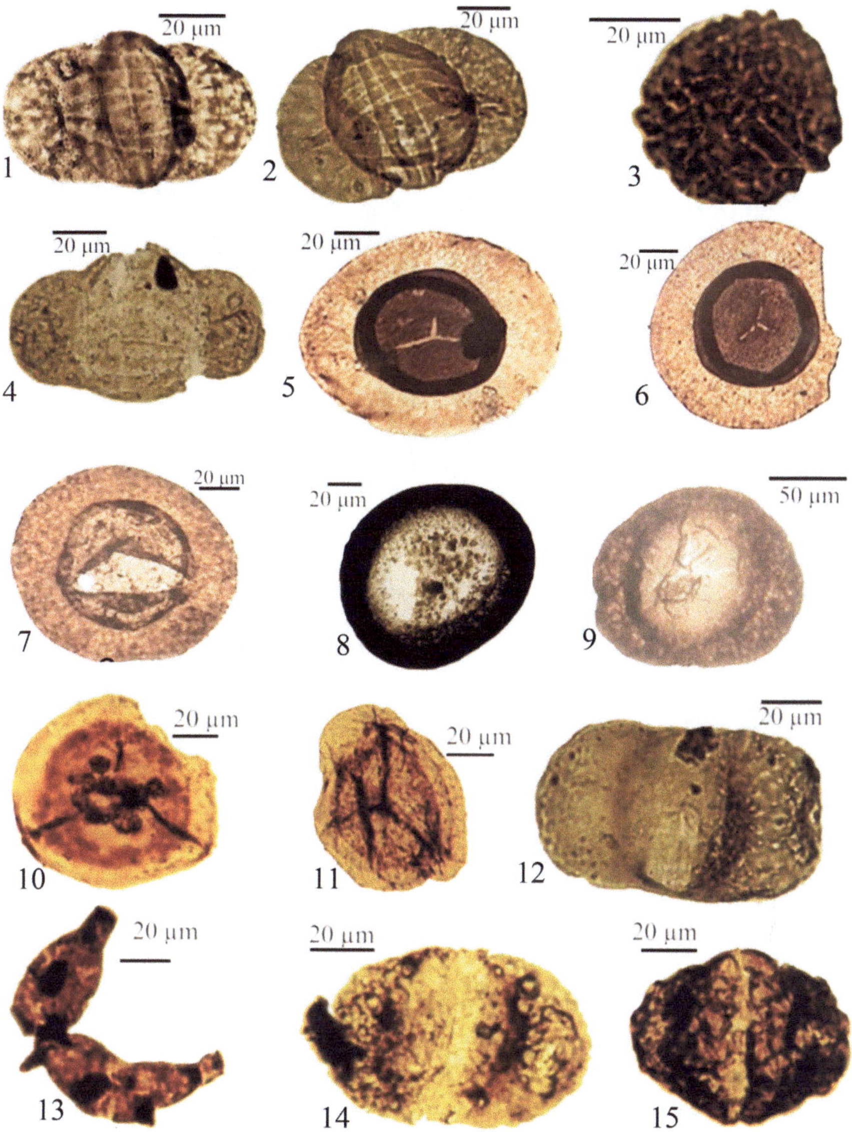

Plate 2.12 Selected diagnostic Permian palynomorphs of Central Iran and Azarbaijan (NIOC 2016). 1, 2. *Distriatites insolitus*, Faraghan-Chalisheh Fm., Zirreh-1 Borehole; 3. *Plicatipollenites malabarensis*, Faraghan Fm., Finu-1 Borehole; 4. *Hamiapollenites dettmannae*, Faraghan Fm., Zard Kuh Section.; 5, 6. *Plicatipollenites malabarensis*, Zakin Fm., Finu-1 Borehole; 7. *Potonieisporites novicus,* Faraghan Fm., Finu-1 Borehole; 8. *Cannanoropollis janakii*, Illiya-1 Borehole; 9. *Barakarites rotatus*, Faraghan Fm., Finu-1 Borehole; 10, 11. *Indotriradites mundus*, Faraghan Fm., Darreh-Yas section; 12. *Alisporites nuthallensis*, Faraghan Fm., Illiya-1 Borehole; 13. *Reduviasporonites*, Faraghan Fm., Darreh-Yas section; 14, 15. *Falcisporites stabili*, Faraghan Fm., Darreh-Yas section

Plate 2.13 Selected diagnostic Permian foraminifers of Central Iran and Azarbaijan (NIOC 2016).1. *Paraleeina* sp. cf. *P. postkraffti*, Jamal Group, Central Iran; 2. *Iranella bella*, Jamal Group, Central Iran; 3. *Paraleeina postkraffti*, Jamal Group, Central Iran; 4. *Cancellina cutalensis*, Jamal Group, Central Iran; 5, 6. *Cancellina* ex gr. *cutalensis*, Jamal Group, Central Iran; 7. *Verbeekina* sp. aff. *V. heimi*, Jamal Group, Central Iran; 8, 9. *Eopolydiexodina persica*, Surmaq Fm., Central Iran; 10, 11. *Afghanella schencki*, Surmaq Fm., Central Iran; 12. *Sumatrina annae*, Surmaq Fm., Central Iran; 13. *Chusenella abichi*, Abadeh Fm., Central Iran; 14. *Dunbarula mathieui*, Abadeh Fm., Central Iran; 15. *Dunbarula nana*, Abadeh Fm., Central Iran; 16, 20. *Reichelina pulchra*, Jamal Group, Central Iran; 17. *Reichelina* sp., Jamal Group, Central Iran; 18. *Paraglobivalvulinoides septulifer*, Jamal Group, Central Iran; 19, 21. *Colaniella* ex gr. *parva*, Jamal Group, Central Iran; 22. *Chenella* sp., Jamal Group, Central Iran

References

Abbasi, M. (2009). *Sequence stratigraphy of the Upper Dalan and Kangan formations in some boreholes of the South Pars Oil Field, an explanation for the reservoir quality variation.*

Afshar-Harb, A. (1994). *The geology of Kopeh-Dagh, Geological book codification project*. Geological Survey of Iran, 274.

Aghanabati, A. (1975). Etude geologique de la region de Kalmard These Doctorate thesis, No AD 11/623 Gronoble, France.

Aghanabati, A. (2004). Geology of Iran. *Geological Survey of Iran, Report* (in Persian).

Aghanabati, A. (2008). Stratigraphic Lexicon of Iran. In: Devonian to Permian. *Geological Survey of Iran, Report* (in Persian).

Aharipour, R. (1996). Microfacies and depositional environment of the Upper Tirgan formation and Sarcheshmeh, Pesteligh and Chehelkaman formations of the Northeastern Gonbad-e-Kavus, western Kopeh-Dagh (Unpublished MSc thesis). Faculty of Sciences, Kharazmi University.

Ahmadi, W. (2011). Depositional environments of the Kangan Formation in the A and B boreholes of South Pars Field based on the lithofacies. In *The 1st symposium of geology*. Geological Survey of Iran.

Akbari-e-Reshvanlou, M. (1992). *Palynostratigraphical and paleoecological studies of the Sardar Formation, Shotori Mountains* (Unpublished M.Sc. thesis). Faculty of Earth Sciences, Shahid Beheshti University.

Alavi, M., & Bolourchi, M. H. (1973). Explanatory text of the Maku quadrangle map 1:250,000, Geological Survey of Iran, Geological Quadrangle, Tehran, A1, p.

Alavi-Naeini, M. (1972). *Etude géologique de la région de Djam*. Geological Survey of Iran, Report 23, 293.

Alavi-Naeini, M. (1993). Paleozoic stratigraphy of Iran. In *Treatise on the geology of Iran* (Vol. 5, pp. 1–492).

Alavi-Naeini, M. (2009). *An introduction on the stratigraphy of Iran*. Geological Survey of Iran, 507.

Allenbach, P. (1966). *Geologie und Petrographie des Damavand und seiner Umgebung (Zentral-Elburz), Iran* (Doctoral dissertation).

Ameri, H .(2009). *Paleontology and biostratigraphy of the Kuhbanan Formation of the Kerman area* (Unpublished Ph.D. thesis). Faculty of Sciences, Shahid Bahonar University.

Amin-Rasouli, H. (1999). *Microfacies, depositional environment and sequence stratigraphy of the shaly and quartzarenite of the Lalun Formation, and Member I and II of Milla Formation* (MSc thesis). Kharazmi University.

Amir Sardari, D. (1998). *Microfacies and depositional environments of the Members 3, 4 and 5 of Milla Formation, Eastern Alborz* (M.Sc. thesis). Faculty of Sciences, Kharazmi University.

Amiri-Bahman, F. (2009). *Microfaceis, diagensis and depositional environments of Mobarak Formation, Mighan and Khoshyeilagh areas* (Unpublished M.Sc. thesis). Faculty of Earth Sciences, Shahid Beheshti University.

Arefifard, S. (1997). *Microstratigraphy and micropaleontology of the Dorud and Ruteh Formation of the Central Alborz* (Unpublished M.Sc. thesis). Faculty of Earth Sciences, Shahid Beheshti University.

Asadi, G. (1997). *Spore biostratigraphy of the Muli and Ilanghare Formations, northeastern Maku, western Azarbaijan* (M.Sc. thesis). Faculty of Sciences, University of Tehran.

Ashouri, A. (2006). Middle Devonian-Mississippian conodont faunas from the Khoshyeilagh formation, Alborz Mountains, north Iran. *Journal of Sciences, Islamic Republic of Iran, 17*, 53–65.

Ashouri, A., & Najjarzadeh, M. T. (2000). Conodont biostratigraphy of the Devonian rocks of the Kuh-Zou Section, southwestern Bojnourd. In *The 4th symposium of geology*, Geological Society of Iran.

Assereto, R. (1963). The Paleozoic formations in Central Elburz (Iran). *Rivista Italiana di Paleontologia e Stratigrafia, 69*(4), 503–543.

Assereto, R. (1966). The Jurassic Shemshak formation in central Elburz, Iran. *Paleontology, 72*(4), 1133–1182.

Baghbani, D. (1996a). The study of complete Permian Section of Northeastern Shahreza. In *The 9th Symposium of Earth Science*, Geological Survey of Iran.

Baghbani, D. (1996b). *The study of Qom basin and the evaluation of its hydrocarbon potential*, National Iranian Oil Company, Report 1838.

Bahrami, A., Königshof, P., Boncheva, I., Tabatabaei, M. S., Yazdi, M., & Safari, Z. (2015). Middle Devonian (Givetian) conodonts from the northern margin of Gondwana (Soh and Natanz regions, north-west Isfahan, Central Iran): Biostratigraphy and palaeoenvironmental implications. *Palaeobiodiversity and Palaeoenvironments, 95*(4), 555–577.

Bahrami, A., Königshof, P., Boncheva, I., Yazdi, M., Khalaji, M., & Zarei, E. (2017). Conodont biostratigraphy of the Kesheh and Dizlu sections, and the age range of the Bahram Formation in central Iran. *Palaeobiodiversity and Palaeoenvironments*, 1–15.

Bakhtiari-Fard, A. (1992). *Palynostratigraphical and lithostratigraphical studies of the Niur Formation in the type section, Ozbak-Kuh* (Unpublished MSc thesis). Faculty of Earth Sciences, Shahid Beheshti University.

Balazadeh, A. (1983). *Evaluation of the organic-rich beds of Kangan and Dalan formations, Salman oil field* (Unpublished M.Sc. thesis). Faculty of Science, University of Tehran.

Barham, S. (2006). *Study of diagenesis processes of the upper Dalan formation of the Kuh-e-Surmeh and SPD Borehole, South Pars Oil Field* (Unpublished M.Sc. thesis). Faculty of Sciences, Kharazmi University.

Bayet-Goll, A. (2009). *Depositional environment and diagenesis of the Shirgesht Formation, Kuh-e-Asheghan and Kuh-e-Rahdar Sections* (Unpublished M.Sc. thesis). Faculty of Earth Sciences, Shahid Beheshti University.

Bayet-Goll, A., Geyer, G., Wilmsen, M., Mahboubi, A., & Moussavi-Harami, R. (2014). Facies architecture, depositional environments, and sequence stratigraphy of the Middle Cambrian Fasham and Deh-Sufiyan Formations in the central Alborz, Iran. *Facies, 60*(3), 815–841.

Berberian, M., & King, G. C. P. (1981). Towards a paleogeography and tectonic evolution of Iran. *Canadian Journal of Earth Sciences, 18*(2), 210–265.

Birounro, M, Ahmadzadeh-Heravi, M., & Vosoughi-Abdini, M. (2003). Introduction of conodont zones of Khoshyeilagh Formation, Members Mighan and Tilabad. In *The 22th symposium of Earth Science*, Geological Survey of Iran.

Bozorgnia, F. (1973). Paleozoic foraminiferal biostratigraphy of central and east Alborz Mountains, Iran, National Iranian Oil Company, Geological Laboratories Publication, Tehran 4, 185.

Brenckle, P. L., Gaetani, M., Angiolini, L., & Bahrammanesh, M. (2009). Refinements in biostratigraphy, chronostratigraphy, and paleogeography of the Mississippian (Mississippian) Mobarak formation, Alborz Mountains, Iran. *GeoArabia, 14*(3), 43–78.

Bruton, D. L., Wright, A. J., & Hamedi, M. A. (2004). Ordovician trilobites of Iran. *Palaeontographica, Abteilung, A271*, 111–149.

Clark, G. C., Davies, R., Hamzepour, B., & Jones, C. R. (with contributions by Ghorashi, M., Hamdi, B., Navai, N.). (1975a). Explanatory text of the Bandar-e-Pahlavi quadrangle map. *Geological Survey of Iran, Tehran, D3*, 1–198.

Clark, G. C., Davies, R., Hamzehpour, B., Jones, C. R., Ghorashizadeh, M., Hamidi, B., et al. (1975b). Bandar-e-Pahlavi, Geological Quadrangle map of Iran; 1:250,000; Geological Survey of Iran, No D3.

Davydov, V. I., & Arefifard, S. (2013). Middle Permian (Guadalupian) fusulinid taxonomy and biostratigraphy of the mid-latitude Dalan Basin, Zagros, Iran and their applications in paleoclimate dynamics and paleogeography. *GeoArabia, 18*(2), 17–62.

Eftekharnezhad, J., & Behroozi, A. (1991a). Geodynamic significance of recent discoveries of ophiolites and Late Paleozoic rocks in NE Iran (including Kopeh-Dagh). *Abhandlungen der Geologischen Bundesanstalt, 38*, 89–100.

Eftekharnezhad, J., & Behroozi, A. (1991b). New findings on the Ophiolitic and upper Paleozoic rocks of the northeastern Iran (including Kopeh-Dagh) and their geodynamic significance. *Journal of Earth Sciences, (Geological Survey of Iran)* (in Persian).

Fahim, Z. (2009). *Corals of the Mobarak formation in the Toye-Darvar section, eastern Alborz, southwestern Damghan* (Unpublished M.Sc. thesis). Paiam-Nour University of Tehran.

Gaetani, M., Angiolini, L., Ueno, K., Nicora, A., Stephenson, M. H., Sciunnach, D., Rettori, R., Price, G. D., & Sabouri, J. (2009). Pennsylvanian–Early Triassic stratigraphy in the Alborz Mountains (Iran). *Geological Society, London, Special Publications, 312*(1), 79–128.

Gansser, A., & Huber, H. (1962). Geological observations in central Elburz Iran. *Schweizerische mineralogische und petographische mitt, 42*(4), 583–630.

Ghaderi, A., Agha, N. S. A., Hamdi, B., & Miller, J. F. (2008). Biostratigraphy of the First and Second members of type section of the Shirgesht Formation in north of Tabas with special emphasis on conodonts. *Earth Science Journal of Geological Survey of Iran* 150–163.

Ghaderi, A., Leda, L., Schobben, M., Korn, D., & Ashouri, A. R. (2014). High-resolution stratigraphy of the Changhsingian (Late Permian) successions of NW Iran and the Transcaucasus based on lithological features, conodonts and ammonoids, Mitteilungen aus dem Museum für Naturkunde in Berlin. *Fossil Record, 17*(1), 41.

Geyer, G., Bayet-Goll, A., Wilmsen, M., Mahboubi, A., & Moussavi-Harami, R. (2014). Lithostratigraphic revision of the middle Cambrian ('Series' 3) and upper Cambrian (Furongian) in northern and central Iran. *Newsletters on Stratigraphy, 47*(1), 21–59.

Ghavidel-Syooki, M. (1986). Palynological studies and age determination of the Faraghan formation in the Gahkum Mountain and its temporal correlation with that of Faraghan Mountain. *Scientific Journal of Tehran University, 10*(1), 11–28.

Ghavidel-Syooki, M. (1988). *Palynostratigraphy and paleoecology of the Faraghan formation of southeastern Iran*. Michigan State University. 279.

Ghavidel-Syooki, M. (1990a). Achritarchs and chitinozoans of the Milla, Ilbeyk, Zardkuh and Faraghan formations in the Zardkuh area, and its correlation with Paleozoic sequence of the Chalisheh section and Darang borehole. In *The symposium of Diapirism; special implication on Iran*, pp. 141–201.

Ghavidel-Syooki, M. (1990b). Palynological study of Mila, Ilebek, Zard-kuh & Faraghan Fm. at Zard Kuh, Chal-i-sheh areas & Darang well No 1. In Zagros basin, Southern Iran, GR N, 1766.

Ghavidel-Syooki, M. (1992). Palynostratigraphy and paleogeography of the Padeha Formation and Khoshyeilagh Formation, in Kuh-e Ozom at Kopeh Dagh region, northeastern Iran. *TR N, 966*.

Ghavidel-Syooki, M. (1993). Palynological study of Palaeozoic sediments of the Chal-iSheh area, southwestern Iran. *Iranian Journal of Science, 4*, 32–46.

Ghavidel-Syooki, M. et al. (1994). Lower Paleozoic rocks of the Tang-e-Zakin, Faraghan Mountain, and Introduction of the Siahoo and Sarchahahn Formation in the Zagros Basin. *Earth Science Quarterly*, 1–21.

Ghavidel-Syooki, M. (1995). Srchahan formation. *Earth Science Quarterly, 15*(16), 74–89.

Ghavidel-Syooki, M. (1998). Upper Paleozoic deposits of the Zagros basin and introduction of the Zakin formation in the Faraghan Mountain. *Earth Science Quarterly, 29*(30), 54–73.

Ghavidel-Syooki, M. (2006). Palynostratigraphy and Paleogeography of the Cambro Ordovician strata in Southwest of Shahrud city (Kuh-e-Kharbash, near Deh-Molla), Central Alborz Range, Northern Iran. *Review of Palaeobotany and Palynology, 139*, 81–95.

Ghavidel-Syooki, M., Hassanzadeh, J., & Vecoli, M. (2011a). Palynology and isotope geochronology of the Upper Ordovician-Silurian successions (Ghelli and Soltan Maidan Formations) in the Khoshyeilagh area, Eastern Alborz Range, Northern Iran; stratigraphic and Palaeogeographic implications. *Review of Palaeobotany and Palynology, 164*, 251–271.

Ghavidel-Syooki, M., Javier Álvaro, J., Popov, L., Ghobadi Pour, M., Ehsani, M. H., & Suyarkova, A. (2011b). Stratigraphic evidence for the Hirnantian (latest Ordovician) glaciation in the Zagros Mountains, Iran. *Palaeogeography, Palaeoclimatology, Palaeoecology, 307*, 1–16.

Ghavidel-Syooki, M., & Owens, B. (2007). Palynostratigraphy and palaeogeography of the Padeha, Khoshyeilagh and Mobarak formations in the eastern Alborz Range (Kopeh-Dagh region), North-eastern Iran. *Revue de Micropaleontology, 50*(1), 129–144.
Ghavidel-Syooki, M., & Vecoli, M. (2008). Palynostratigraphy of Middle Cambrian to Lowermost Ordovician strata sequences in Zagros Mountains, southern: Regional stratigraphic implications, and palaeobiogeographic significance. *Review of Palaeobotany and Palynology, 150*, 97–114.
Ghavidel-Syooki, M., & Winchester-Seeto, T. (2002). Biostratigraphy andpalaeogeography of Late Ordovician chitinozoans from the Northeastern Alborz Range, Iran. *Review of Palaeobotany and Palynology, 118*, 77–99.
Gholamalian, H., Ghorbani, M., Sajadi, S. H. (2009). Famennian conodonts from Kal-e-Sardar section, eastern Tabas, central Iran. *Rivista Italiana di Paleontologia e Stratigrafia (Research in Paleontology and Stratigraphy), 115*(2).
Gholamalian, H., & Kebriaei, M. R. (2008). Late Devonian conodonts from the Hojedk section, Kerman Province, southeastern Iran. *Rivista Italiana di Paleontologia e Stratigrafia (Research in Paleontology and Stratigraphy), 114*(2).
Ghorbani, M. (1999). *Petrological analysis and metallogeny of the tertiary-quaternary magmatic rocks*. Geological Survey of Iran, Report p. 263.
Glaus, M. (1964). *Trias und Oberperm im zentralen Elburs (Persien)*. Geologisches Institut der Eidg, Technischen Hochschule und der Universität Zürich.
Glaus, M. (1965). *Die Geologie des Gebietes nördlich des Kandevan-Passes (Zentral Elburz), Iran* (Diss) Mitteilungen aus dem Geologischen Institut der Eidgenössischen, Technische Hochschule und Universität Zürich, N S, 48, 1–165.
Habibi, T., Corradini, C., & Yazdi, M. (2008). Conodont biostratigraphy of the Upper Devonian-Mississippian Shahmirzad section, central Alborz, Iran. *Geobios, 41*(6), 763–777.
Haghipour, A. (1971). *Hercynian movements in Iran*. Geological Survey of Iran.
Hairapetian, V., Pour, M. G., Popov, L. E., & Modzalevskaya, T. L. (2012). Stegocornu and associated brachiopods from the Silurian (Llandovery) of Central Iran. *Estonian Journal of Earth Sciences, 61*(2).
Hairapetian, V., Pour, M. G., Popov, L. E., Männik, P., & Miller, C. G. (2017). Silurian stratigraphy of Central Iran–an update. *Acta Geologica Polonica, 67*(2), 201–233.
Hamdi, B. (1975). Conodont studies on the Upper Silurian–Lower Devonian of the Kolur area, east Azarbayjan. *Geological Survey of Iran, Geological Quadrangle, D3*, 153–171.
Hamdi, B. (1989). *Stratigraphy and paleontology of the late Precambrian in the Alborz mountains, North Iran*. Geological Survey of Iran, Report 59.
Hamdi, B. (1995). *Precambrian and Cambrian sedimentary rocks of Iran*. Treatise on the Geology of Iran, Ministry of Mines and Metals.
Hamdi, B., Golshani, F. (1983). *Preliminary investigation on the lower Cambrian and Percambrian-Cambrian boundary of North Iran, Tehran*. Geological Survey of Iran.
Hamedi, M., & Alavi-Naeini, M. (1984). *A revision of the Dezu and Rizu 'Series', and introduction of a new Banestan formation of the Kuhbanan area, Kerman*.
Hasani, M. (1999). *Geological study of Plio-Quaternary of the Tehran area* (Unpublished Ph.D. thesis). Faculty of Earth Sciences, Shahid Beheshti University.
Heydari, E., Hassanzadeh, J., Wade, W. J., & Ghazi, A. M. (2003). Permian-Triassic boundary interval in the Abadeh section of Iran with implications for mass extinction: Part 1–Sedimentology. *Palaeogeography, Palaeoclimatology, Palaeoecology, 193*(3), 405–423.
Hoseini, B. (2011). *Petrography and petrogenesis of metamorphic rocks assigned to the Paleozoic of the Sanandaj-Sirjan Zone and their connection with mineralization* (Unpublished Ph.D. thesis). Faculty of Earth Sciences, Shahid Beheshti University.
Hoseini-e Barzi, M. (1996). The geological provenance of the Lalun sandstone of the type section (Unpublished MSc thesis). Faculty of Sciences, University of Tehran.
Hosseini, S. R. (2011). *Microfacies and depositional environment of the Asmari Formation of the Masjed-Soleyman Oil Field* (Unpublished MSc thesis). Bu-Ali Sina University.

Huckriede, R., Kürsten, M., Venzlaff, H. (1962). *Geotektonische Kartenskizze des Gebietes zwischen Kerman und Sagand, (Iran)*, Bundesanstalt für Bodenforschung.

Hushmandzadeh, A., Nabavi, M. H., Hamdi, B. (1988). Precambrian-lower Cambrian rocks of Iran. In *The Symposium of Reservoir Evaluation and Ore Potential of the Yazd Province*, Geological Survey of Iran.

ICS (2015). http://www.stratigraphy.org/index.php/ics-chart-timescale.

Jafarizadeh, H. (1997). *Biostratigraphy, lithostratigraphy and depositional environments of the Qom Formation* (Unpublished M.Sc. thesis). Faculty of Earth Sciences, Shahid Beheshti University.

Jahangir, H., Pour, M. G., Holmer, L. E., Popov, L. E., Ashuri, A. R., Rushton, A., et al. (2015). Biostratigraphy of the Cambrian–Ordovician boundary beds at Kopeh-Dagh, Iran. *Stratigraphy, 2*(12), 40–47.

Jenny, J. (1977). Géologie et stratigraphie de l'Elbourz oriental entre Allabad et Shahrud, Iran (unpublished). Thèse Université Genève, 1820.

Kalantari, M. (2006). *Microbiostratigraphy, microfacies and depositional environment of the Carboniferous deposits, Mobarak Formation, in the northern and northwestern Damghan* (Unpublished M.Sc. thesis). Faculty of Sciences, Ferdowsi Mashhad University.

Kavousi, M. A. (2005). *Microfaceis, temestites and depositional environment of the Milla Formation's Member C in the High Zagros.*

Kazeruni, N. (2009). *Conodont biostratigraphy of the Mobarak Formation in the Geirud section, northeastern Tehran* (Unpublished M.Sc. thesis). Faculty of Earth Sciences, Shahid Beheshti University.

Kebriaei-Zadeh, M., & Gholamalian, H. (2003). *Brachiopods of the Geirud Formation in the Sarkhon Oil Field (Bandar Abbas) and type section of Jahrom Formation* (Unpublished M.Sc. thesis). Faculty of Earth Sciences, Shahid Beheshti University.

Kohansal-Langaroudi, A. (2009). *Biostratigraphy of Ruteh formation in the Harijan section, Chalus Road, with special implications on conodont* (Unpublished M.Sc. thesis). Faculty of Earth Sciences, Shahid Beheshti University.

Kushan, B. (1973). *Stratigraphie un trilobiten fauna in der Mila-formation (mittelkambrium-tremadoc) im Alborz-gebireg (N-Iran).*

Lasemi, Y., & Mossadegh, H. (2005). Depositional environment and sequence stratigraphy of the mud stacks of Lower Mobarak Formation, northeastern Shahmirzad, Semnan. *Scientific Journal of Tehran University.*

Leven, E. J., & Gorgij, M. N. (2005). The Pennsylvanian-Permian of the Central and East Iran: Anark, Ozbak-Kuh and Shirgesht Areas. *Permophiles, 46,* 20–21.

Leven, E. J., & Gorgij, M. N. (2011a). Fusulinids and stratigraphy of the Carboniferous and Permian in Iran. *Stratigraphy and Geological Correlation, 19*(7), 687–776.

Leven, E. J., & Gorgij, M. N. (2011b). First record of Gzhelian and Asselian Fusulinids from the Vazhnan Formation (Sanandaj-Sirjan zone of Iran). *Stratigraphy and Geological Correlation, 19*(5), 486–501.

Lotfpour, M. (2003). *Porosity and permeability of the Upper Dalan formation in the South Pars Oil Field, based on the geophysical logs* (Unpublished M.Sc. thesis). Faculty of Sciences, University of Tehran.

Mahdavi Feiz-Abadi, A., Ashouri, A., & Khaksar, K. (2001). Paleobiogeography of the Member A of Gachal formation based on the coral fauna. In *The 20th Symposium of Geology, Geological Survey of Iran.*

Mehrban, N. (1983). *Lithostratigraphy and biostraigraphy of the Permian deposits in the Marsal section, Sangroud area, western Alborz* (Unpublished M.Sc. thesis). Faculty of Sciences, University of Tehran.

Mehrjerdi, H. (1995). *Palynostratigraphy and paleobiogeography of the Lower Paleozoic succession, Derenjal, Shirgesht and Niur Formations) in the Shirgesht area, Tabas* (Unpublished M.Sc. thesis). Faculty of Earth Sciences, Shahid Beheshti University.

Mirshahani, M., Adabi, M. H. (2003). Petrographical and geochemistrical features of dolomite beds of Barut Formation, Central Alborz. *Scientific Journal of Tehran University.*

Mohammadi, G.H. (1993). *Lithology and geochemical studies of Mobarak Formation in the Mobrakabad area* (Unpublished M.Sc. thesis). Faculty of Sciences, University of Tehran.

Mossadegh, H. (2000). *Microfossils, microfacies, depositional environment and stratigraphy of the Mississippian Mobarak Formation of the Central Alborz* (Unpublished Ph.D. thesis). Faculty of Sciences, Kharazmi University.

Motiei, H, (1993). *Stratigraphy of Zagros*. Geological Book Codification Project, Geological Survey of Iran.

Nabavi, M. H. (1976). *An introduction to geology of Iran*. Geological Survey of Iran, Tehran.

Najafi, M. (2001). Microbiostratigraphy of the Mobarak Formation in the Abnik Section (Unpublished M.Sc. thesis). Faculty of Sciences, Kharazmi University.

Najafi-Neisani, R. (2004). *Palynostratigraphy of the Zakin Formation in the Gahkum Mountain, Norhtern Bandar Abbas* (Unpublished M.Sc. thesis). Faculty of Earth Sciences, Shahid Beheshti University.

NIOC. (2016). *Paleontology and biozonation of paleozoic sediments of Central Iran and Zagros Basins*. National Iranian Oil Company, Tehran (no 89230).

Okhravi, R. (1991). *The study of depositional environment and diagensis of the Dorud Formation, Haraz Section, Central Alborz* (Unpublished M.Sc. thesis). Faculty of Sciences, University of Tehran.

Partoazar, H. (1995). *Permian of Iran, geological book codification project*. Geological Survey of Iran, No 222, 340.

Rahimi, B. (2002). *Structural studies of the Alborz Range in the North Damghan* (Unpublished Ph.D. thesis). Faculty of Earth Sciences, Shahid Beheshti University.

Rastegar, G. H. (1995). *Brachiopods and the location of the cephalopod-bearing strata of the Shishtu Formation* (Unpublished M.Sc. thesis). Faculty of Sciences, Ferdowsi Mashhad University.

Reitlinger, E. A. (1965). *Development of foraminifera during the late Permian and early Triassic Epochs in Transcaucasian territory*.

Roozgar, R. (2008). *Palynostratigraphy of Padeha and Bahram formations of the Garik Section, Eastern Zarand* (Unpublished M.Sc. thesis). Faculty of Earth Sciences, Shahid Beheshti University.

Ruttner, A., Nabavi., M. H., & Hajian, J. (1968). *Geology of the Shirgesht Area (Tabas area, East Iran)*. Geological Survey of Iran, Report 4, pp. 5–133.

Ruttner, A., & Stöcklin, J. (1966). Forward, with key map, stratigraphy summary and description of fossil-localities. *Geological Survey of Iran Report, 6,* 2–6.

Sabouri, J., Mahmoudi, M., Sabbaghiyan, H., Aria-Nasab, M. (2014). The first evidences of Carboniferous (Middle-?Late Mississippian) sediments in Zagros Basin, Iran and its palaeogeographic implications. In *Abstract Volume and Proceedings of the Third International Symposium of the IGCP-589, Development of the Asian Tethyan Realm: Genesis, Process and Outcomes*, pp. 29–30.

Saidi, M. (2010). The geological provenance of sandstone part of Padeha Formation, Simeh Kuh section, Central Iran; implication on the diagenesis effects. *Earth Science Quarterly, 78*(20).

Sardar Abadi, M., Kulagina, E. I., Voeten, D. F., Boulvain, F., & Da Silva, A. C. (2017). Sedimentologic and paleoclimatic reconstructions of carbonate factory evolution in the Alborz Basin (northern Iran) indicate a global response to Mississippian (Tournaisian) glaciations. *Sedimentary Geology, 348,* 19–36.

Seger, F. E. (1977) *Zur Geologie des Nord-lamut Gebietes* (Zentral-Elburz, Iran) [microform]/von Felix Subjects. Geology-Iran-Elburz Mountains.

Setudehnia, A. (1972). Paleozoic sequence at Zard Kuh and Kuh-e Dina IOOC Report, No 1196 (Unpub). Original survey was done by Setudehnia and Kherapir in 1971 but Reported by Setudehnia.

Setudehnia, A. (1975). The Paleozoic sequence in Southwest Iran and adjacent areas. *Journal of Petroleum Geology, 1,* 3–34.

Shadan, M. (2008). *Depositional environment and diagenesis of the Khan Formation of the Chahrouf and Cheshemeh-Bakhsi Sections, Tabas area* (Unpublished M.Sc. thesis). Faculty of Earth Sciences, Shahid Beheshti University.

Shafiei, N. (2008). *Palynostratigraphy of Padeha and Bahram formations of the Shamsabad Section* (Unpublished M.Sc. thesis). Faculty of Earth Sciences, Shahid Beheshti University.

Shemirani, A. (2002a). *The handbook of Palaeozoic of Iran*. Faculty of Earth Sciences, Shahid Beheshti University.

Shokravi, N. (1997). *Lithostratigraphy and biostratigraphy of Nesen formation in the Central Alborz, with a special implication on the Permo-Triassic Boundary* (Unpublished M.Sc. thesis). Faculty of Sciences, Kharazmi University.

Siasi-e-Torbati, A. (2003). *Biostratigraphy of Devonian Khoshyeilagh Formation of the Badeleh-Kuh Section, Northern Damghan; with special implication on conodonts* (Unpublished M.Sc. thesis). Faculty of Earth Sciences, Shahid Beheshti University.

Steiger, R. (1966). Die Geologie der West-Firuzkuh-area (Zentralelburz/Iran), Mitteilungen aus dem Geologischen Institut der Eidgen6ssischen Technischen Hochschule und der Universitiit Zurich, 68, 145 (in German).

Stepanov, D. L., Golshani, F., & Stöcklin, J. (1969). Upper Permian and Permian-Triassic boundary in North Iran. *Geological Survey of Iran, 12,* 71.

Stöcklin, J. (1960). General geology, East Alborz. *Geological Survey of Iran, 52*(2).

Stöcklin, J. (1961). Lagoon formation and salt domes in East Iran. *Bulletin of the Iranian Petroleum Institute, 3,* 29–46.

Stöcklin, J. (1962). Note on the geology of the Iran-Pakistan border area. *Geological Survey of Iran, Geological Note, 1,* 15.

Stöcklin, J. (1974). Northern Iran: Alborz Mountains. *Geological Society, London, Special Publications, 4*(1), 213–234.

Stöcklin, J., & Setudehnia, A. (1991). Stratigraphic Lexicon of Iran. Tehran, Iran: *Geological Survey of Iran, Report* no. 18.

Stöcklin, J., Eftekharnezhad, J., & Hushmandzadeh, A. (1965). *Geology of the Shotori Range (Tabas area, East Iran)*. Geological Survey of Iran, Report 3, 1–69.

Stocklin, J., Ruttner, A., & Nabavi, M. H. (1964a). *New data on the Lower Paleozoic and Precambrian of North Iran*. Geological Survey of Iran, Report 1, 13.

Stöcklin, J., Ruttner, A., & Nabavi, M. H. (1964b). *New data on the lower Paleozoic and Precambrian of North Iran*. Geological Survey of Iran, Report 1, 29.

Szabo, F., & Kheradpir, A. (1978). Permian and Triassic stratigraphy, Zagros Basin, South-West Iran. *Journal of Petroleum Geology, 1*(2), 57–82.

Tabatabaei, G. (2003). *Geochemistry of the Dalan formation, late Permian, and Kangan formation (early Triassic) of the Salman Field* (Unpublished M.Sc. thesis). Faculty of Sciences, Amirkabir University.

Taraz, H. (1972). Geology de la Surmaq- Deh bid, Central Iran Geological Survey of Iran report 37.

Taraz, H. (1973). Correlation of uppermost Permian in Iran, central Asia, and south China. *AAPG Bulletin, 57*(6), 1117–1133.

Taraz, H. (1974). *Geology of Surmaq-Deh bid area, Abadeh region, Central Iran*. Geological Survey of Iran, Report 47, 147.

Taraz, H., Golshani, F., Nakazawa, K., Shimizu, D., Bando, Y., Ishii, K.I., et al. (1981). *The Permian and the lower Triassic systems in Abadeh region, Central Iran*.

Thiele, O., Alavi, M., Assefi, R., Hushmandzadeh, A., Seyed-Emami, K., & Zahedi, M. (1968). 1: 250000 Golpaygan quadrangle map. Geological Survey of Iran, Report.

Vachard, D., & Arefifard, S. (2015). Foraminifers and algae of the late Tournaisian–early Viséan boundary interval (MFZ8–9) in the Gachal Formation (Central Iran). *Revue de Micropaléontologie, 58*(3), 85–216.

Wendt, J., Kaufmann, B., Belka, Z., Farsan, N., & Karimi Bavandpur, A. (2002). Devonian/Mississippian stratigraphy, facies patterns and palaeogeography of Iran, Part I: Southeastern Iran. *Acta Geologica Polonica, 52,* 129–168.

Yamini, A. (1995). *Cephalopods of cephalopod-bearing beds of Shishtu I Member in the Shotori Mountains, Tabas area* (Unpublished M.Sc. thesis). Faculty of Sciences, Ferdowsi Mashhad University.

Yazdi, M. (1999). Late Devonian-carboniferous conodonts from eastern Iran. *Rivista Italiana di Paleontologia e Stratigrafia, 105*(2), 167–200.

Zaman, S. H. (2001). *Paleontology of sperifirids, rynchonellids and biostratigraphy of the Devonian rocks of the Lower Khoshyeilagh Formation* (Unpublished M.Sc. thesis). Faculty of Earth, University of Tehran.

Zandkarimi, K., Najafian, B., Bahrammanesh, M., & Vachard, D. (2014a). Permian Foraminiferal biozonation in the Alborz mountains at Valiabad Section (Iran). *Permophiles, 60,* 10–16.

Zandkarimi, K., Najafian, B., Vachard, D., Bahrammanesh, M., & Vaziri, S. H. (2014b). Latest Tournaisian–late Viséan foraminiferal biozonation (MFZ8–MFZ14) of the Valiabad area, northwestern Alborz (Iran): Geological implications. *Geological Journal, 51*(1), 125–142.

Zandkarimi, K., Vachard, D., Cózar, P., Najafian, B., Hamdi, B., & Mossadegh, H. (2017a). New data on the Late Viséan-Late Serpukhovian foraminifers of northern Alborz, Iran (biostratigraphic implications). *Revue de Micropaléontologie, 60*(2), 257–278.

Zandkarimi, K., Vachard, D., Najafian, B., Hamdi, B., & Mosaddegh, H. (2017b). Viséan-Serpukhovian (Mississippian) archaediscoid foraminifers of the northern Alborz, Iran. *Neues Jahrbuch für Geologie und Paläontologie-Abhandlungen, 286*(1), 105–123.

Zandkarimi, K., Vachard, D., Mosaddegh, H. & Ehteshami-Moeinabadi, M. (2018). Mississippian lithofacies and foraminiferal biozonation of the Alborz Mountains, Iran: Implications for regional geology. *Geological Journal* (doi: https://doi.org/10.1002/gj.3242).

Chapter 3
Mesozoic

The Mesozoic era consists of three periods the Triassic, Jurassic and Cretaceous spanning in total 186 Myr (Fig. 3.1).

The Mesozoic era consists of three periods the Triassic, Jurassic and Cretaceous spanning in total 186 Myr (Fig. 3.1).

The Paleozoic stratigraphies of the various regions of Iran show many similarities owing to their deposition under the same marine platform conditions. However, the paleogeography of the regions considerably changed at the beginning of Mesozoic when the Zagros basin was detached from Central Iran and Alborz, resulting in lithological similarities with southern and west southern Iran's neighboring territories (Fig. 3.2). Consequently, the Mesozoic and younger sediments of the Kopeh-Dagh differ from those of Alborz and Central Iran.

3.1 Triassic

In general terms, the Lower Triassic strata of Iran were deposited on an epicontinental shelf (e.g. Alavi 2004). The lowermost Triassic units can be traced in the areas where deposition across the Permian-Triassic boundary was continuous, such as Julfa (northwestern Iran; Tichert et al. 1973), Abadeh (southern and Central Iran; Taraz et al. 1981) and northern Alborz (Karj-Chalus road; Bozorgnia 1973; Gaetani et al. 2009; Zandkarimi et al. 2014, 2018).

In the other parts of Iran where there is no evidence of lowermost Triassic rocks, the Lower-Middle Triassic Elika[1] Formation disconformably overlies either the Nesen Formation or the Ruteh Formation (Gaetani et al. 2009). The Upper Triassic rocks of Northern Iran are less well-exposed due to a sedimentary gap or to the effects of subsequent erosion (Aghanabati 2004). The red-colored shales and sandstones

[1] Worth mentioning is that Glaus (1964) mistakenly named the Lower Triassic rocks of Alborz the "Elikah Formation" while the true name of the village around of which the formation is defined is in fact "Elika".

M. Ghorbani, *Lithostratigraphy of Iran*, Springer Geology,
https://doi.org/10.1007/978-3-030-04963-8_3

Fig. 3.1 International chronostratigraphical chart of Mezozoic (ICS 2015)

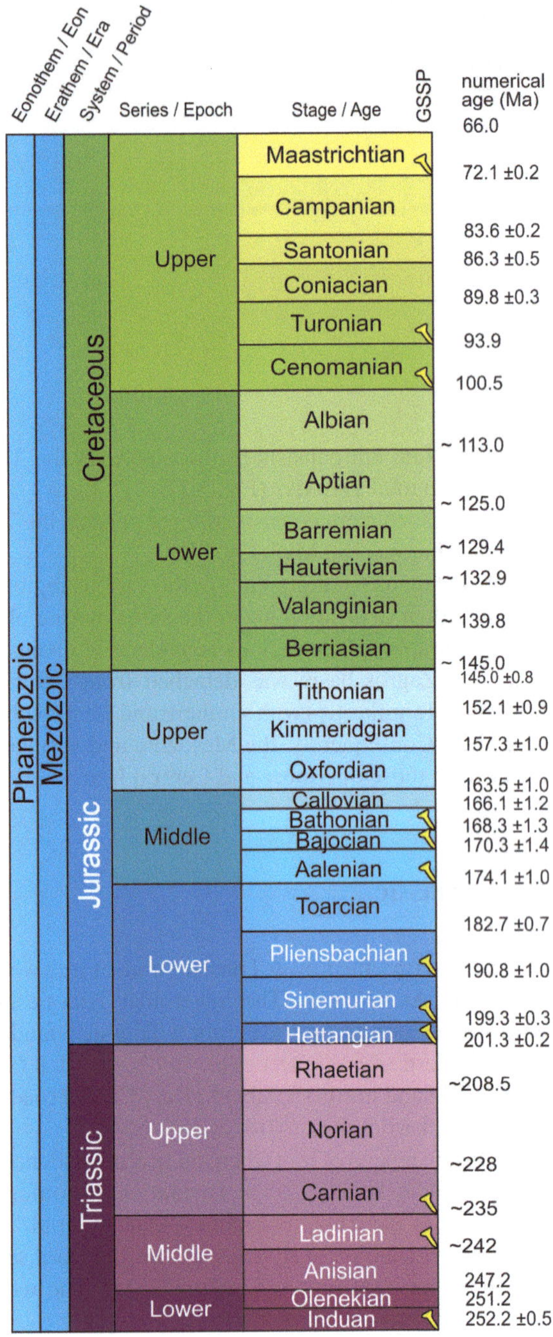

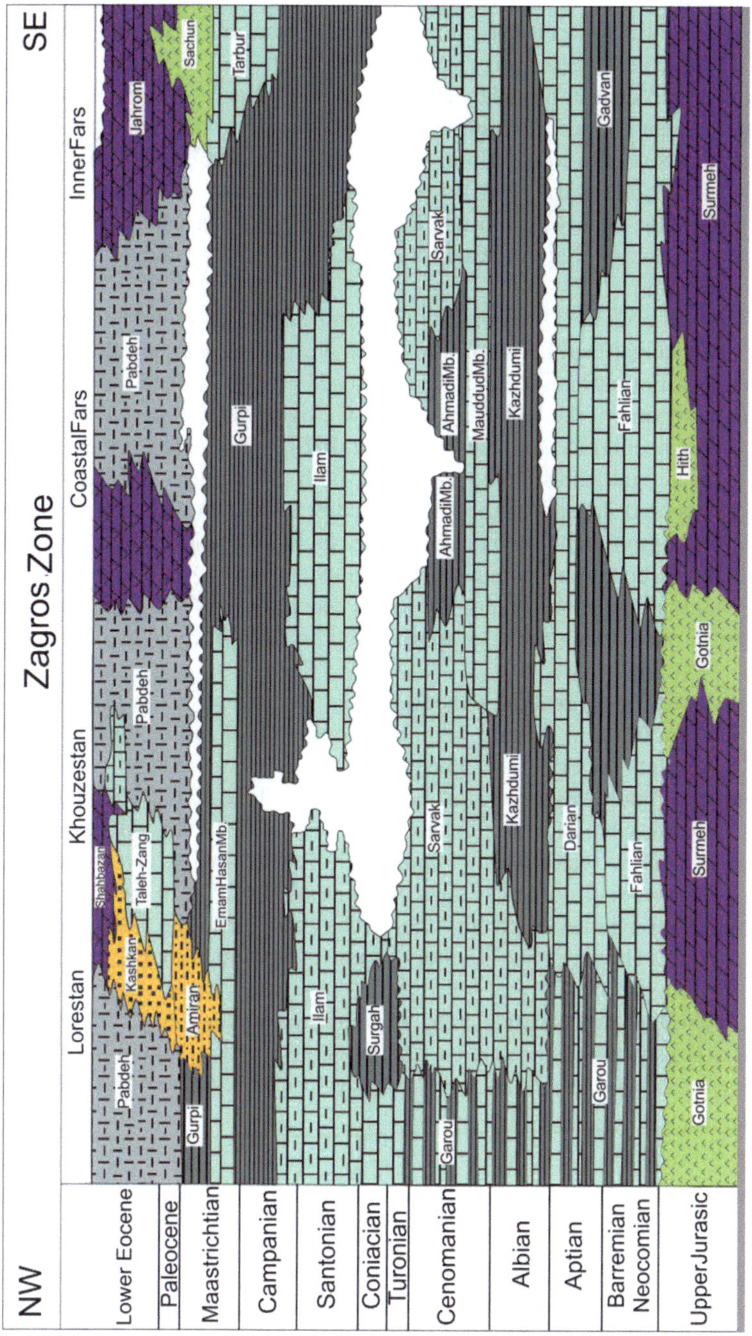

Fig. 3.2 A generalized scheme of the Mesozoic units of Southwestern Iran and Southern Persian Gulf Countries, modified after Motiei (1993)

interbedded with dolomites and limestones of the Sorkh Shale Formation, constitute the Lower Triassic rocks of east Central Iran. This succession is disconformably underlain and overlain by the cream-colored Permian Jamal and light-colored Shotori Formations, respectively (Stöcklin et al. 1965).

The Middle-Late Triassic boundary coincides with the Early Cimmerian Orogeny. This structural event divided the Iranian basin into three new basins, Zagros (in the southern and east southern Iran), Alborz (Northern Iran) and Central Iran.

In the Alborz, Central Iran and Eastern Iran the Triassic epicontinental shelf sediments include the mainly Dolomitic Elika (Alborz), shaly-carbonate Sorkh Shale (Central Iran), Dolomitic Shotori (Central Iran) and the shaly Nayband Formation (East of Central Iran).

In the Zagros, the Khaneh Kat Formation is composed of shallow water carbonate sediments, and Kangan and Dashtak formations are of carbonate-evaporite facies. The Khaneh Kat Formation is generally composed of fine-grained, silica-bearing dolomite in most parts of the High Zagros and of clayey carbonate, shale and limestone in the type section (Szabo and Kheradpir 1978). The same formation in the High Zagros was regarded as the equivalent of both the Kangan and Dashtak formations of the Folded Zagros (Szabo and Kheradpir 1978). The Kangan Formation is lithologically composed of three distinct facies including clean carbonate, argillaceous shale and carbonate-evaporate facies. The evaporate Dashtak Formation also includes with inhomogeneous lithological facies encompasses six member including the Aghar shale, the A evaporite, C evaporite, Sefidar Dolomite and the D evaporite Members.

As already mentioned, generally, the Mesozoic formations of the Zagros significantly differ from those of Central Iran and Alborz, rather can be correlated with their time equivalents in northern Arabia and eastern Iraq (Fig. 3.2).

Overlying the Middle Permian Ruteh Formation, the Triassic rocks of Azarbaijan are sporadically distributed in the Takab and Shahin-Dezh. This succession was deposited in a shallow water (Alavi-Naini et al. 1982). Having low thickness in the area, the Triassic rocks of Azarbaijan are composed of thin to medium-bedded limestone and thick-bedded dolomite lithologically similar to the Elika Formation of Central Alborz. However, its lithology is not consistent throughout the entire Azarbaijan region.

In the Kopeh-Dagh and Nakhlak areas, in addition to sedimentary carbonate succession, there are some volcanoclastic rocks, possibly reaching 300 m in thickness. In the following account, the Triassic rocks of the various regions of Iran are briefly mentioned.

3.1.1 Alborz

Triassic rocks are widely distributed in the Alborz Mountains and in some areas are of considerable thickness. The dominant lithology is shallow water carbonate (dolomite and limestone), referred to as the "Elika Formation" or its equivalents.

Elika Formation

This formation encompasses two members the Lower Limestone (Vermiculate Limestone) of Early Triassic age and the "Upper Dolomite Member" or "Main Dolomitic Member" of Middle Triassic age (Glaus 1964).

The Lower Member (95 m thick) includes thin-bedded laminated limestone to shaly limestone with marly limestone and rarely dolomitic limestone in some beds. The yellow to pinkish or grey-green is the dominant color of this member. The stratification surfaces of some beds are crowded with worm traces referred to as the "Vermiculate Limestone" (Glaus 1964). Some conglomerate intercalations are observed in the type section.

The upper member is 200 m thick and is composed of dolomitic limestone and dolomite. Middle Triassic age was proposed for this unit based mainly on its foraminifer and conodont content. Some white colored limestone beds overlying this member are referred to as the Veresk Limestone (Nabavi 1976; Alavi-Naeini 2009). Of Late Triassic age (Carnian to Norian), the so called Verask Limestone is exposed in the eastern Alborz (Seyed-Emami 2003), at Ayeneh-Varzan, Veresk Bridge and at Shahmirzad, where its thickness varies from 60 to 80 m (Alavi-Naeini 2009). Some researchers (e.g. Shemirani 2002b) reported that Alborz Mountains were exposed in the Late Triassic. Nevertheless, sections in some areas contain coal-bearing beds (i.e. the beds of the Shemshak Group) that are correlatives of the Upper Triassic Nayband Formation. The Russian geologists proposed the name "Tazareh Formation" for these coal bearing beds (Bragin et al. 1976). The Elika Formation is exposed in the major parts of Iran including eastern, central and western Alborz and most of Azarbaijan.

Shemshak Group

The Shemshak Formation was first introduced as a formation by Assereto (1966) but this was subsequently elevated to Group status (e.g. Fursich et al. 2009), consisting of several formations of Upper Triassic-Jurassic age (Fig. 3.3). The Triassic formations of this group are as follows:

Ekrasar Formation

The type section of the Ekrasar Formation was introduced by Bragin et al. (1976) on the eastern slopes of Ekrasar Village. It is predominantly monotonous, dark-grey argillaceous siltstone and siltstone, locally also of blackish shaly argillaceous siltstone with much pyrite. Occasionally, thin siltstone and fine-grained ripple- or parallel-bedded sandstone beds are intercalated, arranged in thickening-upwards parasequences. In the basal part of the formation, well-bedded, silty micritic limestones are locally developed in the northern Alborz. These are usually reddish, the colour stemming apparently from bauxitic clays reworked from the subaerially exposed Elika Platform further south (Furisch et al. 2009).

The Ekrasar Formation is underlain and overlain by the Elika and Lalehband formations with sharp and gradational contacts, respectively (Fursich et al. 2009). A late Carnian–Norian age can be assigned to this formation based on its ammonite fossil content (Seyed-Emami et al. 2009).

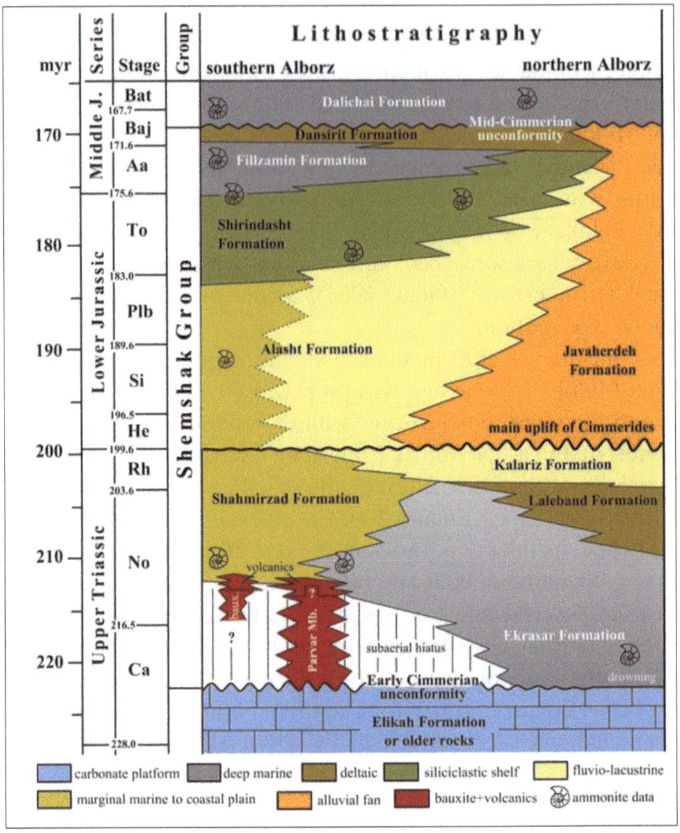

Fig. 3.3 Revised lithostratigraphic scheme of the Shemshak Group (Fursich et al. 2009)

Laleband Formation

The Laleband Formation was first erected for rocks exposed in southeastern Laleband Village, north Damghan (Fursich et al. 2009). The Laleband Formation is mainly composed of regular alternations of bedded sandstones and argillaceous siltstones. The sandstones are mainly fine- to medium-grained, commonly sharp-based and graded. In some cases, the sandstones have erosional bases. Sedimentary structures include planar lamination, ripple bedding and some large-scale trough cross-bedding. It is underlain and overlain by the Ekrasar and Kalariz formations with gradual and sharp boundaries, respectively. Based on rare bivalve occurrences a late Triassic age has been suggested (Fursich et al. 2009).

Shahmirzad Formation

This formation was introduced around Shahmirzad to accommodate a succession of siliciclastic sediments. In some localities volcanic and volcaniclastic rocks, bauxite and, rarely, carbonates occur. The Shahmirzad Formation is developed in most parts of the southern Alborz Mountains, where it interfingers laterally with the Ekrasar and Kalariz formations. The lower boundary with the Elika Formation is completely sharp and the upper boundary with either the Kalariz or the Alasht Formation is gradational (Fig. 3.3). A Norian–Rhaetian, late Triassic, age has been suggested based on bivalves (Fursich et al. 2009).

Kalariz Formation

The Kalariz Formation was introduced by Bragin et al. (1976) as a "suite" and later redefined by Fursich et al. (2009) as a formation. The type section is located near the Tarazeh Coal Mine, Kalariz Village, eastern Damghan. It extends across much of the Alborz Range but is absent in the Shahmirzad area, where it is replaced by marginal marine, coastal-plain sediments of the Shahmirzad Formation (Fursich et al. 2009). The Kalariz Formation consists of grey argillaceous–silty fine-grained sandstones with thin intercalations of ripple-bedded fine-grained sandstones, argillaceous silt with ferruginous concretions, carbonaceous silt and coal alternate with fine-grained, large-scale trough cross-bedded sandstones (Fursich et al. 2009).

3.1.2 Central Iran

Triassic deposits of diverse facies are widely distributed in Central Iran especially in eastern parts. These rocks are of the same lithology in all of Central Iran except for the Anarak area.

Generally, a long-term erosional phase, from Middle Permian to Middle Triassic, occurred in most parts of Iran. This resulted in the deposition of red colored clastic sediments referred to as the "Sorkh Shale Formation".

Sorkh Shale Formation

The Sorkh Shale Formation rests on the Jamal Formation, grading upward to red colored shaly limestone of the Sorkh shale Formation. In some areas, the underlying formation was reported to have gradual contact with limestones of the Jamal Formation. The upper boundary with the Shotori Formation is also gradual (sedimentary continuity) consisting of a few meters of yellow to green shaly limestone and dolomite (Stöcklin et al. 1965).

The formation is composed of clayey and limey shale with dark red color, which is characteristic of the formation, and of few pink to yellow colored limestone intercalations. Some ripple mark structures can be observed on the stratification surface (Stöcklin et al. 1965).

The Sorkh Shale Formation is widely distributed from its type area of the Shotori Mountain through southern area of Nayband and Lakar-kuh in the northern Kerman. In the northern part of Central Iran, it can be traced to the Gardou area of the northern Shirgesht (Stöcklin et al. 1965).

Westward from the type section to the Soh area of Kashan, the equivalent of the Sorkh Shale Formation, mainly consist of 50 m-thick, red-colored, sandstone and conglomerate with red marly shale intercalations, which has been named the "Abyaneh Formation" (Aghanabati 2004).

The Sorkh Shale Formation of the Toroud area contains 100–120 m yellowish red limestones. The lithological composition of the Sorkh Shale Formation of the Yazd area tends to show more calcareous content in comparison to type area of the formation in the Shotori Mountains. Its thickness may reach 95 m in the Bafgh and Biabanak areas. Being the time equivalent of the Elika Formation, Sorkh Shale Formation has been assigned to the Triassic age based on the fossil contents and stratigraphical position (NIOC).

Shotori Formation

The Middle Triassic Shotori Formation, gradationally covering the Sorkh Shale Formation, is mainly composed of carbonate and dolomitic rock units. The dolomitic unit is considered to be the main unit of the Formation. The type section of the Shotori Formation contains 220 m well-bedded dolomite and limestone. A white colored limestone unit "Spahak limestone" overlies the dolomitic unit which is of a few meters thick (30–150 m) and of limited distribution. Its occurrence does not depend on the underlying dolomitic unit and cannot be seen everywhere the dolomitic unit is observable. The Shotori Formation is exposed in most of the mountains of Central Iran, except for the Anarak area where Triassic rock of other facies are developed (Stöcklin et al. 1965).

Shemshak Group

The Upper Triassic-Lower Jurassic Shemshak Group of Central Iran consists of post Eo-Cimmerian orogeny sediments divided into several formations (Aghanabati 1977). Herein, the Upper Triassic Nayband Formation is described and other formations will be discussed in the Jurassic section.

Nayband Formation

This formation is the product of Upper Triassic marine deposition in Central Iran. The Nayband Formation at its type section, southwest of the Nayband Mountain, unconformably overlies the Middle Triassic karstified carbonates of the Shotori Formation and, in turn is overlain by the Lower Jurassic Ab-Haji Formation (Seyed-Emami 2003). The Nayband Formation (ca. 2200 m thick) is subdivided, in ascending order, into the Gelkan, Bidestan, Howz-e Sheikh, and Howz-e Khan members (Bronnimann et al. 1971). Moving northward, to the south Shotori Mountains (Kamar Macheh Kuh), the Nayband Formation changes lithologically and its thickness is greatly reduced (Stöcklin et al. 1965). The following members have been established (Bronnimann et al. 1971).

Gelkan Member (915 m-thick). This is composed of intensely weathered pencil shales, light gray siltstones and thin sandstones. Sedimentary structures such as ripple marks and cross-bedding are present in this member. The Gelkan Member contains the bivalve *Megalodon* and also plant fossils in some horizons.

Bidestan Member (450 m-thick). This consists of shale, siltstone, sandstone with intercalations of brownish, thin-bedded sandy limestones and marl bearing Coelantaria fossil (e.g. *Heterastridum* sp.). In general, this member is marked by pinkish-to cream color which distinguishes it from other members of the Nayband Formation. A Late Triassic age (Norian) has been proposed for this Member based on the occurence of *Heterastridum*.

Howz-e Sheikh Member (365 m-thick). This consists of pencil shales changing upward into siltstones and brownish sandstones. The most common fossils in this member are bivalves and hydrozoans. This Member has been assigned to the Late Triassic (Norian) based on its lithostratigraphical position.

Howz-e Khan Member This (465 m-thick) consists of cliff-forming reefal limestone with intercalations of shale and sandstone. This member contains corals, hydrozoans, brachiopods, bivalves, gastropods and some echinoderms, indicating a Late Triassic (Norian) age.

Anarak-Nakhlak Area

This succession differs paleontologically and lithologically from that of Central Iran. The thick Triassic sequence (2549 m-thick) of the Anarak-Nakhlak area, northern Naein, shows remarkable lithological resemblance to that of northern Afghanistan. This complete and thick sequence that was named the "Nakhlak Group" is subdivided into the following formations:

1. Alam Formation, 873 m-thick, of Olenekian to Anisian age (Early to Middle Triassic).
2. Baghrogh Formation, 1251 m-thick, of Ladinian age (Middle Triassic).
3. Ashin Formation, 425 m-thick, of Carnian age (Late Triassic).

Almost all of the Middle Triassic sediments of Central Iran were deposited in an epicontinental sea and consist of dolomite and limestone accompanied by evaporite sediments, especially gypsum.

The Late Triassic is also characterized by several thousands of meters of clastic and lagoonal sediments. This high thickness is unique in the Triassic of Iran.

The most important Triassic rocks of Central Iran are as follow:

The Sorkh Shale and the Shotori and Nayband formations of the east of Central Iran, Kerman and Yazd. The Nakhlak Group including the Alam, Baghrogh and Ashin formations.

Abadeh Area

The Abadeh area appears to contain the most complete and interesting Triassic sections known till now in Iran. Taraz (1972) subdivided the Triassic deposits of the Abadeh area into three main groups:

A Lower Triassic group including chrome-bearing hieroglyphic limestone, oolitic limestone and marl (683 m-thick).

A Middle Triassic group consists of well-bedded dark grey to light grey dolomite, gradually covering the latter group (780 m-thickness).

An Upper Triassic group including reefal limestones, sandy limestones and limey sandstones and shales.

3.1.3 *Kopeh-Dagh*

The Triassic rock exposed in the Kopeh-Dagh are limited to the Jajarm area, in the eastern mountains of the Gorgan and in the erosional window of Aghdarband (Baud et al. 1991). The sedimentary facies of the later are completely different to the two other areas. In the Jajarm area and eastern Alborz, Triassic rock facies resemble to those of Alborz Mountain, i.e. Elika Formation.

The Triassic lithostratigraphic units of Kopeh-Dagh are as follows:

Elika Formation: Lithologically this resembles exposures in the Alborz Mountains.

Ghareh-Gheytan Formation: shale, sandy shale, sandstone and conglomerate.

Aghdarband Group: The Triassic succession of Aghdarband is about 1500 m-thick and consists of the following units (Ruttner 1991):

1. Sefid Kuh Formation (200 m): a sequence of light-coloured and bioturbated limestones with few conodonts assigned to the Late Scythian (late Olenekian) (Baud et al. 1991);
2. Nazarkardeh Formation (20–50 m), a succession of dark, cherty and thin-bedded-limestones of early Anisian age;
3. Sina Formation (400–700 m), a turbiditic and siliciclastic sequence with many volcanoclasts and some conglomeratic intercalations with conodonts, echinoderms, pelecypods and ammonoids assigned to the Late Ladinian and ?Early Carnian;
4. Miankuhi Formation (250 m): a rather monotonous, brownish sequence of shales, siltstones and fine-grained sandstones.

It is noteworthy that paleogeographically the fauna seems to be restricted to a narrow strip at the southern edge of Laurasia (Krystyn and Tatzreiter 1991).

The Aghdarband Group is underlain and overlain by Permian rocks and basal conglomerate of the Lower Jurassic Kashafroud Formation, respectively. Both contacts are angular unconformities.

3.1.4 *Zagros*

The Triassic sediments of southern Iran can be into two groups:

1. Carbonate and evaporite sediments comprising the Kangan Formation and the overlying Dashtak Formation. Three facies were determined in the Kangan Formation (Szabo and Kheradpir 1978) as follows:

1. Clean carbonate facies;
2. Basal argillaceous and shaly facies; and
3. Carbonate-evaporate facies.

The Dashtak Formation consists of a succession of Upper Triassic shaly and evaporite rocks.

2. The Khaneh Kat Formation includes carbonate sediments without any evaporite beds. The type section of the formation consists of 364 m of medium to thin-bedded, dark grey fine-grained dolomite in the lower part; and 122 m breccia and brown massive recrystallized dolomite (Szabo and Kheradpir 1978).

Due to the facies variation, the Triassic rocks of Zagros have been divided into 2 facies (1) carbonate facies without certain evaporite, i.e. the Khaneh Kat Formation and (2) carbonate and evaporite facies, of which the latter was subdivided into two completely distinct formations in ascending stratigraphic order as follows:

A: the Kangan Formation
B: Dashtak Formation.

The geographical distribution of the facies of the Zagros area and their relationships are summarized in the Table 3.1 and lithological features of the Triassic rocks of Iran are shown in the Table 3.2. Also, the selected stratigraphically-diagnostic fossils of Iranian Triassic rocks are illustrated in Plates 3.1 and 3.2.

3.2 Jurassic

The Jurassic Period is geochronologically divided into three Epochs, the Early Jurassic, the Middle and the Late Jurassic. The Lower Jurassic rocks of Iran are of more varied depositional environments compared to the Upper Triassic (Seyed-Emami 2003).

The Early Cimmerian Orogeny which occurred during the late Middle Triassic is considered to be one of the most important events that affected the tectonic evolution of Iran, causing significant changes to patterns of paleobiogeography and sedimentation (e.g. Aghanabati 2004). In Central Iran and the Alborz, the Upper Triassic, Lower and Middle Shemshak Group is composed of shale and sandstone facies ranging from several meters to more than three thousand meters in thickness (Wilmsen et al. 2009). The occurrence of plant fossil remains and coal layers indicate that some parts of this succession were formed in lagoonal or continental-related environments (Wilmsen et al. 2009). The existence of multiple marine ammonite-bearing limestone and sandstone intercalations hints at frequent and short-lived rises in sea level indicating temporary replacement of continental environments by open marine conditions. Consequently, it seems that these sediments were mainly formed in tectonically unstable conditions (Aghanabati 1975).

Table 3.1 Geographical distribution of Triassic sediments of Iran (Alavi-Naeini 2009)

Alborz	Upper Triassic		Paland Fm.		
	Lower and Middle Triassic		Elika Fm.	Veresk limestone Mb.	
				Lower dolomitic unit	
				Lower limestone unit	
Central Iran	Eastern part	Upper Tri-assic	Nayband Fm.	TRn7	
				Ghadir Mb.	TRn6
					TRn5
				Howz- Khan Mb.	
				Hoz-shiekh Mb.	
				Bidestan Mb.	
				Gelkan Mb.	
	Western Part		Shotori Fm.	Spahak limestone	
				Dolomite Mb.	
			Sorkh Shale Fm.		
	Southwestern part		Nakhlak Gr.	Ashin Fm.	
				Baghrogh Fm.	
				Alam Fm.	
			Abadeh Fm.	Upper Tiassic Gr.	
				Middle Triassic Gr.	
				Lower Triassic Gr.	
Sanandaj-Sirjan	Eghlid area				
Zagros	Folded Zagros zone		Dashtak Fm.	D Evaporite Mb.	
				Sepidar Dolomite	
				C Evaporite Mb.	
				B Evaporite Mb.	
				A Evaporite Mb.	
				Aghar shale Mb.	
			Kangan Fm.	Evaporite carbonate facies	
				Basal Shale and clayey facies	

(continued)

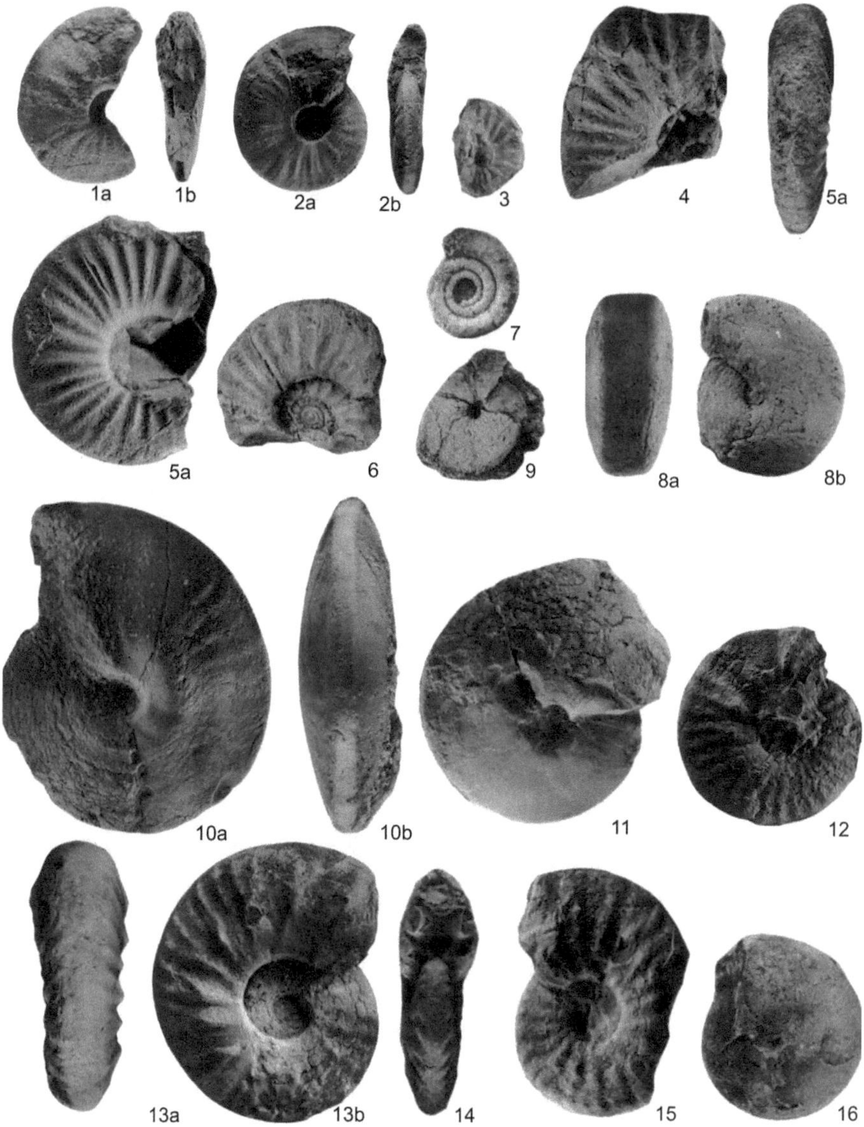

Plate 3.1 Selected Triassic ammonoids of the Aghdarband, northeastern Iran (Krystyan and Tatzreitern 1991). 1a, 1b. *Nicomedites osma*; 2a, 2b. *Pseudohollandites ?marmarensi*; 3. *Pseudohollandites eurasiatic*; 4–6 *Leiophyllites suessi*; 7. *Leiophyllites suessi*; 8. *Procladiscites* sp. cf. *P. propontic*; 9–11. *Semibeyrichites ruttneri*; 12–15. *Aghdarbandites ismidicus*; 16. *Romanites simionescui*

Table 3.1 (continued)

Alborz	Upper Triassic	Paland Fm.		
	Lower and Middle Triassic	Elika Fm.	Veresk limestone Mb.	
			Lower dolomitic unit	
			Lower limestone unit	
Kopeh-Dagh	Aghdarband area	Aghdarband Gr.	Ghaleh-Gabri Shaly Fm	
			Mian-Kuhi Shaly Mb.	Brown homogenous shale without tuffic and volcanic composition
				Conglomerate and volcanic pebbles
				Coal bearing beds
			Sina Volcanic Fm.	Upper Shale.
				Lower Sandstone Mb.
			Nazar-Kardeh Fm.	
			Calcareous Sefid-Kuh Fm.	
		Ghareh-Gheytan Fm.		

Aghanabati (1998) referred the Norian to Lower Bajocian sediments as the "Shemshak Group", believing that this unit is of consistent and homogeneous lithological facies. However, the formation nomenclature differs from place to place. It can be traced in most parts of the Alborz, Central Iran, Kopeh-Dagh and Sanandaj-Sirjan.

The lower boundary of this group with Middle Triassic rocks was reported to be gradational in some areas including Sari, Kashan and Tafresh, whereas its top also is associated with structural events (Aghanabati 1998).

The Upper Triassic rocks of the Shemshak Group in the Northern Iran (Ramsar, Glandroud and Tazareh) Central Iran (Tabas, Ghadir, Nehbandan and Kerman) and northeastern Iran (Aghdarband) contain economic coal deposits. Also, the Lias and Dogger rocks of the same group in the Southern Alborz (Tazareh, Shahroud), Central Iran (Tabas and Northern Kerman) and Southern Azarbaijan (southwestern Zanjan and northern Shahin-Dezh) also contain extractable coal layers. The Kopeh-Dagh is the only area of the Iranian plate where the Jurassic succession, Kashafroud Formation, is barren of coal.

The shale and sandstone of the Kashafroud Formation are source rocks of gas reservoirs. There are several gas reservoirs in the limestones of the Mozduran Formation and in the clastic deposits of the Shurijeh Formation (Kopeh-Dagh).

Table 3.2 Lithostratigraphic features of Triassic rocks of Iran

Rock unit	Chronostratigraphic affinity	Lower/upper boundary	Lithology and thickness at type section	Note (references, comments)
Triassic (Alborz and Azarbaijan)				
Elika Fm.	Early to Middle Triassic	Lb. discontinuous but not sharp over cherty limestone of Nesen Fm. Ub. Shemshak Fm. with disconformity	Includes two or in some areas three members: the lower Mb. with variable thickness consists of thin-bedded limestones and marly limestones accompanied by some marly beds and thin-bedded dolomite (informally referred to vermiculate limestones); Middle Mb. consists of thick-bedded limestone and dolomites referred to Elika Dolomite; Upper Mb. consists of white massive fine-grained limestones named Veresk limestones Thickness: 295 m	Book: Shahrabi (1998, p. 42), Aghanabati (2004, p. 201), Aghanabati (2009, p. 7), Alavi-Naeini (2009, p. 284) Thesis: Tahmasbi (1997), Rabani et al. (1993), Gholi-nataj (2000), Pashaei (2009) Article: Gholi-nataj (2001), Rabani et al. 1995) Jahani (2000) revised the Elika Fm. whose thickness measured 346 m in the new revision
Parvar Fm.	Late Triassic	Lb. Veresk Mb. of Elika Fm. Ub. Tazareh Mb.	Consists of claystone rich in iron oxide (laterite); tuff and black limestone Thickness. 25.80 m	Book: Aghanabati (1998, p. 75), Aghanabati (2004, p. 204), Aghanabati (2009, p. 134)

(continued)

Table 3.2 (continued)

Rock unit	Chronostratigraphic affinity	Lower/upper boundary	Lithology and thickness at type section	Note (references, comments)
Tazareh Fm.	Late Triassic	Lb. Elika Fm. with disconformity contact Ub. Alasht Mb. with conformable contact	Includes following members Akrasar Mb. consisting of grey siltstone and argillite accompanied by some successions of coal bearing sediments Alah-band Mb. consisting of siltstone, argillite and sandstoneKalariz Mb. including sandstone, argillaceous siltstone with 11 beds of coal; and Alasht Mb. (Lias)	Book: Aghanabati (1998, p. 120), Aghanabati (2004, p. 204), Aghanabati (2009, p. 140.)
Triassic (Central Iran, Eastern Iran and Sanandaj-Sirjan)				
Sorkh shale Fm.	Early Triassic	Lb. Jamal Fm. with disconformable contact Ub. grey dolomite of Shotori Fm. with continuous contact	Includes red colored clayey-limey shale accompanied by regular successions of light limestone and crème-grey dolomite. Carbonate intercalations are abundant in the upper part Thickness: 122 m	Book: Shahrabi (1999, p. 71), Aghanabati (2004, p. 205), Aghanabati (2009, p. 189), Alavi-Naeini (2009, p. 287)
Shotori Fm.	Middle Triassic	Lb. Sorkh shale Fm. with gradual and continuous contact Ub. transgressive Upper Triassic strata	Condensed well-bedded fine grain grey dolomite Thickness: 820 m	Book: Shahrabi (1999, p. 85), Aghanabati (2004, p. 206), Aghanabati (2009, p. 222), Alavi-Naeini (2009, p. 288) Article: Pirouzi (2007)

(continued)

Table 3.2 (continued)

Rock unit	Chronostratigraphic affinity	Lower/upper boundary	Lithology and thickness at type section	Note (references, comments)
Darkuh Fm.	Late Triassic	Lb. Spahak Mb. of Shotori Fm. Ub. Ab-Haji Fm.	Includes three members viz. Kuhbanan coal-bearing limestone; colored marlstone of Eshkli-Kuh and Behabad Gypsum Thickness: 51 m	Book: Aghanabati (2009, p. 377)
Alam Fm.	Early to Middle Triassic	Lb. shelf carbonate beds Ub. turbidity clastic facies	Sandy limestone with some sandstone layers, conglomerate, purple shale and a sandstone unit with plant fossil remnants at the uppermost part of Alam Fm. Thickness: 873 m	Book: Shahrabi (1999, p. 116), Aghanabati (2004, p. 221), Aghanabati (2009, p. 441), Alavi-Naeini (2009, p. 259) The first Fm. of Nakhlak Gr.
Baghrogh Fm.	Middle Triassic (late Anisian to Middle Ladinian)	Lb. Alam Fm. with disconformable contact Ub. Ashin Fm. with disconformable contact	Includes coarse-grained clastic sediments, conglomerate, sandstone and shale Thickness: 1251 m	Book: Shahrabi (1999, p. 118), Aghanabati (2004, p. 221), Aghanabati (2009, p. 482), Alavi-Naeini (2009, p. 259) Thesis: Ahmadi (2011) The second Fm. of Nakhlak Gr.

(continued)

Table 3.2 (continued)

Rock unit	Chronostratigraphic affinity	Lower/upper boundary	Lithology and thickness at type section	Note (references, comments)
Nayband Fm.	Late Triassic	Lb. Shotori dolomite or Spahak limestone with sharp and angular unconformity Ub. younger successions with conformable contact without lithological difference	Includes four formally introduced members: Gelkan Mb.: strongly weathered pencil shale (915 m-thick); Bidestan Mb.: shale, siltstone and some sandstone beds (450 m-thick); Hoz-Sheikh Mb: a homogenous succession of pencil shale (365 m-thick); Hoz-khan Mb.: Reefal limestone, mostly coral bearing and cliff-forming (465 m-thick). It is also divided into six informally introduced members including coal bearing shale and, sandstone Mb., coral limestone mb. (75 m), red sandstone mb. (76 m), Ghadir mb. (540 m of a succession of grey siltstone, sandstone Mb., argillite and coal beds); Rajab Mb. (slightly metamorphosed argillite and siltstone; Ferdows (400–500 m of a succession of fine feldspar sandstone to quartz sandstone Thickness: 2800 m	Book: Shahrabi (1999, p. 91), Aghanabati (2004, p. 208), Aghanabati (2009, p. 259), Alavi-Naeini (2009, p. 290) Thesis: Ahmadi (2011) The last unit of Shemshak Group of Central Iran

(continued)

Table 3.2 (continued)

Rock unit	Chronostratigraphic affinity	Lower/upper boundary	Lithology and thickness at type section	Note (references, comments)
Ashin Fm.	Middle to Late? Triassic (Late Ladinian to Early Carrnian)	Note recorded	Fine-grained sandstone and shale interbedded with sandy limestone Thickness: 281 m	Book: Shahrabi (1999, p. 120), Aghanabati (2004, p. 221), Aghanabati (2009, p. 494), Alavi-Naeini (2009, p. 295) The third Fm. of Nakhlak Gr.
Triassic (Zagros)				
Kangan Fm.	Early Triassic (Olenekian)	Lb. Dalan Fm. with disconformable contact Ub. Aghar shale Mb. of evaporite Dashtak Fm. with conformable contact	Three facies are recognized 1. Clean carbonate facies including oolitic grainstone, mudstone slightly dolomitized accompanied by some anhydrite; 2. Basal clayey and shaly facies including shale and clayey limestone and dolomite beds; 3. Evaporite- carbonate facies including vermiculate limestone, dolomite and dissolved breccia Thickness: 178 m in type section and 140 m in surface section of Kuh-e-surmeh	Book: Motiei (1993, p. 79), Shahrabi (1999, p. 148), Aghanabati (2004, p. 213), Aghanabati (2009, p. 695), Alavi-Naeini (2009, p. 299) Thesis: Bala-zadeh (1983), Abdi (2004), Lotfpour (2005), Sheikhpour (2005), Khodaei (2007), Abbasi (2009), Ahmadi (2011) Article: Lotfpour et al. (2002), Abdi et al. (2003)

(continued)

Table 3.2 (continued)

Rock unit	Chronostratigraphic affinity	Lower/upper boundary	Lithology and thickness at type section	Note (references, comments)
Dashtak Fm.	Middle to Late Triassic	Lb. Either Dashtak Fm. or Kangan Fm. with conformable contact Ub. Neyriz Fm. with angular unconformity	Dashtak Fm. includes some evaporite successions subdivided into two formal and four informal members: Aghar Shale Mb. (formal); An evaporite Mb. including limestone, clayey dolomite and a succession of limestone, dolomite, anhydrite and shale (informal); B evaporite Mb. consisting of anhydrite and dolomite (informal); C evaporite Mb. (informal); Sepidar Dolomite (formal); D evaporite mb. including dolomite, anhydrite and shale (informal) Thickness: 814 m	Book: Motiei (1993, p. 84), Shahrabi (1999, p. 152), Aghanabati (2004, p. 215), Aghanabati (2009, p. 716), Alavi-Naeini (2009, p. 301) The first Fm. of Kazeroun Gr. Kheradpir and Szabo (1978) assigned an Early to Middle Triassic age to the formation
Khaneh Kat Fm.	Early to Middle Triassic	Lb. is not clear in the type section, however in the Oshtoran-kuh area the formation disconformably overlies the carbonate beds of Dalan Fm. Ub. Lias shale and dolomite (Neyriz Fm.) with disconformable contact	Include clayey and shaly carbonate in its lower part; and limestone and dolomite in the upper part. The type section of the formation is mainly dolomite Thickness: 364 m	Book: Motiei (1993, p. 91), Shahrabi (1999, p. 159), Aghanabati (2009, p. 685)

(continued)

Table 3.2 (continued)

Rock unit	Chronostratigraphic affinity	Lower/upper boundary	Lithology and thickness at type section	Note (references, comments)
Triassic (Kopeh-Dagh)				
Ghareh-Gheytan Fm.	Early Triassic (Early Olenekian)	Lb. covered in the Sefid-kuh section, disturbed by fault in the Sina section and Permian ophiolite rocks in the reference section Ub. Sefidkuh Fm. with low angular unconformity	Shale, sandstone and conglomerate Thickness: about 1000 m	Book: Afshar-Harb (1994, p. 53), Shahrabi (1999, p. 174), Aghanabati (2004, p. 217), Aghanabati (2009, p. 521), Alavi-Naeini (2009, p. 303)
Sefidkuh Fm.	Early Triassic (Late Olenekian)	Lb. Ghareh-Gheytan Fm. Ub. either Nazarkrdeh Fm. or Sina Fm., both with disconformable contact	Yellow colored thick-bedded limestone with light grey to blue grey thin-bedded limestone bearing some nodules and worm traces Thickness: 200 m	Book: Afshar-Harb (1994, p. 55), Shahrabi (1999, p. 178), Aghanabati (2004, p. 218), Aghanabati (2009, p. 549), Alavi-Naeini (2009, p. 304)
Nazarkardeh Fm.	Middle Triassic (Early Anisian)	Lb. Sefidkuh Fm. Ub. Sina Fm. with gradual contact	Includes shale and silty marl, tuffic sandstone, glauconite -nodular limestone, sandy marl and bioclastic limestone Thickness: 20–50 m	Book: Afshar-Harb (1994, p. 56), Shahrabi (1999, p. 183), Aghanabati (2004, p. 218), Aghanabati (2009, p. 569), Alavi-Naeini (2009, p. 305)

(continued)

Table 3.2 (continued)

Rock unit	Chronostratigraphic affinity	Lower/upper boundary	Lithology and thickness at type section	Note (references, comments)
Sina Fm.	Middle Triassic (Late Ladinian)	Lb. Nazarkardh Fm. with disconformable contact Ub. Miankuhi Fm. with disconformable contact	Includes two members, viz. 1. Lower Sandstone Mb. consisting of a succession of tuffic sandstone and green-red purple thin-bedded tuffic shale, limey shale, tuffic limestone (270–350 m-thick) and 2. Upper Shale Mb. consisting of tuffic shale accompanied by thin-bedded, fine-grained sandstone and highly fossiliferous limey shale. The basal part contains a key beds referred to "Poor Marl" consists of 20–25 m highly fossiliferous marl differentiated from other parts by light green to pink color (300–500 m-thick) Thickness: 400–700 m	Book: Afshar-Harb (1994, p. 56), Shahrabi (1999, p. 186), Aghanabati (2004, p. 218), Aghanabati (2009, p. 593), Alavi-Naeini (2009, p. 305)

(continued)

Table 3.2 (continued)

Rock unit	Chronostratigraphic affinity	Lower/upper boundary	Lithology and thickness at type section	Note (references, comments)
Mian-kuhi Fm.	Late Triassic (Norian)	Lb. Sina Fm. with disconformable contact Ub. alluvial and losses sediments	Include a homogenous succession of shale with fine-grained sandstone and siltstone	Book: Afshar-Harb (1994, p. 59), Aghanabati (1998, p. 328) Shahrabi (1999, p. 20), Aghanabati (2004, p. 219), Aghanabati (2009, p. 644), Alavi-Naeini (2009, p. 306)
Ghaleh-Gabri Fm.	Late Triassic (Rhaetian)	Lb. Sefid-kuh Fm. with sharp angular unconformity Ub. Kashafroud Fm. with disconformable and disconcordant contact	A succession of dark shale, sandstone and coal bearing beds Thickness: 10–20 m	Book: Aghanabati (1998, p. 332), Aghanabati (2004, p. 220), Aghanabati (2009, p. 660), Alavi-Naeini (2009, p. 306)

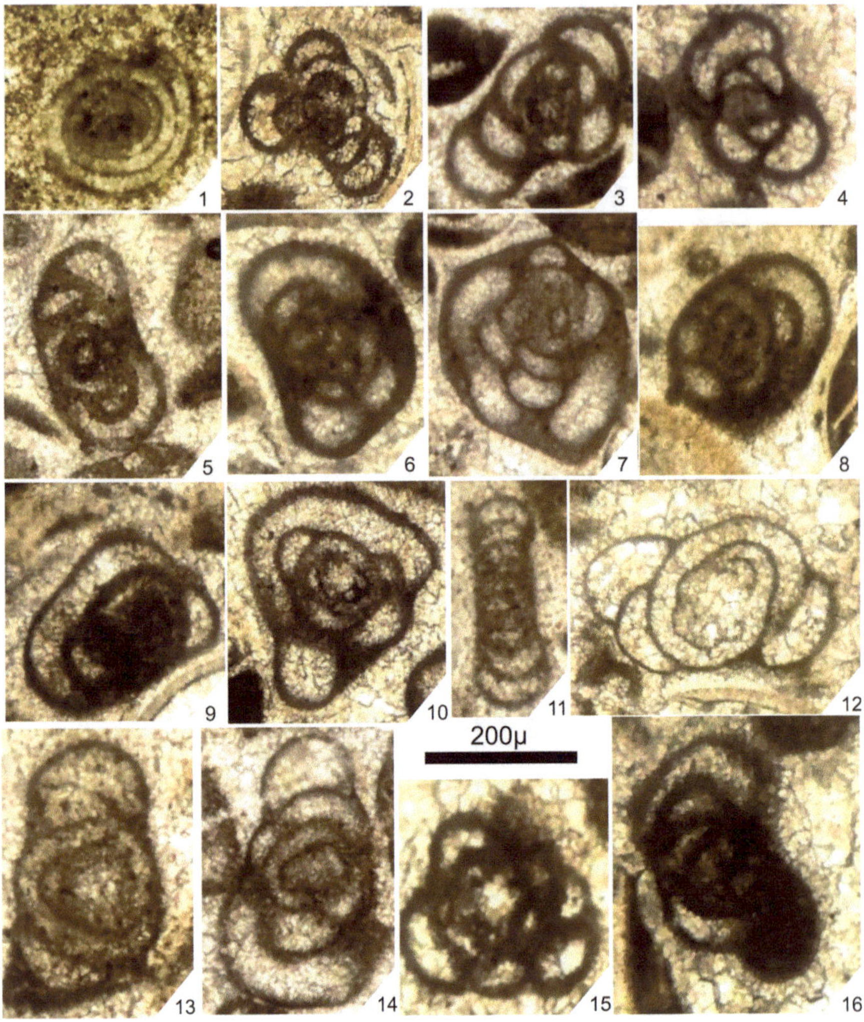

Plate 3.2 Selected Triassic foraminifers of the Kuhbanan area, Central Iran (NIOC). 1. *Ammodiscus incertus*; 2–5. *Agathammina austroalpina*; 6–10. *Hoyenella inconstans*; 11. *Pilamminell* ex gr. *semiplana*; 12–14. *Hoyenella inconstans*; 15, 16. *Hoyenella* sp.

3.2.1 Alborz

In all parts of Northern Iran, Alborz, Kopeh-Dagh, and Central Iran, the Shemshak Group starts with conglomerate and ammonite bearing marl and limestone and gradually or discontinuously passes upwards into reef-forming and pelagic limestone. These beds eventually changed to red clastic or evaporite sediments whose deposition triggered by a Late Jurassic-Early Cretaceous marine regression. Marly-limey

Fig. 3.4 A view of Upper Jurassic? Lower Cretacaceous? Rocks at northern Ravar

sediments of lower Magu Group are referred to as the "Dalichai Formation" in the Alborz Mountains; the Baghamshah Formation in Central Iran and the Chamanbid Formation in the Kopeh-Dagh. Other names applied to the Upper Jurassic rocks include the Lar Formation (Alborz), the Mozduran Formation (Kopeh-Dagh) and Esfandiar and the Ghaleh-Dokhtar formations and Pecten-bearing Limestone (Central Iran; Aghanabati 2004).

Generally, these rocks are of Oxfordian-Kimmeridgian age but are probably younger than Tithonian or even Berriasian in some regions (northern Alborz, eastern Alborz, Sanandaj-Sirjan and some parts of southern Alborz). The Upper Jurassic successions are composed of clastic or evaporitic sediments, especially in Central Iran (Ruttner et al. 1968) and the Kopeh-Dagh Mountains. The Upper Jurassic of the Kerman area is mainly composed of red conglomerate, sandstone and shale apparently indicating deposition in a continental environment (Huckriede et al. 1962) (Figs. 3.3 and 3.4) .

These rocks are less widely distributed in the Alborz Mountains and can be found only in the Damavand area, where they consist of evaporite-volcanic deposits named the "Damavand Gypsum and Melaphyre"(Allenbach 1966). These rocks which can be traced through wide areas of the Kopeh-Dagh are represented by gypsiferous clastic sediments of the Shurijeh Formation, and provide evidence for the onset of the Late Cimmerian Tectonic Event. The Upper Jurassic-Lower Cretaceous clastic and evaporitic sediments crop out mainly in the area located between Tabas and Kerman, and include the Garedu red beds, the Bidou Formation, the Continental Red Formation and the Salty-gypsiferous Formation of the Ravar area. Of note is the fact that Middle and Upper Jurassic marly-calcareous sediments are not observed in western Central Iran (Yazd, Esfahan) and Western Alborz. In these areas, the Middle-Upper Triassic sandstones and shales of the Shemshak Group are overlain by transgressive Lower Cretaceous sediments of Barremian to Aptian age.

Shemshak Group

As already mentioned in the Triassic chapter, this group includes both Upper Triassic and Lower Jurassic rocks, of which the Upper Triassic formations were described in the Triassic section. In the following account, the Jurassic formations of the group are described.

Alasht Formation

This name was first proposed by Bragin et al. (1976) as a "suite" but this was subsequently changed by Fursich et al. (2009) to the Alasht Formation. This formation is generally composed of alternating packages of fine-grained and medium—to coarse-grained siliciclastics, separated by finer-grained interbeds. It is underlain by either the Kalariz Formation or the Shahmirzad Formation with sharp contact and is overlain by Shirindasht Formation. The plant fossils of the Formation are not age diagnostic to suggest any age. However, based on the Triassic character of the plant association of the underlying Kalariz Formation and on the Toarcian ammonites occurring at the base of the overlying Shirindasht Formation, the Alasht Formation is assigned an Early Triassic (Hettangian/Sinemurian–Pliensbachian, most probably also Toarcian) age. This formation represents fluvial–lacustrine environments, with the widespread development of coal swamps (Fursich et al. 2009).

Shirindasht Formation

This formation corresponds to the "Shirindasht Suite" of Bragin et al. (1976) which was later changed to formation by Fursich et al. (2009). The type section is located in the Tazareh area, northeastern Damghan, and is characterized by small-scale alternations of heavily bioturbated dark-grey–olive-grey silt-fine-sandy silt and fine-grained sandstones. At Shahmirzad and Sharif-Abad, this formation is characterized by highly bioturbated argillaceous silt with scattered, often highly fossiliferous limestone concretions, which are occasionally reworked into pebble layers, and some intercalations of hummocky cross-stratified or parallel-laminated sandstones (Fursich et al. 2009). It is considered to represent a storm-dominated shale. The underlying unit in most of the area is the Alasht Formation and overlying it is the Fillzamin Formation. A Toaracian to early Aalenian age is suggested based on the available ammonite evidence (Fusrich et al. 2009).

Fillzamin Formation

The type section of this formation is located in the Fillzamin area, south Central Alborz (Fursich et al. 2009). This formation corresponds to the upper part of the "Shirinidasht Suite" of Bragin et al. (1976) and to the lower part of the Diktash Formation of Nabavi and Seyed-Emami (1977). According to Fursich et al. (2009), their new "Fillazamin Formation" is necessary because the existing lithostratigraphic schemes overlooked the distinct facies changes occurring at the base and top of the formation, and because of the characteristic, very uniform, facies of the formation is unique within the Jurassic part of the Shemshak Group. The formation is a monotonous succession of indurated, dark-grey-slightly greenish, highly bioturbated, partly fine sandy, and argillaceous silt with occasional limestone concretions in its lower part. At Sharif-Abad, the sediment is marly and ferruginous concretions are common in the lower part (Fursich et al. 2009). The Fillzamin Formation is underlain and overlain by the Shirindasht and the Dansirit formations with gradual and sharp contacts, respectively. Ammonite evidence indicates an Aalenian to Bajocian age for the formation (Fursich et al. 2009).

Dansirit Formation

The Dansirit Formation was first defined by Bragin et al. (1976) as the "Dansirit suite" and which was later changed by Fursich et al. (2009) to the "Dansirit Formation". It consists of large-scale cross-bedded sandstone packages, between which silt carbonaceous silt and coal layers are intercalated. It is underlain and overlain by the Fillzamin and Dalichai formations, respectively. According to Fursich et al. (2009), ammonites found in the underlying Fillzamin Formation and the overlying Dalichai Formation constrain the age of the Dansirit Formation to Early Bajocian.

Javaherdeh Formation

This formation was first defined by Bragin et al. (1976) for a conglomeratic unit mostly exposed in the Ramsar area, northern Alborz. It is mostly composed of polymict conglomerates consisting mainly of well-rounded pebbles of quartzites, milky quartz, volcanics, metamorphics and rarely granite (Fursich et al. 2009). In most places, the base of the formation is recorded to be the Fillzamin Formation with an angular unconformity and the top has not yet been observed. (Fursich et al. 2009). A Hettangian–Early Bajocian age was inferred based on its stratigraphic context.

These thick, coarse conglomerates represent alluvial-fan–braided-river deposits, and are evidence of uplift and erosion of the Cimmerian Mountain Range to the north, situated in the southern part of the present-day Caspian Sea (Fursich et al. 2009).

Dalichai Formation

This formation is mostly composed of cliff forming limestone and, marly limestone, disconformably overlying the Shemshak Formation (NICS 1962; Unpublished data). It is be observed in the areas where the Shemshak Formation is exposed. Time-equivalents of the Upper Bajocian-Callovian Dalichai Formation of different facies are exposed in other parts of Iran, such as the Chamanbid Formation of Kopeh-Dagh, the Baghamshah Formation of Central Iran and the Sargelu Formation of Zagros.

Lar Formation

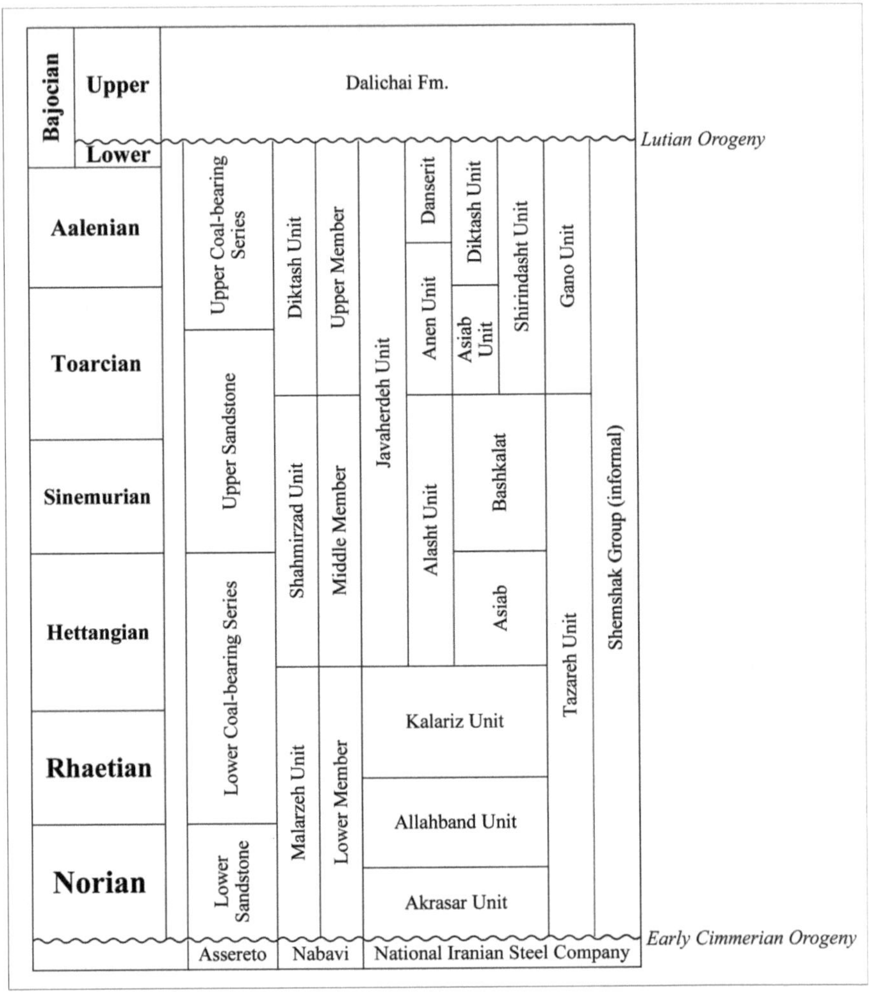

Fig. 3.5 Lithostratigraphic units of the Late Triassic to early Bajocian of northern Iran and Alborz Mountains. Modified after Alavi-Naeini (2009)

A massive, cliff-forming and thick-bedded unit overlying the Dalichai Formation is differentiated from the Dalichai Formation by its massive bedding, light color and the presence of cherty nodules in the upper part. The Lar Formation is widely distributed in the Alborz, Azarbaijan and the northern part of Sanandaj-Sirjan. Its equivalents can be observed in the Zagros (Surmeh Formation), Kopeh-Dagh (Mozduran Formation), Shotori Mountain (Esfandiar Limestone) and northern Khorasan (Ghaleh-Dokhtar Formation) (Fig. 3.5).

In addition to aforementioned formations, additional formations have been reported from the Jurassic of the Alborz. For example, equivalents of the Shemshak

Formation but with some facies variations. Also, equivalents of Lar Formation exists that are informally referred to as the Abnak and Shal formations, being the time equivalent and facies variants of the Lar Formation. The former is composed of black micritic limestone and dolomite, while the latter consists of brown to green sandstone and thin-bedded sparry limestone.

3.2.2 *Zagros*

The Jurassic rocks, referred to as the "Neyriz Formation" (Lower Jurassic) and the "Surmeh Formation" (Middle to Upper Jurassic), are exposed in the Zagros zone with two distinct facies in areas such as Fars, Bandar Abbas, the Dezful Embayment and Lorestan. The Jurassic rocks of Zagros differ from other parta of Iran. However, they are not of same facies, everywhere, in the all of the Zagros and show some facies variations within this area (Meitei 1991).

The Jurassic-Upper Triassic rocks of the High Zagros are composed of limestone and radiolarite indicating a relatively deeper depositional environment. In the Neyriz area, these rocks are called the "Pichakun (or Pishakun) Formation" consisting of an assemblage of black marl, silica limestone and radiolarite that have been assigned a Late Triassic-Jurassic age. In the Kermanshah area, time equivalent rocks comprise the Kermanshah Radiolarite, Bistoon Limestone and the Sahneh-Harsin ophiolite that were deposited in the environments similar to those of Neyriz area. The rocks of both areas are characterized by over-thrusting flysch feature.

These rocks were probably deposited in the deeper marginal environments of northeastern Zagros (High Zagros). However, they were presumably first formed in the Sanandaj-Sirjan zone and were then over-thrusted onto the high Zagros.

Neyriz Formation

In the Fars area and southern part of the Dezful embayment, the Neyriz Formation is composed of siltstone, limestone and dolomite. It is dated as early Jurassic in age. However, a Late Triassic age has been proposed for the same formation in the Neyriz area based on fossil assemblages. This formation shows that facies variation in the Zagros zone is diachronous, being the time equivalent of some Jurassic formations in Central Iran (Aghanabati 2004).

Being of different facies in comparison to other portions of the Zagros zone, the Lower Jurassic rocks of Lorestan and the northern Dezful Embayment are shallow water and evaporitic sediments closely resembling to their time equivalents in Iraq (Afshar-Harb 2004). Towards the Late Jurassic, a tectonic phase (exposure) occurred in the southern Fars, Lorestan and the northern Dezful Embayment that produced an angular unconformity at the base of the Lower Cretaceous sediments. However, in the area around Shiraz, the deep-water carbonate deposition of the Late Jurassic continuously persisted without break through the Early Cretaceous.

Lower Jurassic sedimentation in Lorestan and the northern Dezful Embayment started with dolomite and continued with Middle Jurassic, deep-water shale and thin-bedded limestone of the Sargelu Formation. After a break in deposition at the beginning of Late Jurassic time, the area was submerged again leading to sedimentation of the Najmeh Formation. Finally the area was affected by a marine regression that occurred at the end of the Late Jurassic, which resulted in the deposition of Gutnieh (Ghatnieh) Anhydrite. The southern Fars, Lorestan and the Northern Dezful Embayment were exposed in the Late Jurassic, while the Northern Fars experience deep-water carbonate deposition that continued into the Early Cretaceous and is represented by Tintinnid Biofacies (Motiei 1993) (Fig. 3.2).

3.2.3 *Central Iran*

Shemshak Group

The Shemshak Group of Central Iran is similar to that of the Alborz Mountains. However, the base of the formation has been dated as Late Triassic and has been compared to the basal parts of the Nayband Formation. Generally, the Lower Jurassic flood plain and lagoonal facies of Central Iran are referred to as the Ab-Haji Formation. The Toarcian-Bajocian sediments in this area have been assigned to the "Baghamshah, Ghaleh-Dokhtar and Esfandiar formations, the gypsiferous pecten-bearing limestone, the calcareous Tar Formation, the Magu Gypsum, the Garadu red beds and Bidou unit".

Alavi-Naeini (1972) considered this group to as a formation and believed in attribution of Shemshak "Formation" to a succession overlying the Nayband Formation. He suggested the name "Shemshak Formation" for a sequence confined by two tectonic phases of Early Cimmerian and Middle Cimmerian (Lutian), whether in Central Iran or in the Alborz and also introduced some members and beds can be introduced.

However, the currently accepted rank for this unit is actually "Group" (Aghanabati 1977, 2004; Seyed-Emami 2003; Wilmsen et al. 2009), consisting of the Nayband, Ab-Haji, Badamu and Hojedk Formations (Fig. 3.6). The Upper Triassic Nayband Formation was described in the Triassic section, other formations are described herein.

Ab-Haji Formation

The Ab-Haji Formation was first described by Aghanabati (1977) in the Kerman, Tabas, and Mazinou areas and the Lut Block. The name was given to a succession of shale and sandstone overlying and overlain by the Nayband and Badamu formations, respectively. In some areas, the Ab-Haji, Hojedk and Badamu formations are of the same lithological facies and, in consequence, the differentiation of those formations is difficult (Alavi-Naeini 2009). The Ab-Haji Formation has not been studied in detail, but appears to be largely non-marine (Wilmsen et al. 2009). In terms of depositional setting and chronostratigraphy, the Ab-Haji Formation is comparable to the Alasht Formation of the Shemshak Group in the Alborz Mountains of northern Iran (Fursich et al. 2009).

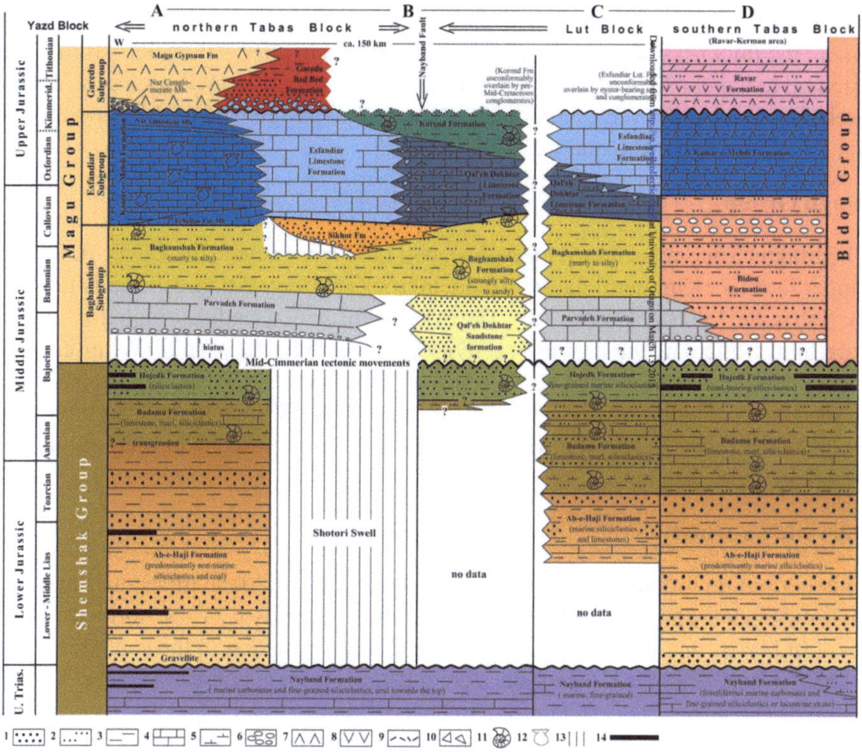

Fig. 3.6 Revised lithostratigraphic scheme of Jurassic system in Central Iran (Wilmsen et al. 2009)

Badamu Formation

This formation consists of relatively condensed, ammonite-rich, and dark, often oolitic limestones, marls and siliciclastic rocks. The type section is located near Titu village, 17 km NNE of Zarand, NW of Kerman (southern Tabas Block). It is believed that the Badamu formations is important for recognition of the underlying Ab-Haji and overlying Hojedk formation. The Badamu Formation is fairly fossiliferous, and mainly yields ammonites, bivalves, corals and serpulids (Wilmsen et al. 2009). Abundant ammonites indicate an upper Toaracian to earliest Bajocian age (Seyed-Emami et al. 2004).

Hojedk Formation

The Hojedk Formation is a siliciclastic unit, first defined for the higher coal-bearing beds overlying the Badamu Formation. This Formation is composed of an alternation of sandstone and sandy to argillaceous shale with dull green-grey to brown-green colours. "Flaky" sandstones, composed of reworked shale flakes, are particularly characteristic and contrast with the more quartzitic sandstones of the otherwise similar Shemshak Formation (Stöcklin and Setudehnia 1991). Noteworthy are the rapid lateral facies and thickness changes in the northern Tabas Block at Parvadeh Mine, where the thickness of the formation is reduced within a few kilometers from a few hundred to a mere 30 m of marginal marine siliciclastic rocks (Wilmsen et al. 2009). Well preserved and diverse ammonites allow an age assignment of the Formation to the Bajocian (Seyed-Emami et al. 2004).

Magu Group

The Magu Group was erected by Aghanabati (1977) to include the Upper Bajocian-Upper Jurassic (Kimmeridgian–? Tithonian) rocks of Central Iran, especially those of the Tabas and Lut Blocks (Fig. 3.6). The existing formations indicate high lithological variability, reflecting the increased activity of syn-sedimentary faults (Wilmsen et al. 2009). This group was named "Bidou Group" in the southern Tabas Block, consisting of the Bidou (equivalent to the Baghamshah Fm.), the Kamar-Mahdi and the Ravar Formation (equivalent to the Magu Gypsum). The constituent formations of this group are described below.

Parvadeh Formation

This Formation is composed of condensed, oncolitic–microbial limestones and siliciclastic rocks. The Parvadeh Formation is mostly exposed in the northern Tabas and Lut blocks and laterally changes to the Bidou Formation in the southern Tabas Block. It was deposited across the Tabas Block, also onlapping onto the Shotori Swell (Aghanabati 1977). According to Seyed-Emami et al. (1991), the formation is of Late Bajocian to early-middle Bathonian age based on ammonite occurrences.

Baghamshah Formation

This formation consists of shale, light-green shaly marl, sandstone and limestone accompanied by minor quantity of gypsum, salt, coal and clayey, iron bearing rocks. It is exposed in the Tabas, Semnan, southern Miami, Ferdows, northern Kerman and Naibandan areas. The Baghamshah Formation can be correlated with the Dalichai Formation of Alborz Mountain and the Chamanbid Formation of Kopeh-Dagh. This formation which was dated as late Bathonian-early Callovian (Seyed-Emami et al. 1997) may extend locally up into the Middle Callovian (Seyed-Emami et al. 2002).

Esfandiar Formation

The Esfandiar Formation is mostly composed of sandstone in the lower part and light colored massive, reefal limestone overlying the Baghamshah Formation in the type section.

It is geographically restricted to the western and northern slopes of the Shotori Mountain. The reefal Esfandiar Formation can be correlated with Lar Formation of Alborz, Mozduran Formation of Kopeh-Dagh and upper parts of Surmeh Formation of Zagros (Aghanabati 2001). The age of the Esfandiar Limestone Formation is early Callovian–(? early) Kimmeridgian (e.g. Stöcklin et al. 1965).

Ghaleh-Dokhtar Formation

The type section of the Ghaleh-Dokhtar Formation is located in the Boshruyeh area, and is composed of following three members (Aghanabati 2001).

1. Lower Sandstone Member with grey to brown color
2. Middle Shale Member consisting of marly shale, green to brown sandstone and siltstone
3. Middle Limestone Member of light color.

Kamar-Mahdi Formation

The Kamar-Mahdi Formation is a very widespread and carbonate-dominated unit that crops out in the Tabas Block. It is laterally equivalent to the Ghaleh-Dokhtar Formation. Wilmsen et al. (2003) divided it into two members a basal Echellon Limestone Member and the capping, cliffforming Nar Limestone Member. It was deposited in a widespread shelf lagoon on the western side of the Esfandiar Platform (Wilmsen et al. 2009).

In the northern Tabas Block this formation is overlain by a gypsiferous unit, the "Magu Gypsum" which laterally grades into a conglomeratic and sandy unit, the "Garadu Red Beds".

The temporal-spatial relation of the Mesozoic rock units are given in Fig. 3.6 (Fig. 3.7).

The lithostratigraphic features of the Jurassic rock units of Iran can be availed of Table. 3.3 and selected diagnostic fauna of Jurassic rocks of Iran are illustrated in Plates. 3.3 and 3.4.

3.3 Cretaceous

The Cretaceous rocks are widely distributed through most of Iran and exhibit considerable facies variations, so much that that some Iranian geologists have referred to the Cretaceous of Iran as "the time of High Facies Change". The Cretaceous-Jurassic transition is continuous in the some areas of Iran and Upper Jurassic deposits gradually grade to the Lower Cretaceous ones. These regions exprienced continuous marine condition (e.g. Razmjooei et al. 2014) but in other regions, there is an unconformity between the contact of the uppermost Jurassic (or older) and lower Cretaceous rocks, indicated by red sandstone and conglomerate beds, are overlain by Aptian-Albian strata deposited during main early Cretaceous transgression. These beds can be observed in most parts of Iran, except Zagros and the flysch basins of eastern Iran and Makran. Several names were assigned to the red rocks in the different areas, such as the Sangestan Formation of Kerman and Esfahan, the Noghreh Formation of the Anarak area, and the Shurijeh Formation of Kopeh-Dagh.

The exact age of the red conglomerates at the base of the Cretaceous succession has not been determined due to the absence of diagnostic fossil. Nevertheless, a Neocomian age has been proposed based on their stratigraphic position. Overlying

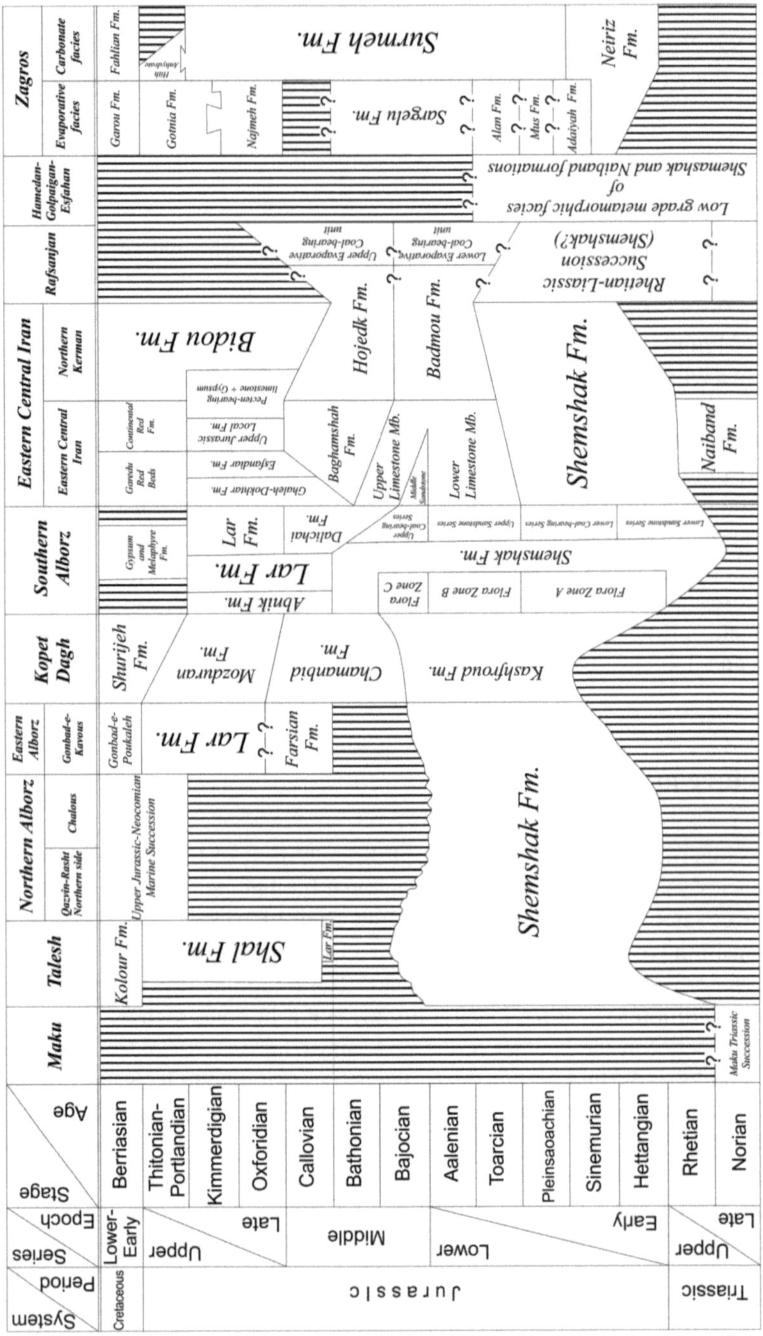

Fig. 3.7 Temporal-spacial distribution of former Jurassic units. Modified after Shemirani (1991)

Table 3.3 Lithological features of Jurassic formations of Iran

Rock unit	Chronostratigraphic affinity	Lower/Upper Boundary	Lithology and thickness at type section	Note (references, comments)
Jurassic (Alborz -Azarbaijan)				
Shemshak Gr.	Late Triassic-Jurassic	Lb. Geirud Fm. with disconformable boundary Ub. Dalichai Fm. with gradual contact	Four lithological units including Lower Sandstone Mb., Lower coaly shale, Upper sandstone and Upper Coaly Shale Thickness: 1027 m	Book: Aghanabati (1998, p. 31), Aghanabati (2004, p. 231), Alavi-Naeini (2009, p. 309). Thesis: Vaez-Javadi (2001), Safar-Khanlou (2000), Hosein-Zadeh (2001), Hoseyni (2001), Jamshidi (2010) Article: Vaez-Javadi et al. (2003), Ghasemi-Nejad et al. (2001), Fursich et al. (2009)
Javaherdeh Fm.	Early to Middle Jurassic	Lb. Kalariz Fm. disconformable boundary Ub. Dalichai Fm. with disconformable contact	Includes pebbly conglomerate with abundant rock fragments accompanied by some successions of siltstone and sandstone	Book: Aghanabati (1998, p. 152), Aghanabati (2004, p. 234). Article: Fursich et al. (2009)
Farsian Fm.	Middle Jurassic	Lb. Shemshak Gr. with conformable contact Ub. Lar Fm. with gradual contact	A succession of red thin-bedded limestone Thickness: 234	Book: Aghanabati (1998, p. 385), Aghanabati (2004, p. 235)

(continued)

Table 3.3 (continued)

Rock unit	Chronostratigraphic affinity	Lower/Upper Boundary	Lithology and thickness at type section	Note (references, comments)
Dalichai Fm.	Middle Jurassic (Bajocian-Callovian)	Lb. Shemshak Fm. with disconformable boundary Ub. Lar Fm. with gradual contact	Includes marl, thin-bedded marly limestone with slight sparry limestone and marly shale intercalations Thickness: 107 m	Book: Aghanabati (1998, p. 367), Aghanabati (2004, p. 236) Alavi-Naeini (2009, p. 311). Thesis: Abarghani (2000), Naderi-Roshanavand (2002), Falahpour (2002), Hashemi-Yazdi, (2008), Boroumand (2009) Article: Shafie-zadeh (2005)
Ganou Fm.	Early to Middle Jurassic	Lb. Alasht Mb. Ub. Dalichai Fm.	Includes members consisting of Shirindasht Mb. (sandstone), Anen Mb. (sandstone, grainstone, conglomerate and siltstone), and Dansirit Mb. (sandstone, conglomerate, grainstone, argillite and siltstone)	Book: Aghanabati (1998, p. 140), Aghanabati (2004, p. 234)
Lar Fm.	Late Jurassic-Early Cretaceous	Lb. Alasht Mb. Ub. Dalichai Fm.	Include light brown to white, thin-bedded limestone accompanied by nodular chert in the lower part and layered chert in the massive limestone of the upper part Thickness: 250–350 m	Book: Aghanabati (1998, p. 391), Aghanabati (2004, p. 237) Alavi-Naeini (2009, p. 312) Thesis: Abarghani (2000), Naderi-Roshanavandi (2002) Article: Lasemi (2000)
Abnik Fm.	Middle –Late Jurassic (Callovian-Kimmeridgian)	Lb. coal-bearing rocks of Shemshak Fm. with conformable but diachronous contact Ub. Kharsang Mb. of Tiz-Kuh Fm.	A homogenous succession of black micritic, dolomitic and piceous limestone Thickness: 115	Book: Aghanabati (1998, p. 435), Aghanabati (2004, p. 238) Alavi-Naeini (2009, p. 315)

(continued)

Table 3.3 (continued)

Rock unit	Chronostratigraphic affinity	Lower/Upper Boundary	Lithology and thickness at type section	Note (references, comments)
Shal Fm.	Middle Jurassic (Callovian) to Early Cretaceous (Berriasian)	Lb. coal bearing rocks of Shemshak Fm. Ub. micritic limestone of Early Cretaceous	Includes two informal members of Lower Mb. consisting of brown to green medium to thick-bedded sandstone with some pebbly layers in the lower part and reddish green to brown, slightly recrystallized, thin-bedded sparry limestone in the upper member Thickness: 61 m	Book: Aghanabati (1998, p. 438), Aghanabati (2004, p. 239) Alavi-Naeini (2009, p. 315) Article: Azimi et al. (2008)
Damavand gypsum and melaphyre Fm.	Late Jurassic-Early Cretaceous	Lb. Lar Fm. Ub. Tiz Kuh Fm.	A succession of colored marl, gypsum and augite-olivine bearing diabase Thickness: 143 m	Book: Aghanabati (1998, p. 444), Aghanabati (2004, p. 240)
Jurassic (Central Iran, Eastern Iran and Sanandaj-Sirjan)				
Badamu Fm.	Early to late Jurassic (Toracian-Middle Bajocian)	Lb. Ab-Haji Fm. Ub. Hojedk Fm.	Includes dark grey sandy oolitic limestone Thickness:163 m	Book: Aghanabati (1998, p. 236), Aghanabati (2004, p. 243) Thesis: Negarestani (1998), Kabir (1981) Article: Nir-Abadi et al. (2003), Wilmsen et al. (2009)
Ab-Haji Fm.	Early Jurassic	Lb. Upper Triassic Nayband Fm. Ub. Toracian -Bajocian Badamu Fm.	Includes a basal quartz sandstone unit and shaly and sandstone unit Thickness: 485	Book: Aghanabati (1998, p. 223), Aghanabati (2004, p. 242), Alavi-Naeini (2009, p. 318) Article: Wilmsen et al. (2009) Report: Aghanabati (1977), Khasi (2009)

(continued)

Table 3.3 (continued)

Rock unit	Chronostratigraphic affinity	Lower/Upper Boundary	Lithology and thickness at type section	Note (references, comments)
Parvadeh Fm.	Middle Jurassic (Bajocian -Bathonian)	Lb. Sandstone of Hojedk Fm. Ub. Baghamshah Fm. with either sharp or conformable contact	Except for the basal clastic bed, it is mainly composed of homogenous grey limestone Thickness: 46 m	Book: Aghanabati (1998, p. 454), Aghanabati (2004, p. 248) Report: Aghanabati (1977) Article: Wilmsen et al. (2009) The first unit of Magu Gr.
Hojedk Fm.	Middle Jurassic (Middle Bajocian-Bathonian)	Lb. Badamu Fm. with gradual contact Ub. Hojedk Fm. with disconformable contact resulting from Middle Cimmerian tectonic event	The Hojedk Fm. lithologically is composed of quartz sandstone, arkosic sandstone grey to brown-green shale characterized by red colored clayey and hematitic nodules and various coaly lenses (roughly 30 beds, whose so called "D Bed" located at the 150–200 m of the basal part is of the economic significance	Book: Aghanabati (1998, p. 253), Aghanabati (2004, p. 245) Thesis: Soheyli (2002) Article: Mirzaei-Ataabadi (2003), Wilmsen et al. (2009)
Baghamshah Fm.	Middle Jurassic (Late Bathonian to Late Callovian)	Lb. Parvadeh Fm. or its clastic equivalents with sharp but conformable contact Ub. Ghaleh-Dokhtar Fm. with disconformable contact resulted from the sedimentary cessation	It is mostly composed of shale, light green slightly shaly marl and some limestone and sandstone beds accompanied by very rare salt, coal and clayey-ironic nodules. The outcrops of the formation is covered by rock fragments and is characterized by foothills resulted from the marly constituents especially in the lower part Thickness:496 m	Book: Aghanabati (1998, p. 477), Aghanabati (2004, p. 250), Alavi-Naeini (2009, p. 319) Article: Wilmsen et al. (2009) Second unit of Magu Gr.

(continued)

Table 3.3 (continued)

Rock unit	Chronostratigraphic affinity	Lower/Upper Boundary	Lithology and thickness at type section	Note (references, comments)
Ghaleh-Dokhtar Fm.	Late Jurassic (Callovian-Kimmeridgian)	Lb. covered in the type section, however, Stöcklin et al. (1965) believes that the formation is underlain by the Baghamshah Fm. with gradual and continuous contact, however, field observation results show that the basal sandstone, pebbly and cobble content indicate an unconformity in the lower contact of the Ghaleh-Dokhtar Ub. is tectonically disturbed	Being lithologically inhomogeneous succession, the formation is divided into three distinct members including 1. Lower sandstone Mb. (grey or brown quartz sandstone characterized by cross bedding) 2. Middle shale Mb. (green to brown marly, sandy and silty shale with sandstone and limestone intercalations), 3. Upper limestone Mb. (, thin-bedded limestone and shaley limestone characterized by the light and sometimes brown colors due to dolomitization Thickness: 974 m	Book: Aghanabati (1998, p. 495), Aghanabati (2004, p. 251), Alavi-Naeini (2009, p. 320) Article: Wilmsen et al. (2009)

(continued)

Table 3.3 (continued)

Rock unit	Chronostratigraphic affinity	Lower/Upper Boundary	Lithology and thickness at type section	Note (references, comments)
Esfandiar Fm.	Late Jurassic (Callovian-Oxfordian)	Lb. Baghamshah Fm. with sharp and conformable contact Ub. red bedded clastic deposits of Garadu Red Beds in the Northern Shotori Mountain and Shirgesht area	Light colored massive cliff forming Reefal limestone Thickness: 690 m	Book: Aghanabati (1998, p. 509), Aghanabati (2004, p. 252), Alavi-Naeini (2009, p. 321) Article: Wilmsen et al. (2009)
Pecten bearing limestone	Late Jurassic	Lb. Baghamshah with conformable contact Ub. Nar Fm.	A homogenous succession of greenish or crème well-bedded limestone forming low foothill covered by a sheet of limestone fragments Thickness: 364	Book: Aghanabati (1998, p. 515), Aghanabati (2004, p. 253)
Nar Fm.	Late Jurassic	Lb. Pecten Bearing Limestone Ub. in some areas is red marly, conglomerate deposits (Garadu equivalent) and in somewhere is Upper Jurassic-Lower Cretaceous gypsum deposits of Magu Gypsum	It is lithologically divided into three members including 1. Lower Limestone Mb. (massive, fine-grained, slightly recrystallized limestone); 2. Middle Marly Limestone (grey to light green thin-bedded limestone interbedded with marly laminates); Upper Limestone Mb. (same as the lowermost member) Thickness: 60–90 m	Book: Aghanabati (1998, p. 521), Aghanabati (2004, p. 255)

(continued)

Table 3.3 (continued)

Rock unit	Chronostratigraphic affinity	Lower/Upper Boundary	Lithology and thickness at type section	Note (references, comments)
Garadu Red Beds	Late Jurassic	Lb. Upper Jurassic or older rocks (Ghaleh-Dokhtar, Baghamshah or Nar formations) with either erosional or angular unconformity	A relatively homogeneous succession of conglomerate, purple limey sandstone and dark red siltstone interbedded with blue grey limestone Thickness: 100–500 m	Book: Aghanabati (1998, p. 532), Aghanabati (2004, p. 256) Article: Wilmsen et al. (2009)
Red Continental Formation	Late Jurassic-Early Cretaceous	Lb. Upper Jurassic salt deposits with gradual contact Ub. coarse-grained sandstone covered by Cretaceous orbitolina bearing limestone and gypsum bearing marl, with disconformable contact	A thick succession of dark red to blackish well-bedded sandstone Thickness: more than 500 m	Book: Aghanabati (1998, p. 531)

(continued)

Table 3.3 (continued)

Rock unit	Chronostratigraphic affinity	Lower/Upper Boundary	Lithology and thickness at type section	Note (references, comments)
Bidou Fm.	Middle to Late Jurassic	Lb. Parvadeh Fm. Ub. Kamar Mehdi Fm.	The Bidou Fm. is composed of three members including 1. Lower conglomerate (Lower Part); 2. Red to green sandstone and marl (middle part) and 3. Bidou limestone consisting of marly limestone in lower parts and red marl and sandstone in upper parts	Book: Aghanabati (1998, p. 540), Aghanabati (2004, p. 257) Huckriede et al. (1962) redefined the unit to include the colored clastic beds occurring in the below the Bidou Limestone as well as such clastic beds found higher in the section but outside the Bidou syncline till Lower Cretaceous sequence. So the Bidou Formation can be ranked as a group
Jurassic (Zagros)				
Neyriz Fm.	Early Jurassic (Lias)	Lb. Khaneh-Kat with conformable contact Ub. Surmeh with gradual contact	The lower part of the formation is composed of thin-bedded dolomite, greenish shale; the middle part of brown weathering color dolomite, sandy-silty dolomite; and upper part of thin-bedded clayey limestone to shale Thickness: 290 m	Book: Motiei (1993, p. 105); Aghanabati (1998, p. 663), Aghanabati (2004, p. 265), Alavi-Naeini (2009, p. 324). Second Fm. of Kazerun Gr.
Adaiyah Fm.	Late Lias	Lb. Butmah Fm. with gradual and conformable contact Ub. Mus Fm. with gradual and conformable contact	It is mainly composed of well-bedded anhydrite occasionally accompanied by black shale, green marl and anhydrite	Book: Motiei (1993, p. 110), Aghanabati (1998, p. 654), Aghanabati (2004, p. 263), Alavi-Naeini (2009, p. 326). The proposed age is only based on stratigraphic position

(continued)

Table 3.3 (continued)

Rock unit	Chronostratigraphic affinity	Lower/Upper Boundary	Lithology and thickness at type section	Note (references, comments)
Alan Fm.	Late Lias	Lb. Mus Fm. with gradual contact Ub. Sargelu Fm. with gradual and conformable contact	includes layered anhydrite and some brown limestone beds	Book: Motiei (1993, p.114), Aghanabati (1998, p. 655), Aghanabati (2004, p. 263), Alavi-Naeini (2009, p. 326)
Mus Fm.	Lias	Lb. Adaiyah Fm. with gradual and conformable contact Ub. anhydrite bearing Alan Fm. with gradual and conformable contact	Includes grey to pale recrystallized limestone, recrystallized marly limestone in the lower part; grey brown, medium grained, recrystallized limestone with fine marly limestone, brown limey shale, fine anhydrite intercalations and autogenic quartz Thickness: 56 m	Book: Motiei (1993, p. 112), Aghanabati (1998, p. 655), Aghanabati (2004, p. 263), Alavi-Naeini (2009, p. 326)
Gotnia (Ghatnieh) Fm.	Late Jurassic	Lb. Najmeh Fm. with conformable and gradual contact Ub. Makhol Fm. with conformable and gradual contact	Well-bedded anhydrite interbedded with brown limey shale, thin-bedded bitumen bearing shale, recrystallized limestone rarely	Book: Motiei (1993, p. 121), Aghanabati (1998, p. 659), Aghanabati (2004, p. 264), Alavi-Naeini (2009, p. 327)
Najmeh Fm.	Late Jurassic	Lb. Sargelou Fm. with disconformable contact Ub. Garadu Fm. with disconformable contact	A succession of limestone, oolitic limestone, dolomite and thin-bedded anhydrite Thickness: 341 m	Book: Motiei (1993, p. 118), Aghanabati (1998, p. 658), Aghanabati (2004, p. 263), Alavi-Naeini (2009, p. 327)

(continued)

Table 3.3 (continued)

Rock unit	Chronostratigraphic affinity	Lower/Upper Boundary	Lithology and thickness at type section	Note (references, comments)
Sargelu Fm.	Lias to Bathonian	Lb. Seh-Kanian Fm. with gradual and conformable contact Ub. Naklakan Fm. with gradual and conformable contact	Black, thin-bedded fetid bitumen bearing limestone, dolomitic limestone, laminated black shale and block cherty beds in the upper part Thickness: 115	Book: Motiei (1993, p. 115), Aghanabati (1998, p. 656), Aghanabati (2004, p. 263), Alavi-Naeini (2009, p. 326) The proposed age is only based on stratigraphic position
Hith Fm.	Late Jurassic (Tithonian)	Lb. Surmeh Fm. Ub. Fahlian Fm.	Anhydrite with carbonate intercalation	Book: Motiei (1993, p. 131), Aghanabati (1998, p. 678), Aghanabati (2004, p. 266), Alavi-Naeini (2009, p. 325)
Surmeh Fm.	Middle to Late Jurassic	Lb. Clayey limestone and shaly beds of Neyriz Fm. Ub. Hith Fm. in the Coastal Fars and southern part of Dezful embayment, and Fahlian Fm.in the Internal Fars	Dolomitic limestone, clayey limestone, marl and marlstone, limestone and dolomitic limestone. Drilled wells of southern Dezful embayment contains a clean carbonate succession informally introduced as the Mond mb. Thickness: 672 m	Book: Motiei (1993, p.127), Aghanabati (1998, p. 669), Aghanabati (2004, p. 266), Alavi-Naeini (2009, p. 324) Thesis: Mirshahani (2011), Sadeghi (2009), Yosef-Pour (2004), Mahmoudi (2006), Lotfi (2006) The first Fm. of Khami Gr.

(continued)

Table 3.3 (continued)

Rock unit	Chronostratigraphic affinity	Lower/Upper Boundary	Lithology and thickness at type section	Note (references, comments)
Jurassic (Kopeh-Dagh)				
Chamanbid Fm.	Late Bajocian-Oxfordian	Lb. Shale beds of Bash-Kalateh Fm. in the type section and in the areas where the Bash-Kalateh Fm. does not outcropped, the Lb. is conglomerate and clastic beds which are older than Lias Ub. Mozduran Fm. With conformable boundary showing sharp lithological changes	Includes dark grey to black pyrite-bearing fine-grained limestone with black shale and marl intercalations Thickness: 1722 m	Book: Afshar-Harb (1994, p.74), Aghanabati (1998, p. 551), Aghanabati (2004, p. 272) Thesis: Salahshour (2004), Shahsavani (2003), Dehghan (2001), Yahyaei (2003)

(continued)

Table 3.3 (continued)

Rock unit	Chronostratigraphic affinity	Lower/Upper Boundary	Lithology and thickness at type section	Note (references, comments)
Kashafroud Fm.	Early Jurassic	Lb. Upper Triassic deposits with angular unconformity and transgressive conglomerate Ub. Middle Jurassic Chamanbid Fm. or Upper Jurassic Mozduran Fm. with conformable contact	Mostly composed of black colored shale, sandstone and some limestone and conglomerate beds	Book: Afshar-Harb (1994, p.66), Aghanabati (1998, p. 334), Aghanabati (2004, p. 270), Alavi-Naeini (2009, p. 328) Thesis: Deh-Bozorgi (2004), Hadad-Tavakoli (2005)
Khanehzou Fm.	Early –late Jurassic	Lb. turbiditic flysch of Kashafroud Fm. Ub. marly- limey deposits of Chamanbid Fm.	A succession of thick-bedded and sometimes cliff-forming limestone	Book: Afshar-Harb (1994, p.72), Aghanabati (1998, p. 561), Aghanabati (2004, p. 273), Alavi-Naeini (2009, p. 329)
Mozduran Fm.	Oxfordian-Kimmeridgian	Lb. Kashafroud with conformable contact Ub. Shurijeh Fm. with disconformable contact. The possibility of gradual contact is more than later contact	Mostly composed of cliff forming, thick-bedded micritic limestone with cherty vein and nodules interbedded with marl and shale Thickness: 420 m	Book: Afshar-Harb (1994, p.82), Aghanabati (1998, p. 564), Aghanabati (2004, p. 273), Alavi-Naeini (2009, p. 329) Thesis: Ghazanphari (1991) Article: Lasemi (1989, 1994)

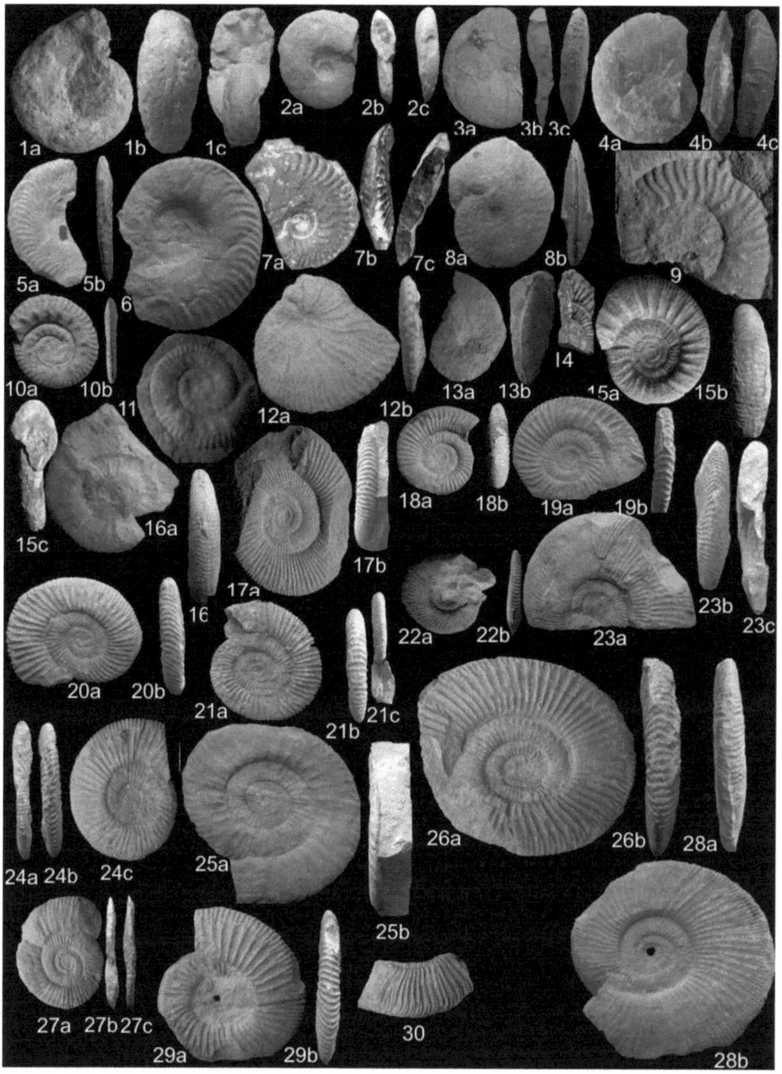

Plate 3.3 Selected Jurassic ammonites of Binalud Zone, eastern Iran (Dabaghi et al. 2012).1a–1c: *Phylloceras* sp.; 2a–2c: *Callipylloceras* sp.; 3a–3c: *Holcophylloceras* sp.; 4a–4c: *Ptycophylloceras* sp.; 5a–5b: *Glochiceras* (*Lingulaticeras*) *semicostatum*; 6. *Glochiceras* (*Lingulaticeras*) *sculptatum*; 7a–7c: *Prohecticoceras* sp. cf. *P. retrocostatum*; 8a–8b. *Oxycerites* sp.; 10a–10b: *Hecticoceras* (*Rossiensiceras*) sp. cf. *H. metomphalum*; 11: *Hecticoceras* (*Putealiceras*) *gigas*; 12a–12b: *Taramelliceras* sp. cf. *T. dentostriatum*; 13a–13b: *Tramelliceras externnodosum*; 14: *Vermisphinctes* sp.; 15a–15c: *Choffatia (Grossouvria) kontkiewiczi*; 16a–16b: *Homeoplanulites* (*Homeoplanulites*) *difficilis*; 17a–17b: *Perisphinctes* sp. cf. *P. thurmanni*; 18a–18b: *Prososphinctes* sp. cf. *P. mairei*; 19a–19b: *Perisphinctes* (*Dichotomosphinctes*) sp. cf. *P. antecedens*; 20a–20b: *Perisphinctes* (*Dichotomosphinctes*) sp. cf. *P. luciaformis*; 21a–21c: *Perisphinctes* (*Dichotomoceras*) sp. cf. *P. bifurcatus*; 22a–22b: *Sequeirosia* (*Gemmellarites*) *trichoploc*; 23a–23c: *Subdiscosphinctes* sp. cf. *S. lucingae*; 24a–24c: *Orthosphinctes* (*Orthosphinctes*); 25a–25b: *Lithacosphinctes* sp.; 26a–26b: *Orthosphinctes* (*Lithacosphinctes*) sp. cf. *O. evolutes*; 27a–27c: *Orthosphinctes* (*Ardesica*) *schaireri*; 28a–28b: *Ataxioceras* (*Ataxioceras*) *hypselocyclum*; 29a–29b: *Ataxioceras* (*Schneidia*) *guilherandense*; 30: cf. *Sublithacoceras* sp

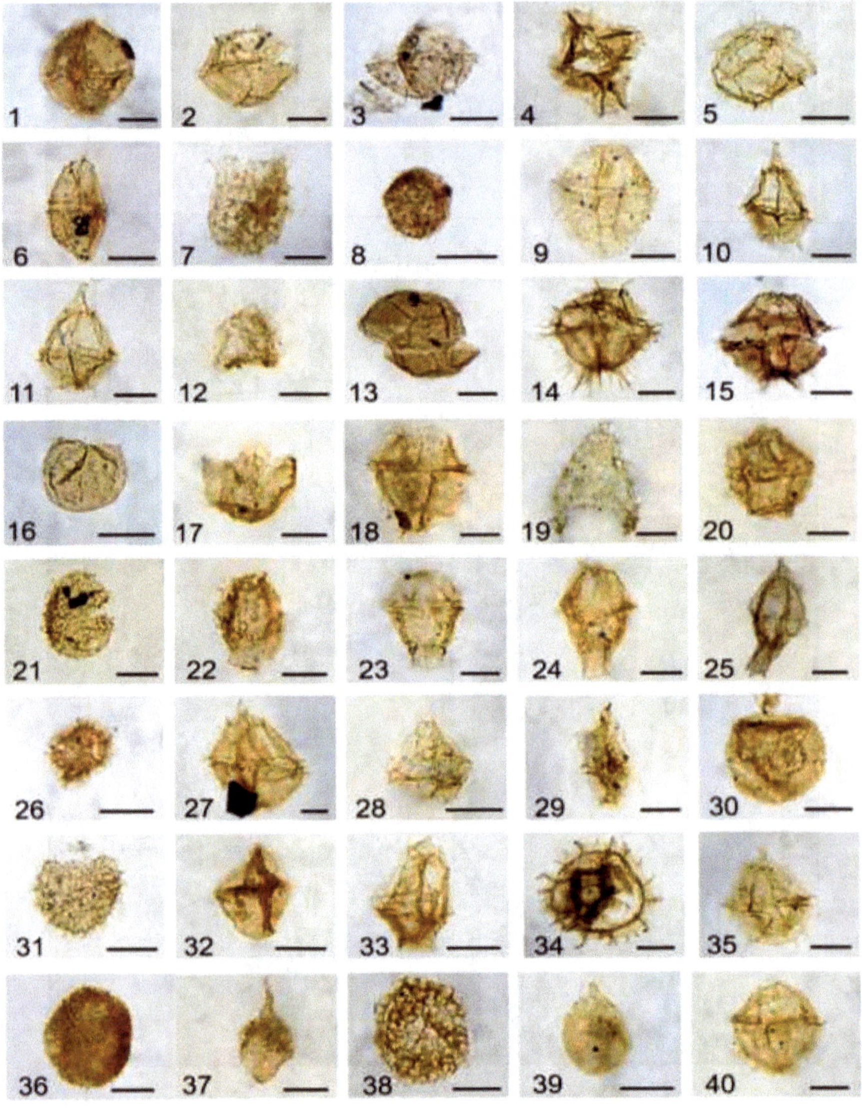

Plate 3.4 Dinoflagellate cysts of eastern Alborz (Mafi et al. 2013). Scale bar: 30 micron. 1. *Cribroperidinium crispum*; 2, 3. *Dichadogonyaulax sellwoodii*; 4. *Ctenidodinium continuum*; 5. *Ctenidodinium tenellum*; 6. *Carpathodinium preda*; 7. *Ellipsoidictyum reticulatum*; 8. *Sentusidinium sparsibarbatum*; 9. *Lithodina caytonensis*; 10, 11. *Gonyaulacysta jurassica*; 12. *Nannoceratopsis raunsgaardii*; 13. *Mendicodinium groenlandicum*; 14, 15. *Ctenidodinium combazii*; 16. *Chytroeisphaeridia chytroeides*; 17. *Escharisphaeridia pocokii*; 18. *Aldorfia aldorfensis*; 19. *Nannoceratopsis pellucida*; 20. *Leptodinium subtile*; 21. *Egmontodinium diminutum*; 22. *Stephanelytron scarburghense*; 23. *Meiourogonyaulax valensii*; 24. *Tubotuberella dangeardii*; 25. *Tubotuberella apatella*; 26. *B. pilosa.*; 27. *Rhynchodiniopsis cladophora*; 28. *Atopodinium prostatum*; 29. *Endoscrinium asymmetricum*; 30, 31. *Sentusidinium rioultii*; 32. *Glossodinium bicuneatum*; 33. *Gonyaulacysta eisenackii*; 34. *Ctenidodinium ornatum*; 35. *Gonyaulacysta centriconnata*; 36. *Tehamadinium konarae*; 37. *Pareodinia prolongata*; 38. *V. ovulum*; 39. *Pareodinia ceratophora*; 40. *Meiourogonyaulax caytonensis*

these beds, the Lower Cretaceous cliff-forming limestone is exposed. This limestone shows insignificant facies variation but has been referred to various formations.

The orbitolina-bearing limestone is the most characteristic Cretaceous unit of Iran. It is named the Tiz-Kuh Formation in the Alborz Mountains; the Tirgan Formation in the Kopeh-Dagh; the Taft Formation in the Yazd; the Shah Kuh formation in the Esfahan; the Parba formation in the Eastern Semnan; and Fahlian-Darian formations in the Zagros. The widely-occurring orbitolina-bearing limestone in the Esfahan area is of the same age in comparison to those of Central Iran, Sanandaj-Sirjan and Central Iran but with minor facies variation in facies. It is here suggested that consider the Tiz-Kuh, the Shah Kuh and the Taft formations, are all assigned to a new "**Esfahan Formation**".

In most of the above areas, Albian black shales and ammonite-bearing marls overlie the Barremian-Aptian rocks with no break in deposition. The former are referred to the Amdovan Formation in the Eastern Semnan; the Darreh-zanjir Formation in the Yazd area (Nabavi 1972); Kazhdomi Formation in the Zagros area (Motiei 1993), and the Sarcheshmeh-Sanganeh formations in the Kopeh-Dagh (Afshar-Harb 1994) and the Hezar-Masjed mountains. A tectonic phase occurred in the Early-Late Cretaceous interval (Albian-Cenomanian boundary), in most of Iran (equivalent to the Austrian Phase). In contrast to the Lower Cretaceous sediments that were deposited in wide interconnected basins, the Upper Cretaceous rocks show high facies variation resulting from sedimentation in small and discrete basins. The Upper Cretaceous rocks of Iran are characterized by so much facies variation and lithological diversity that they have not been named previously.

In general, the Cretaceous succession of Iran is constrained by two tectonic phases; first: the late Cimmerian phase which can be traced in the Lower Cretaceous through most of Iran except for Zagros, Kopeh-Dagh and the northern Alborz and second the Laramide Tectonic Phase that occurred in the latest Cretaceous, i.e. late Maastrichtian.

3.3.1 Central Iran and Sanandaj-Sirjan

At the beginning of Cretaceous, Central Iran and Sanandaj-Sirjan were experiencing continental depositional environment. The Cretaceous marine transgression was diachronous starting in the Hauterivian of the Golpayegan and Kabudarahang, the Barremian of the Esfahan, the early Aptian of Kashan, the late Aptian in the Jandagh (Gaz and Chupanan Mountains), the Cenomanian in the Nakhlak and the Santonian of Jandagh.

The Barremian to Aptian marine transgression resulted in deposition of clastic sediments within a shallow, high-energy basin with some variation from place to place. Thickness of 10–15 m have been reported from Esfahan and Kashan increasing to 400 m in the Zafreh area, 450 m in Jandagh (Kuh-Gaz), 600 m in Yazd and 1400 m in Poshtbadam. The age of this sequence also shows significant variation in these areas, so that the succession at Esfahan-Zafreh is assigned to the Barremian, and at

Golpayegan to Hauterivian, whereas at Kashan and parts of Jandagh, it is Aptian. Deposition of these clastic deposits probably continued into the Albian in the Jandagh area (Kuhegaz).

The Lower Cretaceous orbitolina-bearing rocks, mostly starting with red clastics are some of the most diagnostic rocks in the Central Iran. The wide lateral distribution of these rocks indicates a considerable extent of the Cretaceous Sea over the older rocks resulting in either non-conformities or angular unconformities (Aghanabati 2004). Recently, Wilmsen et al. (2015) revised the Cretaceous lithostratigraphic classification of the Khur area (see Fig. 3.8), which had been previously defined by Aistov et al. (1984).

Chah Palang Formation

The succession starts with conglomerates and sandstones filling depressions in a pronounced palaeo-topography of the basement rocks (metamorphics, granitoids, metasediments) of various ages belonging to the Anarak metamorphic complex and the metamorphic Shemshak Group (Wilmsen et al. 2015). Like the Cretaceous transgressive sequence already described, this formation rests on the older rocks with angular unconformity due to the effects of the Late Cimmerian Tectonic event. A Late Jurassic to early Early Cretaceous age was inferred based on stratigraphic relationships (Aistov et al. 1984).

Noghreh Formation

The Noghreh Formation consists of red sand- and siltstones, gypsum layers, dolostones, oolitic-bioclastic limestones and sandy marls. This formation rests unconformably on basement rocks or conformably on the Chah Palang Formation (Wilmsen et al. 2015). A Late Barremian age was inferred based on the calcareous algae in the upper part of the formation. It is believed that this formation was formed in the coastal sabkha, braided streams, meandering rivers, floodplain and marginal to shallow marine environments (Wilmsen et al. 2015).

Shah Kuh Formation

The Shah Kuh Formation consists of thick-bedded to massive, dark-coloured muddy limestones (wackestone, packstone, rudstone and floatstone) with abundant orbitolinids and rudists. It is in fact a carbonate platform deposit that overlies Upper Jurassic-Barremian continental and marginal marine sediments (the Chah Palang and Noqreh formations) (Wilmsen et al. 2013). A Late Barremian–Early Aptian age is inferred from calcareous algae in the upper Noqreh Formation and Late Aptian ammonoids in the overlying lower Bazyab Formation. This formation was deposited on a shallow-marine, but comparatively low-energy carbonate platform (Wilmsen et al. 2013a).

Bazyab Formation

The Bazyab Formation consists of marls and nodular limestones with intercalated fine-grained sand- and siltstones as well as some sandy bioclastic limestone levels (Wilmsen et al. 2015). The ammonites suggest a Late Aptian age for the lower part

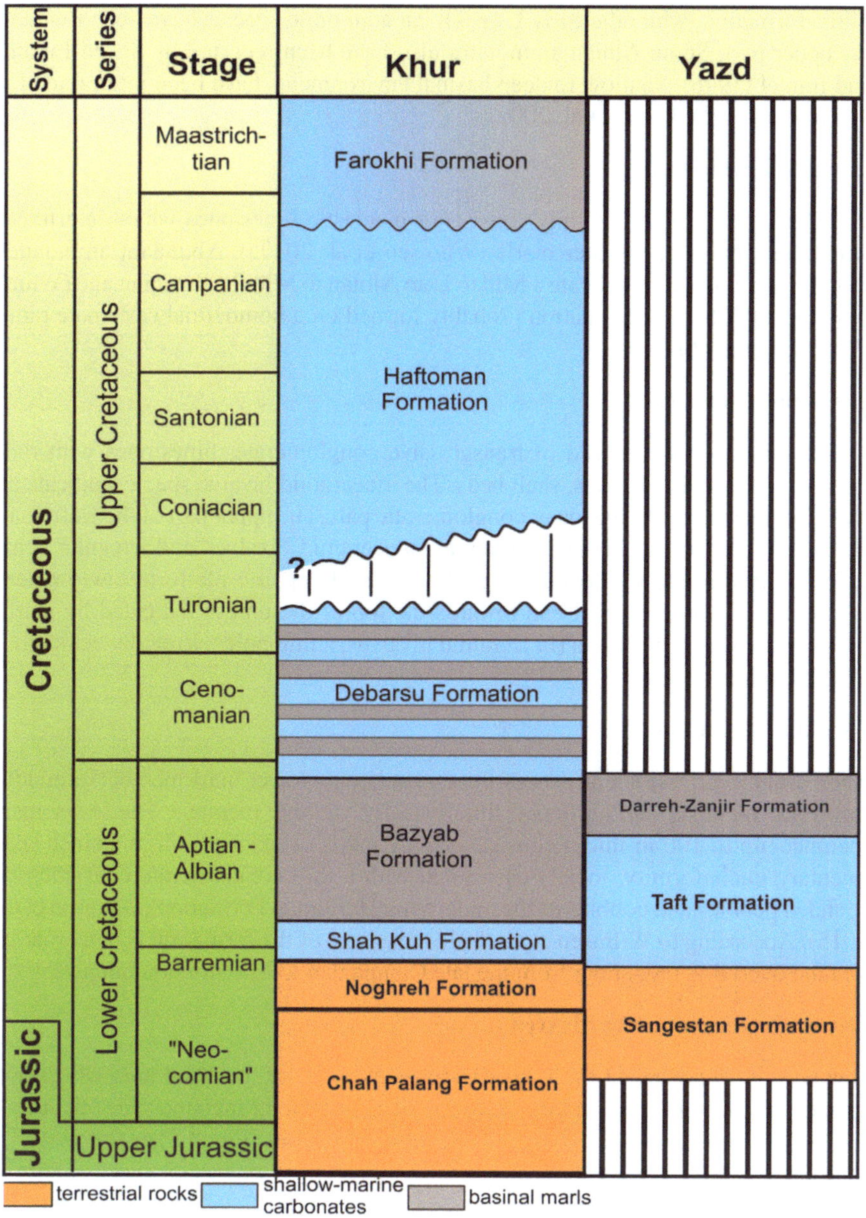

Fig. 3.8 Lithostratigraphy of the Cretaceous rocks of Central Iran, the Khur and the Yazd areas. Modified after Wilmsen et al. 2015)

of the formation, while the early Late Albian ammonite species have been found in the upper part. Some Albian ammonites also have been reported by Seyed-Emami and Immel (1996). Shallow to deep basinal environments have been interpreted for this formation (Wilmsen et al. 2009).

Debarsu Formation

The Debarsu Formation mainly consists of bioclastic limestones which interfinger with somewhat deeper-water marls (Wilmsen et al. 2013a). Abundant ammonites found in the formation indicate a Mid to Late Albian to Middle Turonian age (Wilmsen et al. 2013b). The formation probably formed on a homoclinal carbonate ramp (Wilmsen et al. 2015).

Haftoman Formation

This formation is composed of transgressive conglomerate, limestones with rudist biostromes and rudstones, shell beds. The inoceramid bivalve species indicate an early Coniacian age for the lower conglomeratic part. The upper part of the Haftoman Formation contains abundant ammonites, inoceramid bivalves and irregular echinoids suggest a late Campanian age. A large-scale epeiric platform environment is envisaged. A deepening trend towards the top of the unit is indicated by marly facies and open marine biota (inoceramid bivalves, ammonoids, irregular echinoids) (Wilmsen et al. 2015).

Farokhi Formation

The Farokhi Formation consists of three members, a lower marl member, a middle limestone member, and an upper, lithologically variable member. The lower marl member (up to 100 m-thick) consists of grey marls overlying an iron-stained sedimentary unconformity, locally associated with a thin conglomerate consisting of rounded pebbles and cobbles of the underlying Haftoman Formation (Wilmsen et al. 2015). According to Wilmsen et al. (2015), the base of the formation is diachronous and the overlying units rang from the late Campanian to Maastrichtian in age.

Lower Cretaceous of Yazd Area

Nabavi (1972) described the Lower Cretaceous rocks of the Yazd area which rest unconformably on either the Jurassic Shir-Kuh granite or the metamorphic Shemshak Group. This succession is divided into several formations (Fig. 3.8) as follows:

Sangestan Formation

The Sangestan Formation mainly consists of reddish conglomerates, sandstones and siltstones, evaporates and carbonates. This unit formed in alluvial fans, braided to meandering rivers, coastal plain including restricted marginal marine settings (Wilmsen et al. 2015). The base of formation is Hauterivian and the top, though diachronous, may ranges into the Barremian (Wilmsen et al. 2015) (Fig. 3.9).

Taft Formation

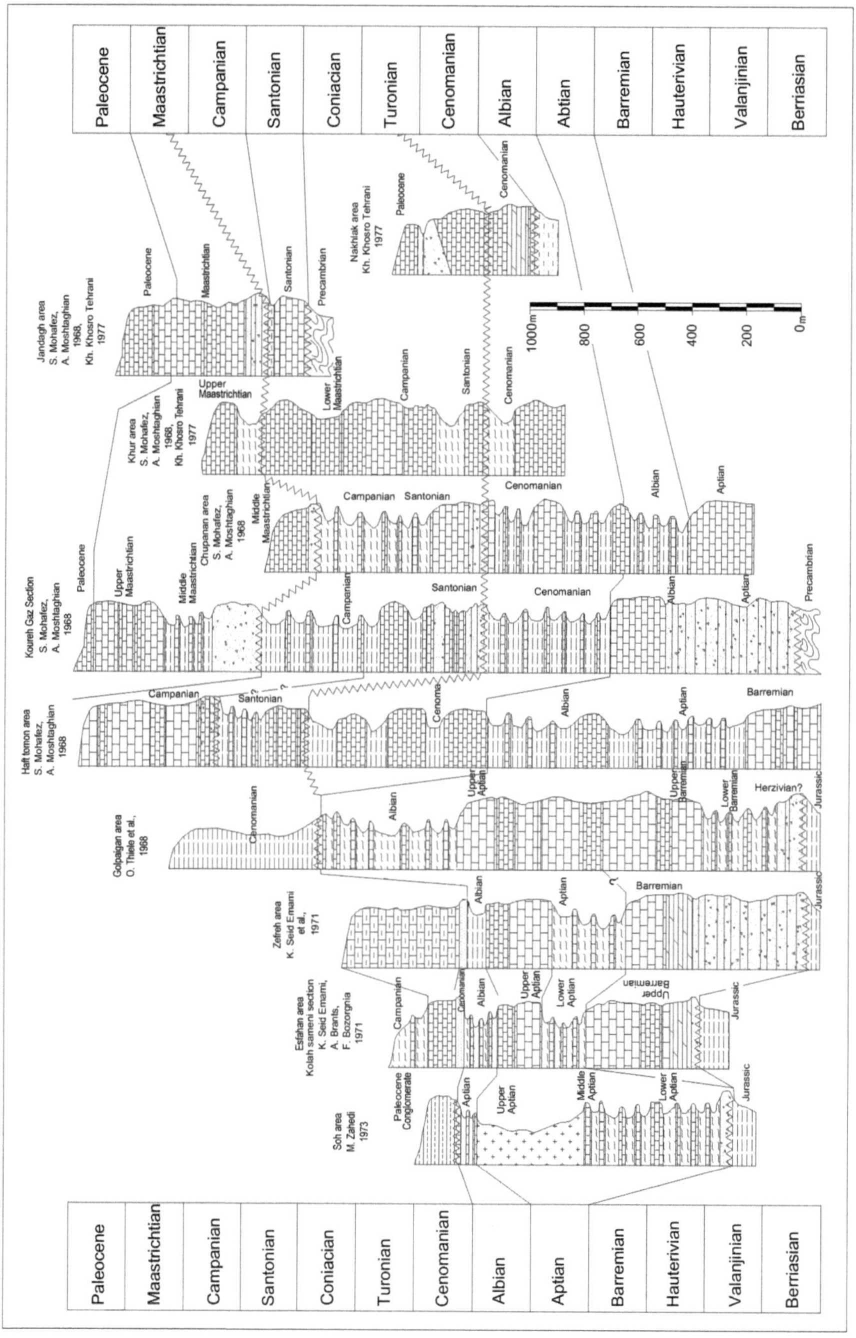

Fig. 3.9 Lithostratigraphical columns of the Cretaceous rock uints of some parts of Sanandaj-Sirjan and Central Iran. Modified after Shemirani 1991

The Taft Formation consist of thick-bedded to massive rudist-, algae- and orbitolinid-bearing shallow-water limestones. The formation attains a maximum thickness of 1 km (in the Darreh-Zereshkeh area) and is part of a large-scale carbonate platform succession that characterizes the major part of Yazd Block (Wilmsen et al. 2013a). The base of the formation has not yielded age diagnostic fossils, nevertheless, absolute age dating suggests a Barremian age. The upper beds of formation were assigned to the mid-late Aptian based on the orbitolinid assemblages (e.g. Schlagintweit and Wilmsen 2014). The Taft Formation rests conformably on the Sangestan Formation and is unconformably overlain by the Darreh-Zanjir Formation.

Darreh-Zanjir Formation

The Darreh-Zanjir Formation is the uppermost Lower Cretaceous rock unit in the Yazd area, where it is composed of deep-waster marls and spiculitic limestones. It contains relatively abundant ammonites that suggest an Albian age (Seyed-Emami and Immel 1995). This formation is economically important because it hosts major lead-zinc deposits (Ghorbani 2013).

3.3.2 *Alborz*

The lithological and paleontological evidence shows that the Cretaceous facies of the northern Alborz differ from that of the southern flanks. Therefore, these two zones are described separately.

Northern Alborz

According to Shemirani (2002b), the late Cimmerian orogeny weakly affected the northern Alborz basin and the marine condition of the late Jurassic persisted through the Cretaceous without any significant change. The Cretaceous-Jurassic boundary is located within continuous sequences in some parts of the Alborz. Due to the different depositional environments in the northern Alborz basin, several facies were developed.

Western northern Alborz

The Cretaceous sediments of this area mostly consists of limestones with sandstone intercalations that accumulated in shoreline, shoal and carbonate ramp environments (Sadeghi 1999). These rocks are exposed in Bandar Anzali-Masoule, southern Rasht and Lahijan. Many lithostratigraphic units have been described in these areas, but not even informal names have been introduced, probably because of the great lithostratigraphic variation observed.

Central Northern Alborz

According to Cartier (1971), the Cretaceous rocks of the Chalus area rest on shales and sandstones of Shemshak Group with a significant temporal gap. The lowermost part of the Cretaceous succession is reported to be oolitic and massive limestones with limey to clayey shale and marls. Cartier (1971) defined the Chalus Formation and described it as consisting of Lower Volcanic, Lower Limestone, Middle Volcanic, Upper Limestone and Upper Volcanic units.

Eastern northern Alborz

Cretaceous rocks in this area have been reported from the Gorgan-Gonbad-Nodeh belt. These consist of mixed siliciclastic-carbonate succession divided into the Pa-Ghaleah, Jami-Shouran and Ghaleh- Mouran formations. For more information, readers are referred to Table 3.4.

Southern Alborz

Cretaceous sediments are widely distributed on the slopes of the southern Alborz. However, these rocks exhibit strong lateral facies variations that have resulted in uncertainties concerning their stratigraphic relationships.

In the southern Alborz, the Mosha-Fasham fault, one of the most important tectonic structures in the region, has played a major role in controlling the deposition and preservation of Cretaceous rocks. Various sedimentary processes influenced deposition on two adjacent blocks resulting in synsedimentary faulting (Sadeghi 1999). Sedimentation was continuous during the Early to Middle Cretaceous, from the Berriasian through to the Maastrichtian, to the west of the fault, in the Sehpaeieh Mountains and the Darbandak area of eastern and southeastern Tehran. To the east, Jurassic sedimentation continued throughout the Cretaceous. During late Neocomian time, the Sehpaeieh Mountains were formed by uplift during the late Cimmerian orogeny. In ascending stratigraphic order, the Cretaceous lithostratigraphic units of the southern Alborz are as follows:

Tiz-Kuh Formation

The Cretaceous succession of the southern Alborz slopes especially in the Central Alborz is characterized by the presence of the cliff-forming Tiz-Kuh Formation. This formation was first described and introduced by Assereto from the Tiz-Kuh Mountain located in southwest Damavand, Central Alborz (Assereto 1964; unpublished data). A formal description of a type section has not yet been published; but, the same author consequently expanded his description of the Tiz-Kuh Formation to include lower and upper units. The former is composed of polygenic conglomerate, limey sandstone and light orbitolina bearing limestone. Some melaphyr and evaporative beds are locally overlain the same formation. In contrast to the lower unit, the upper unit is cliff-forming and forms most of the cliff and mountain peaks of the southern Alborz (Assereto 1966).

The upper Cretaceous deposits of Alborz have not yet been introduced as a formal formation. This succession wss divided into three units C2, C3 and C4 by Steiger (1966) in the Firouzkuh area, and into U2, U3, U4 and U5 by Assereto (1964) in the southern Tehran (Sehpaeieh Mountain). It was also divided by Dellenbach (1964) into four units including K2a, K2b, K2C and K3.

Table 3.4 Lithostratigraphic features of Cretaceous rocks of Iran

Rock unit	Chronostratigraphic affinity	Lower/Upper Boundary	Lithology and thickness at type section	Note (references, comments)
Cretaceous (Alborz-Azarbaijan)				
Tiz-Kuh Fm.	Aptian (Barremian to Early Albian)	Lb. Lar Fm. Ub. Cenomanian successions with angular unconformity	Includes two lithological units Clastic unit is composed of conglomerate, sandstone and yellow sandy limestone in the lower part, and pinkish light, cliff-forming, thick-bedded limestone in the upper part Thickness: 170 m	Book: Aghanabati (2004, p. 289, Alavi-Naeini (2009, p. 335) Thesis: Sadeghi (1999) Assereto (1966) divided the formation into two informal members including Kharsang mb. in the lower part and Hashtar mb. in the upper part
Chalus Fm.	Valanginian to Late Cenomanian	Lb. Berriasian-Valanginian rocks	The formation is composed of 5 members including 1. Lower volcanic; 2. Lower limestone; 3. Middle volcanic; 4. Upper limestone and 5. Upper volcanic Thickness. 1800 m	Book: Aghanabati (2004, p. 298)
Pa-Ghaleah Fm.	Valanginian-Hauterivian	Lb. Lar Fm. with gradual contact Ub. Uncomformable with Polygenetic conglomerate of Aptian- Albian Jamishouran Fm.	It is divided into 3 members as follows: Lower member: Lower Valanginian marl and grey-blue claystone with thin-bedded limestone in the upper part (200–250 m-thick) Middle member: Upper Valanginian-Hauterivian brown clastic limestone with claystone and sandstone intercalations (190 m-thick) Upper member: Upper Hauterivian-Barremian brown dolomite and micritic limestone (120 m-thick) Thickness: till 500 m	Book: Aghanabati (2004, p. 299).

(continued)

Table 3.4 (continued)

Rock unit	Chronostratigraphic affinity	Lower/Upper Boundary	Lithology and thickness at type section	Note (references, comments)
Ghaleh- Mouran Fm.	Late Senonian (Santonian-Campanian)	Lb. Jami-Shouran fm.	The formation is composed of two members including lower member consisting of glauconitic sandstone (10 m-thick) and in some localities sandy biomicritic limestone; and the upper member consisting of nodular limestone and chalk (60–70 m-thick) Thickness: 70–80 m	Book: Aghanabati (2004, p. 299)
Jami-Shouran Fm.	Barremian-Aptian	Lb. Pa-Ghaleh Fm. with parallel unconformity Ub. Upper Cretaceous Ghaleh-Mouran with angular unconformity	The formation starts with a sandstone unit (20–30 m-thick) gradually grading upward into thin-bedded limestone and ruddist-bearing massive reefal limestone (260–270 m-thick) Thickness: 280–300 m	Book: Aghanabati (2004, p. 2004)
Cretaceous (Central Iran, Eastern Iran and Sanandaj-Sirjan)				
Sangestan Fm.	Late Jurassic to Early Cretaceous	Lb. Igneous rocks with nonconformity Ub. Taft Fm.	The formation can be divided into three members as follows: Lower member: a succession of red clastic rocks starting with a conglomerate bed bearing granite fragments of Shirkuh Granite and continuing with sandstone sequence. The conglomerate-sandstone alternating cycle are observed several times. Middle Member: shale and limestone Upper Member: sandstone with red to purple weathering surface and alteration of marl and limestone in the upper part	Book: Aghanabati (2004, p. 310) Article: Wilmsen et al. (2015)

(continued)

Table 3.4 (continued)

Rock unit	Chronostratigraphic affinity	Lower/Upper Boundary	Lithology and thickness at type section	Note (references, comments)
Taft Fm.	Barremian to Aptian	Lb. Sangestan Fm. Ub. Dareh-Zanjir Fm. with continuous contact	The formation is composed of a homogenous succession of grey to dark grey and crème, karstic cliff forming limestone with honeycomb appearance characterized by abundant chert nodules Thickness: 1100 m	Book: Aghanabati (2004, p. 311) Babakhani (1988) informally referred the chertiferous part of formation to "Abkuh" formation having a local uses Article: Wilmsen et al. (2015)
Darreh-Zanjir Fm.	Albian to middle-late Aptian	Lb. Taft Fm. with continuous contact Ub. distributed by thrusting	A homogenous succession of olive green shale	Book: Aghanabati (2004, p. 311). The formation was referred to "Cretaceous green marl" by Huckriede et al. (1962) in the Kerman and to "Bazyab Fm." by Aistov (1984) in the Southern Khur Article: Wilmsen et al. (2015)
Biabanak Fm.	Middle to late Albian	Lb. Noghreh Fm. with conformable contact but laterally grade to Aptian Orbitolina-bearing limestone of either Taft or Shah Kuh formations	The formation is composed of light grey to green, olive, dark to black grey shale with sandy, silty, calcareous and marly shale intercalations and low thickness thin-bedded sandstone. Thickness: 3000 m	Book: Aghanabati (2004, p. 312) Article: Wilmsen et al. (2015)
Mirza Fm.	Early Aptian to middle Albian	Lb. highly eroded shale of Biabanak Fm.	Light to greenish, very soft calcareous shale and slaty shale	Book: Aghanabti (2004)

(continued)

Table 3.4 (continued)

Rock unit	Chronostratigraphic affinity	Lower/Upper Boundary	Lithology and thickness at type section	Note (references, comments)
Noghreh Fm.	Neocomian to Barremian	Lb. the formation evidently overlies the Paleozoic and Mesozoic successions; however in the area where the Upper Jurassic-Lower Cretaceous Chah-Palang formation is observed, these very formations form an apparently continuous succession. Ub. Shah Kuh formation with continuous and gradual contact	Green to red sandstone, siltstone and conglomerate	Book: Aghanabati (2004, p. 315), Alavi-Naeini (2009, p.345 Article: Wilmsen et al. (2015)

(continued)

Table 3.4 (continued)

Rock unit	Chronostratigraphic affinity	Lower/Upper Boundary	Lithology and thickness at type section	Note (references, comments)
Shah Kuh Fm.	Aptian to Albian	Lb. Noghreh with disconformable contact Ub. Bazyab with conformable contact	Grey Orbitolina-bearing limestone accompanied by sandy limestone and marl intercalations	Book: Aghanabati (2004, p. 316), Alavi-Naeini (2009, p.346) Article: Wilmsen et al. (2015) The formation is one of the cliff-forming Cretaceous formations of Central Iran and Sanandaj-Sirjan correlated with Tiz-Kuh formation of Alborz. The Formation was assigned to Baremian-Early Aptian by Seyed Emami (1979)
Bazyab Fm.	Early to middle Albian	Lb. Orbitolina-bearing Shah Kuh formation with conformable contact.	Claystone, clayey marl and 2–3 m-thick succession of limestone and sandy limestone Thickness: 550 m	Book: Aghanabati (2004, p. 316), Alavi-Naeini (2009, p. 347) Article: Wilmsen et al. (2015)
Debarsou Fm.	Cenomanian to Toronian	Lb. Bazyab Formation with conformable boundary. Ub. Haftuman Fm. with angular unconformity	Conglomerate and sandstone, grey bioclastic limestone and marl with calcareous intercalations.	Book: Aghanabati (2004, p. 316), Alavi-Naeini (2009, p. 347) Article: Wilmsen et al. (2015)

(continued)

Table 3.4 (continued)

Rock unit	Chronostratigraphic affinity	Lower/Upper Boundary	Lithology and thickness at type section	Note (references, comments)
Haftuman Fm.	Early Senonian	Lb. Debarsou Fm. with angular unconformity Ub. Choupanan Fm. with disconformable contact	Rudist limestone and some sandy limestone rarely containing some successions of sandstone and marl and/or lens-shaped conglomerate Thickness: 900–925	Book: Aghanabati (2004, p. 317), Alavi-Naeini (2009, p.347) Article: Wilmsen et al. (2015) The formation is one of the cliff-forming Cretaceous formations of Central Iran and Sanandaj-Sirjan comparable with Tiz-Kuh Formation of Alborz
Farrokhi Fm.	Late Senonian to Danian	Lb. Haftuman Fm. with disconformable contact Ub. Choupanan Fm. with disconformable contact	Light to grey limestone with cherty pebbles and layers Thickness: 65–180	Book: Aghanabati (2004, p. 317), Alavi-Naeini (2009, p.348) Article: Wilmsen et al. (2015)
Cretaceous (Zagros)				
Fahlian Fm.	Neocomian to Aptian	Lb. black dolomite of Surmeh Fm. with apparent conformable contact Ub. Marl beds of Gadvan Fm.	Massive limestone grading in the middle part to medium-bedded peloidal, oolitic limestone. The formation tends to show higher clay content than Suremh Fm. Thickness: 365 m	Book: Motiei (1993, p. 154), Aghanabati (2004, p. 322), Alavi-Naeini (2009, p. 355) Thesis: Salehi (2006), Peykani (2005), Najafi (2005), Mohammad-Khani (2003), Vali-pour (2010), Zarea (2003), Khazaei (2003), Karrar-Verdi (2007), Souri (2006), Avarjani (2005), Hashemi-Hoseyni (2006) Article: Najafi et al. (2005)

(continued)

Table 3.4 (continued)

Rock unit	Chronostratigraphic affinity	Lower/Upper Boundary	Lithology and thickness at type section	Note (references, comments)
Gadvan Fm.	Late Neocomian to Aptian	Lb. Fahlian Fm. with gradual and conformable contact Ub. Darian Fm. with gradual and conformable contact	A succession of dark grey limestone and dark grey to brownish green marl. The formation is characterized by a limestone bed named as "Khalij limestone" which is underlain and overlain by shale successions, sometimes called lower Gadvan Shale and upper Gadvan shale, respectively. Thickness: 106.7 m	Book: Motiei (1993, p. 159), Aghanabati (2004, p. 324), Alavi-Naeini (2009, p. 356). Thesis: Kushesh (2008) Article: Shemirani et al. (2001), Movahed (2001).
Darian Fm.	Aptian	Lb. Gadvan Fm. with gradual contact. Ub. Kazhdomi Fm. with conformable contact	Grey to brown, thick-bedded to massive hard limestone Thickness: 286.5 m	Book: Motiei (1993, p. 162), Aghanabati (2004, p. 324), Alavi-Naeini (2009, p. 357). Thesis: Pour-Bagher (2007), Kushesh (2008), Abbasi (2009), Ghalavand (1996), Fakhr-Rahimian (2008), Moradi (2002), Naseri (2002), Ostad-Kalateh (2003), Najafi (2008), Alidadi (2006), Dabbagh (2003) Article: Shemirani et al. (2000)
Garou Fm.	Neocomian to Coniacian	Lb. is not observed in the surface sections; however, in the well section, it overlies the Gotnia (Ghatnieh) Fm. Ub. Bangestan limestone with erosional unconformity	A succession of dark grey limestone with high clay content accompanied by cherty, pyritic bitumen shale Thickness: 894 m	Book: Motiei (1993, p. 166), Aghanabati (2004, p. 325), Alavi-Naeini (2009, p. 358) Thesis: Hekmati-Nia (2004), Ezam-Panah (2011), Akhtari (2006). Article: Angji (2005)

(continued)

Table 3.4 (continued)

Rock unit	Chronostratigraphic affinity	Lower/Upper Boundary	Lithology and thickness at type section	Note (references, comments)
Kazhdomi Fm.	Albian to Cenomanian	Lb. Darian Fm. with disconformable contact accompanied Ub. Saravk limestone with gradual or sharp contact	Grey to black bitumen shale, limestone bearing high clay content with regular weathering, black shale and marl Thickness: 210 m	Book: Motiei (1993, p. 169), Aghanabati (2004, p. 325), Alavi-Naeini (2009, p. 358). Thesis: Norouzi -Chegini (2005), Ghalavand (1996), Khujin (2006) Article: Shemirani et al. (2000)
Sarvak Fm.	Albian to Toronian	Lb. Kazhdomi Fm. with gradual and conformable contact Ub. Gurpi Fm. with sharp contact	Grey fine-grained nodular clayey limestone with black grey marl intercalations in its basal part; light brown hard limestone in the middle part; and thick- to very thick-bedded chalky limestone with irregular weathering profile containing iron oxide and breccia in the upper part. The formation commonly forms a karstified limestone of Zagros Thickness: 821.5 m	Book: Motiei (1993, p. 175), Aghanabati (2004, p. 326), Alavi-Naeini (2009, p. 359). Thesis: Kazemi (2010), Afshin (2007), Afshar (2004), Amir-Zadeh (2000), Eatemadi-Kermani (2010), Shirvani (2006), Amini (2008), Naseri (2005), Dalvand (2006), Ghorbani (2010), Asghari (2005), Ghalavand (2010), Amiri-Bakhtiar (1991), Kamiabi-Shadan (2005), Kalateh (2001), Rostami (2003), Necoeean (2002), Jalilian (1996), Hanzaei (2006), Yazdani (2006). Article: Amir-Zadeh. (2005), Pour-Moatamed et al. (2002)

(continued)

Table 3.4 (continued)

Rock unit	Chronostratigraphic affinity	Lower/Upper Boundary	Lithology and thickness at type section	Note (references, comments)
Ahmadi Shale mb.	Cenomanian	Lb. Modoudlimestone mb. of Sarvak Fm. with conformable contact Ub. Ilam Fm. with erosional unconformity	Grey to green shale with soft erosional profile accompanied by thin-bedded limestone intercalations Thickness: 30–60 m	Book: Motiei (1993, p. 179), Aghanabati (2004, p. 326)
Modoud calcareous mb.	Albian to Toronian	Lb. Kazhdomi Fm. with conformable contact Ub. Ahmadi Shale mb. with conformable contact	Brown thick-bedded to massive limestone. Thickness: 61–122 m	Book: Motiei (1993, p. 180), Aghanabati (2004, p. 326)
Surgah Fm.	Toronian to Early Santonian	Lb. Garo Fm. Ub. Ilam Fm.	Light to black grey pyritic shale showing soft erosional profile interbedding with yellow weathering color, fine-grained limestone Thickness: 175.5 m	Book: Motiei (1993, p. 180), Aghanabati (2004, p. 327), Alavi-Naeini (2009, p. 361) Thesis: Kazemi (2010), Raziani (2010), Rahmani-Zadeh (2009), Farhad (2008), Ghalavand (2009), Akbari (2006)

(continued)

Table 3.4 (continued)

Rock unit	Chronostratigraphic affinity	Lower/Upper Boundary	Lithology and thickness at type section	Note (references, comments)
Laphan Shaly mb.	Early Senonian	Lb. Sarvak Fm. with either gradual or sharp contact. Ub. Ilam formation with sharp and conformable contact	Grey to sometimes greenish grey shale with thin-bedded clayey limestone intercalations Thickness: 84.5 m	Book: Motiei (1993, p. 187), Aghanabati (2004, p. 327)
Ilam Fm.	Santonian-Campanian	Lb. Surgah Fm. Ub. Gurpi Fm.	Fine-grained light to dark grey clayey limestone interbedded with thin and well-bedded shale Thickness: 190 m	Book: Motiei (1993, p. 192), Aghanabati (2004, p. 328), Alavi-Naeini (2009, p. 363) Thesis: Shoshtarian (2011), Hemmat (2009), Kermani (2010), Amini (1999), Mehmandosti (2005), Ghalavand (2009), Razyani (2010), Rahmani-Zadeh (2009), Shadan (2005), Kalateh (2001), Rostami (2002), Bahrami-Samani (1997), Amiri-Bakhtiar (1991), Sipani (2010), Minaei (2002), Ghiasvand (2003)

(continued)

Table 3.4 (continued)

Rock unit	Chronostratigraphic affinity	Lower/Upper Boundary	Lithology and thickness at type section	Note (references, comments)
Gurpi Fm.	Santonian-Paleocene	Lb. Ilam Fm. with erosional disconformity Ub. Pabdeh Fm.	Marl and bluish grey marl with thin-bedded clayey limestone subordinates. Two formal members are introduced viz. “Emam Hasan limestone Mb.” consisting of 144 m-thick fine-grained, thick-bedded grey clayey limestone with condensed grey marl intercalations, and “Seymareh Limestone Mb.” consisting of cliff-forming brown marl occurred as the interbeddings in the shale beds of Gurpi Fm. an informal mb. is also introduced consisting of Neritic limestone named “Mansoori Limestone” Thickness: 320 m	Book: Motiei (1993, p. 197), Aghanabati (2004, p. 328), Alavi-Naeini (2009, p. 364) Thesis: Mahavian-Rad (2009), Falehi (2006), Asle-Shirin (2010), Hovaizavi (1998), Nezhad-Dil (2008), Hooshiar (2008), Gawhari (2010), Rahimi (2010), Nabavi (2009), Badri-Kallou (2008), Salehi (2001), Nawroozi (2001), Fanaei (2010), Zareai (2005), Garoti (2005), Mandani-Zadeh (2005), Saadat-lou (2003), Ghorichaei (2006). Research Project: Farhoodi (1975). Article: Vaziri-Moghadam et al. (2010), Ghasemi-Nejad et al. (2008), Kameli et al. (2008), Soleymani et al. (2008)
Tarbour Fm.	Campanian to Maastrichtian	Lb. Gurpi Fm. with sharp and conformable contact Ub. Sachon Fm. with erosional cotact accompanied with an iron oxide bearing conglomerate	Light to black grey pyritic shale showing soft erosional profile interbedding with yellow weathering color, fine-grained limestone Thickness: 175.5 m	Book: Motiei (1993, p. 206), Aghanabati (2004, p. 329), Alavi-Naeini (2009, p. 365) Thesis: Kazemi (2010), Raziani (2010), Rahmani-Zadeh (2009), Farhad (2008), Ghalavand (2009), Akbari (2006)

(continued)

Table 3.4 (continued)

Rock unit	Chronostratigraphic affinity	Lower/Upper Boundary	Lithology and thickness at type section	Note (references, comments)
Cretaceous (Kopeh-Dagh)				
Shurijeh Fm.	Early Cretaceous	Lb. Mozduran Fm. with gradual contact Ub. Tirgan Fm. with gradual contact and sharp facies change	The formation is mostly composed of clastic sediments, including shale, claystone, sandstone and conglomerate; slightly gypsum; and some carbonate limestone with red brown cement. The formation in the type section is composed of three units as follows: 1. Lower Clastic-Evaporative Mb.: 334 m-thick reddish brown to purple shale, sandstone, limestone and gypsum. 2. Middle Carbonate-Evaporative Mb.:276 m-thick carbonate rocks, gypsum, shale, sandstone and siltstone. 3. Upper Clastic Mb.: 371 m-thick sandstone with reddish brown shale without any carbonate and gypsum beds Thickness: 980 m-thick	Book: Afshar-Harb (1994, p. 99), Aghanabati (1998, p. 591), Aghanabati (2004, p. 331), Alavi-Naeini (2009, p. 366) Thesis: Kohneh-Ghochan (1998), Kamel-Abbasi (1998), Akhavan (2001) Kalantari (2006) assigned the formation to Kimmeridgian to Neocomian based on the paleontological data collected from Hammam-Ghaleh and Zanganlou sections, while Rahaghi (1976) attributed the formation to Tithonian to Neocomian in the Naiba and Jouzak sections
Tirgan Fm.	Barremian-Aptian	Lb. Shurijeh Fm. with sharp facies change Ub. Sarcheshmeh Fm. with sharp facies chang	Thick-bedded oolitic, bioclastic limestone with some marl, calcareous shale and marly shale intercalations Thickness: 780 m-thick	Book: Afshar-Harb (1994, p. 109), Aghanabati (2004, p. 333), Alavi-Naeini (2009, p. 371) Thesis: Mir-Kiani (2006), Khazaei (1999), Sohrabi (1999), Khadaei (1991), Ashouri (2010) Article: Aghadadashi-Abhari (2005).

(continued)

Table 3.4 (continued)

Rock unit	Chronostratigraphic affinity	Lower/Upper Boundary	Lithology and thickness at type section	Note (references, comments)
Sarcheshmeh Fm.	Late Barremian to Middle Aptian	Lb. Tirgan Fm. with conformable contact. Ub. Sanganeh Fm. with conformable contact	It is divided into two informal members including lower member consisting of bluish grey homogenous marl and upper member consisting of dark grey calcareous shale with thin-bedded bioclastic limestone intercalations and a 1 m-thick key bed of bioclastic limestone	Book: Afshar-Harb (1994, p. 124), Aghanabati (2004, p. 334), Alavi-Naeini (2009, p. 373) Thesis: Dehghan (2002), Mahani-pour (2010), Davtalab-Zaraghi (2009) Article: Raisossadat (1995)
Sanganeh Fm.	Albian	Lb. Sarcheshmeh Fm. with conformable contact Ub. Atamir Fm. with conformable contact	It is composed of dark grey to greenish black homogenous shale without any prominent bedding, and some siltstone intercalations Thickness: 750 m-thick	Book: Afshar-Harb (1994, p. 129), Aghanabati (2004, p. 335, Alavi-Naeini (2009, p. 374) Thesis: Mahani-Pour (2010), Musavi-Nia (2008), Raisossadat (1991), Khosravi-Alghar (2007)
Atamir Fm.	Late Aptian-Middle Cenomanian	Lb. Sanganeh Fm. with gradual contact Ub. Abderaz Fm. with erosional unconformity indicating the early Late Cretaceous epeirogenic phase	It is divided into two parts as follows: The lower part: 612 m-thick sandstone with prominent morphology Upper part: a 388 m-thick unit of shale. The formation was measured 100 m in thickness at the type section	Book: Afshar-Harb (1994, p. 137), Aghanabati (2004, p. 336), Alavi-Naeini (2009, p. 375) Thesis: Musavi-Nia (2008), Padyav (1983), Sharafi (2009), Ashouri (2010). Article: Najafi et al. (2010)

(continued)

Table 3.4 (continued)

Rock unit	Chronostratigraphic affinity	Lower/Upper Boundary	Lithology and thickness at type section	Note (references, comments)
Abderaz Fm.	Turonian to Coniacian	Lb. Atamir Fm. with erosional disconformity resulted from erosional phase of early Late Cretaceous. Ub. Ab-Talkh Fm. with gradual contact	It includes two lithologically different facies. In the eastern part of Dargaz-Ghuchan Road, the formation is composed of light grey shale with 3 or 4 successions of gypsiferous limestone. In the western part of the same road, the formation changed to homogenous marly-gypsiferous limestone Thickness: 188 m	Book: Afshar-Harb (1994, p. 145), Aghanabati (2004, p. 336), Alavi-Naeini (2009, p. 376) Thesis: Foroughi (2004), Musavi-Nia (2008), Bahari (2010), Ahsani (2009), Boluri-Moghadam (1996), Abdolshahi (2009), Tavakoli-Nezhad (2007), Zabih-Zavaram (2009) Article: Vahidi-Nia et al. (2002), Hadavi (2009)
Abtalkh Fm.	Middle Santonian-Early Maastrichtian	Lb. Abderaz Fm. with gradual and conformable contact Ub. Neyzar Fm. with conformable and gradual contact	Bluish grey limey shale and some marly beds. In the upper most part there are thin-bedded clayey sandstone and marly-sandy limestone Thickness: 813 m-thick	Book: Afshar-Harb (1994, p. 153), Aghanabati (2004, p. 373), Alavi-Naeini (2009, p. 376) Thesis: Rezaei (2009), Vakili-Baghmisheh (1997) Article: Hadavi (2005)
Neyzar Fm.	Maastrichtian	Lb. Abtalkh Fm. with conformable contact Ub. Kalat Fm. with	Fine to medium grained, thick-bedded glauconitic sandstone and shale with sandy limestone in the upper part Thickness: 318 m-thick	Book: Afshar-Harb (1994, p. 157), Aghanabati (2004, p. 337), Alavi-Naeini (2009, p. 377) Thesis: Musavian (1997), Vakili-Baghmisheh (1997)

(continued)

Table 3.4 (continued)

Rock unit	Chronostratigraphic affinity	Lower/Upper Boundary	Lithology and thickness at type section	Note (references, comments)
Kalat Fm.	Maastrichtian	Lb. Neyzar Fm. with conformable and concordant contact Ub. continental sediments with disconformable contact Thickness: 277 m	5 members can be distinguished in the type section including lower limestone, lower shale, middle limestone, upper shale and upper limestone. the latter mb. is composed of biostorm rudist-bearing limestone	Book: Afshar-Harb (1994, p. 161), Aghanabati (2004, p. 338), Alavi-Naeini (2009, p. 378) Thesis: Bahreh-Mand (1997), Khazaei (1994), Karimian (2009), Vakili-Baghmisheh (1997) Vakil-Baghmisheh (1997) belived that the Maastrichtian fossils are reworked, and based on benthic and planktonic foraminifera, suggested an Eocene age. The proposed age needs more studies and investigations
Nafteh Fm.	Maastrichtian	Lb. Kalat Fm. with gradual and conformable contact Ub. Pesteligh Fm. with gradual and conformable contact	Grey shale sporadically with sandy limestone intercalations Thickness: 60 m-thick	Book: Aghanabati (2004, p. 339).

3.3.3 *Kopeh-Dagh*

In the latest Jurassic, an embayment with an east-west trend formed in the Kopeh-Dagh which resulted in the movements of related blocks, facies variation, thickness changes and sedimentary gaps. Sedimentation in the northern area, located in the deepest part of embayment, continued longer than in the southern area which was relatively shallow and occupied a marginal position. Due to above reasons, sequences on the southern side of the embayment contain more sedimentary gaps than those on the northern side. In ascending stratigraphic order, the Cretaceous units of the Kopeh-Dagh are as follows:

Shurijeh Formation

Shale, reddish brown claystone, greyish brown sandstone, conglomerate, gypsum, and some limestone beds.

Tirgan Formation

Oolitic limestone, clastic sediments, marly limestone, marl and calcareous shale.

Sarcheshmeh Formation

Grey marl and intercalations of shale and thin-bedded limestone

Sanganeh Formation

Shale

Atamir Formation

Shale and sandstone

Abderaz Formation

Shale and chalky limestone

Abtalkh Formation

Sandstone and shale

Neyzar Formation

Sandy limestone

Kalat Formation

Sandy limestone

The Cretaceous sediments of Kopeh-Dagh are characterized by their significant thickness, especially in the northern area, that resulted from the formation of an embayment with an east-west trend in the latest Jurassic (Afshar-Harb 1994). The Lower Cretaceous strata of western Kopeh-Dagh are thicker than those of the eastern Kopeh-Dagh. In contrast, the Upper Cretaceous sequence is thicker in Eastern Kopeh-Dagh. As mentioned above, the succession in the northern Kopeh-Dagh tends to show more continuity than those of northern part that was located at the basin margin.

In general, Lower Cretaceous rocks are more widely distributed than those of the Upper Cretaceous partially due to the latter having been partially removed by erosion. The Cretaceous succession of Kopeh-Dagh comprises the Shurijeh, Zard, Tirgan, Sarcheshmeh, Sanganeh, Atamir, Abderaz, Abtalkh and Kalat formations.

Shurijeh Formation

Exposures of the Shurijeh Formation can be observed around of Bojnourd, where a complete section was described by Bozorgnia (1964). In this section, the formation is 390 m-thick is is composed of shale, siltstone, gypsiferous mudstone and siltstone, conglomerate and some carbonate beds.

To the north of Navia Village and south of Jozak, this formation consists of 540 m of shales, sandstone, thin-bedded conglomerate and gypsum, but on the Ozoun Mountain, it is only 260 m thick. In the latter area, the uppermost bed of the formation has an erosional surface and is overlain by the Maastrichtian Kalat Formation. This indicates that, after deposition of the Shurijeh Formation during Neocomian time, uplift that was more extensive than to the south (e.g. the Kurkhod and Takal-Kuh mountains), and exposed this area to erosion until the early Maastrichtian.

The Shurijeh Formation shows much facies variation. Its lateral equivalent around Kurkhord was referred to as the Zard Formation which was deposited in a marine environment, unlike the Shurijeh Formation which was continental.

Zard Formation

This formation is composed 207 m of micritic-argillitc limestone with high sand content, silty shale and sandstone in the northern Takal-kuh. It is mainly composed of grey sandstone, silty shale and siltstone in the northern Bojnourd, where the sand content tends to decrease upward while the silt content increases downwards. The thickness of formation and its shale and fine-grained limestone content both increases westward. The Zard Formation disappears in the northern areas of Takal-Kuh and is exposed in the southern part of Ishqabad (Afshar-Harb 1994).

Tirgan Formation

The Tirgan Formation is composed of grey well-bedded bioclastic limestone 400 m thick in the eastern Kopeh-Dagh and 1000 m in the west. It has been dated as Aptian in age, mainly based on its foraminiferal content, especially orbitolinids. Interdigitating contacts have been observed between the Tirgan Formation and underlying Shurijeh Formation of the southeastern Kopeh-Dagh, as well as the upper part of the Tirgan Fm and the Sarcheshmeh Formation of northern Kopeh-Dagh. The maximum thickness of the Tirgan Fm has been recorded in the Zaw and Koorkhod areas, reaching 400 m in the latter (Afshar-Harb 1994).

Sarcheshmeh Formation

The Sarcheshmeh Formation is composed of grey to dark grey marls in its lower part and dark shales with thin-bedded limestone in its upper part. The uppermost part of the Sarcheshmeh Formation represents cessation of clastic deposition in the latest Aptian basin and is exposed throughout the Kopeh-Dagh. It contains a thin bed bioclastic limestone bearing terebratulid brachiopods and orbitolinids (Afshar-Harb 1994).

The Sarcheshmeh Formation was deposited in an east-west trending basin with the same lithology developed uniformly. However, the thickness of the formation decreases eastward. Abundant ammonites are found in the formation, especially in its lower part in western Kopeh-Dagh. This suggests the Kopeh-Dagh Basin was of considerable lateral extent and was also connected to the open sea.

Sanganeh Formation

The Sanganeh Formation is composed of dark grey shale in its lower part and shale with siltstone and dark marl intercalations in its upper part. It is 990 m thick in the Hajidagh area and 930 m in northern Takal-Kuh. Its thickness also tends to decrease eastward to 170 m in the Gholaman Anticline where it is composed entirely of shale. The Sanganeh Formation contains diagnostic ammonites, especially in its upper part, that indicate an Albian age.

Atamir Formation

In the type section, at Atamir Village, this formation consists of two parts, a lower sandstone and an upper shale. The widest distribution of this formation can be observed in the northern part of Takal-Kuh, around the northern Maraveh-Tapeh faults. In general, it is dated as late Albian-Cenomanian but a succession assigned to Cenomanian has been described from the Chozak and Artmouli areas. Here, the upper shaly beds contain Cenomanian planktonic foraminifera such as *Hedbergella* and *Rotalipora*.

In latest Cenomanian time, the Kopeh-Dagh area was affected by a marine regression which resulted in the uplift and erosion of the entire area, indicated by an unconformity at the uppermost surface of the Atamir Formation. The Abderaz, or even younger formations rest unconformably on the Atamir Formation (Afshar-Harb 1994).

Abderaz Formation

The Abderaz Formation mainly consists of shale interbedded with fossiliferous limestone bearing abundant inoceramid brachiopods and *Globotruncana*. The fossil contents increases eastward. This formation was deposited during a marine transgression that followed the marine transgression which occurred during deposition of the uppermost Atamir Formation. The Abderaz Fm was assigned to the Turonian-Coniasian to Campanian interval, based on planktonic foraminifers and ammonites (Afshar-Harb 1994).

Abtalkh Formation

The Abtalkh Formation is mainly composed of grey shale, bluish grey to crème limey shale. In its uppermost part, it contains thin-bedded clayey sandstone and marly-sandy limestone. This formation rests on the Abderaz Formation with a gradational boundary with white colored limestone of latter formation grading into grey shale of the Abtalkh Formation. The Abtalkh Formation is overlain by the Neyzar Formation, again with a gradational contact, with shale of former grading into sandstone and siltstone of the latter.

The Abtalkh Formation in eastern Kopeh-Dagh is more widly distributed than in western Kopeh-Dagh, where the 675 m thick Abtalkh Formation is exposed in the northern parts of the Maraveh-Tapeh, Atamir area (Afshar-Harb 1994). The absence of this formation in the southern part of this area was caused either by its non-deposition or by its subsequent removal by erosion.

Kalat Formation

Deposited during the Late Cretaceous marine transgression, the Kalat Formation is relatively widely distributed in the western Kopeh-Dagh. In the southernmost part of Kopeh-Dagh, on the Jajarm Block, on the north slope of Ozum Mountain, the Kalat Formation consists 161 m of buff to yellow colored biomicritic limestone.

This formation was deposited on a shallow platform that included back barrier lagoonal to marginal platform facies. The presence of ammonites and rudists indicates marine depositional environments. In contrast to the other Cretaceous formations of Kopeh-Dagh, the Kalat Formation is thicker and more widely distributed in the southern part of the area, implying more subsidence of the basin in this area (Afshar-Harb 1994).

3.3.4 *Eastern Central Iran*

Unlike other parts of Iran, eastern Iran contains few exposures of Cretaceous rocks. In the east of Central Iran, on the Tabas and north Lut blocks, the Upper Cretaceous is represented by thin deposits, starting with sandstones and in some places conglomerate, with gradual deepening upward, indicated by deposition of dolomitic limestone. The Late Cretaceous of the eastern Lut (i.e. Khash-Nehbandan area) is represented by marine sediments resulting from rifting and subsequent marine basin formation that continued from the Late Cretaceous into the Paleogene (readers are referred to Ghorbani 2014).

3.3.5 *Zagros*

Unlike other parts of Iran where rocks assigned to the Cretaceous System have been assigned to the Early and Late Cretaceous, those in the Zagros Basin have been informally assigned to the Lower, Middle and Upper Cretaceous, with the Lower Cretaceous consisting of the Neocomian to Aptian stages, the Middle Cretaceous comprising the Albian to Turonian and the Upper Cretaceous the Coniacian to Maastrichtian (Motiei 2001).

The Early Cretaceous of Zagros is mainly represented by two distinct facies the first in the Fars area and in the southeastern Dezful Embayment, and the second in the Lorestan area. In the Fars area, the Lower Cretaceous succession starts with shallow water limestone of the Fahlian Formation that continues into the late Neocomian. The Fahlian Formation unconformably overlies the limestone of the Surmeh Formation and rests on the Hith Anhydrite in the southern Fars and in southeastern Khuzestan. As already mentioned, Late Jurassic sedimentation continued into the Neocomian around Shiraz. In the Fars area and Bandar-Abbas, the Barremian-Aptian Gadvan Formation overlies the Fahlian Formation, probably with a sedimentary gap represented by a succession of sandstone, siltstone containing abundant burrowing traces and glauconite, proving a marine regression with subsequent exposure and erosion (Fig. 3.8).

Fahlian Formation

The type section of the calcareous Fahlian Formation is located around village of Fahlian on the southern slope of the Dal Mountain, 90 km east-southeast of Do-Gonbadan and 10 km north of Noorabad-e-Mamasani in the Fars area (Fig. 3.10).

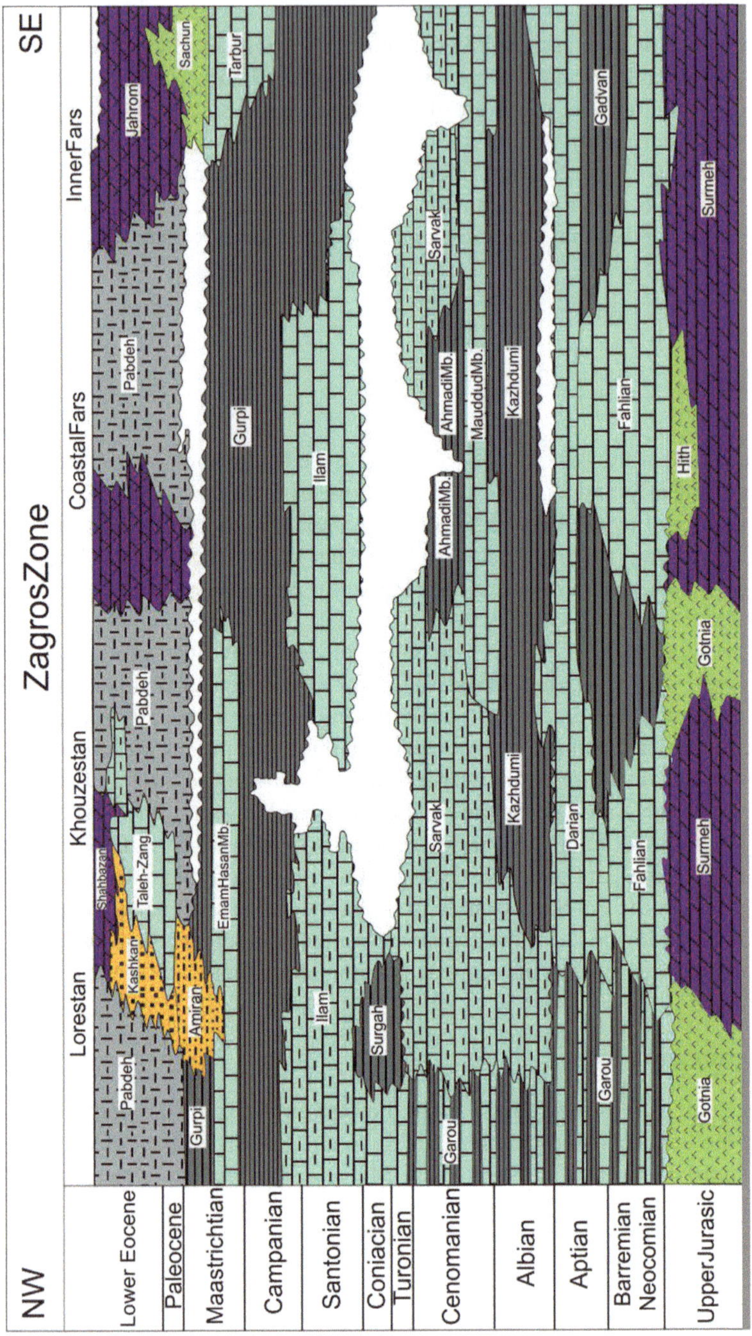

Fig. 3.10 Cretaceous rock units of Zagros zone. Modified after Motiei (1993)

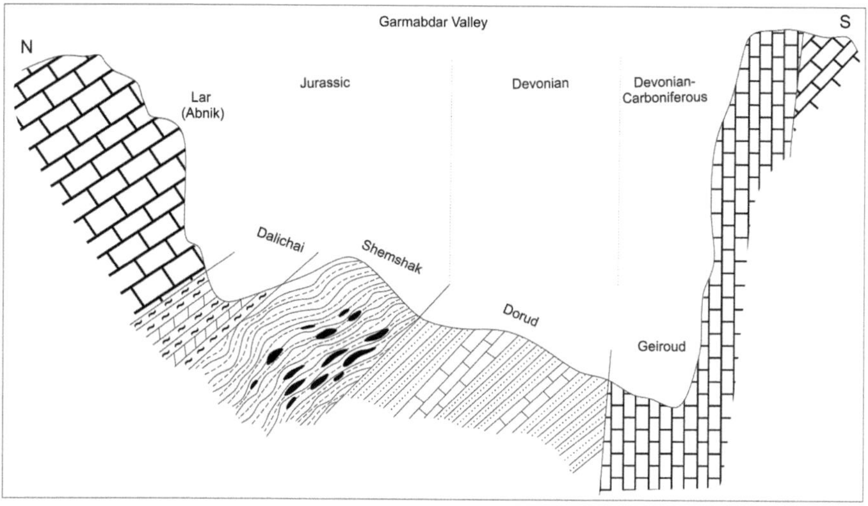

Fig. 3.11 Geological section of Abnik Valley, northeastern Tehran

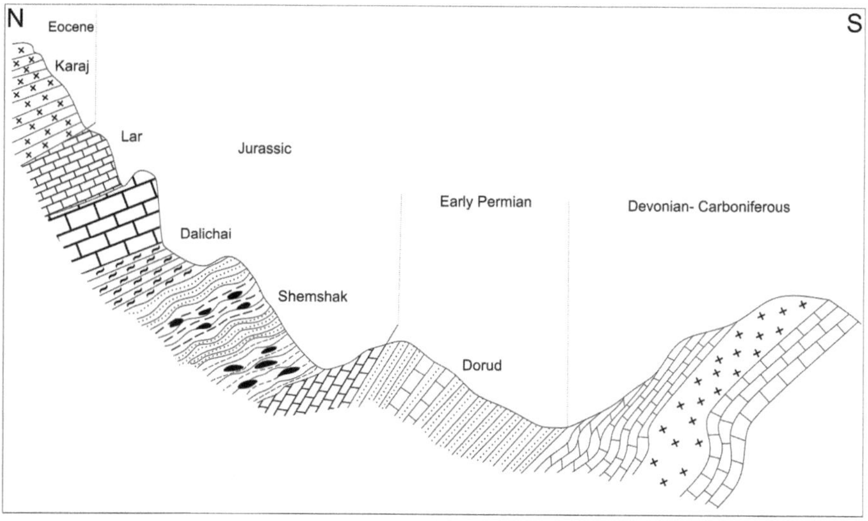

Fig. 3.12 Geological section of Geiroud-Kolunbaste, Shemshak area

The Fahlian Formation consists of grey to brown, cliff-forming massive limestone including medium-bedded pellet-oolitic limestone in its upper part. It has a higher clay content than the underlying Surmeh Formation. In some areas, the formation is underlain by an evaporate-solution breccia which is regarded as lateral equivalent of the Hith Formation.

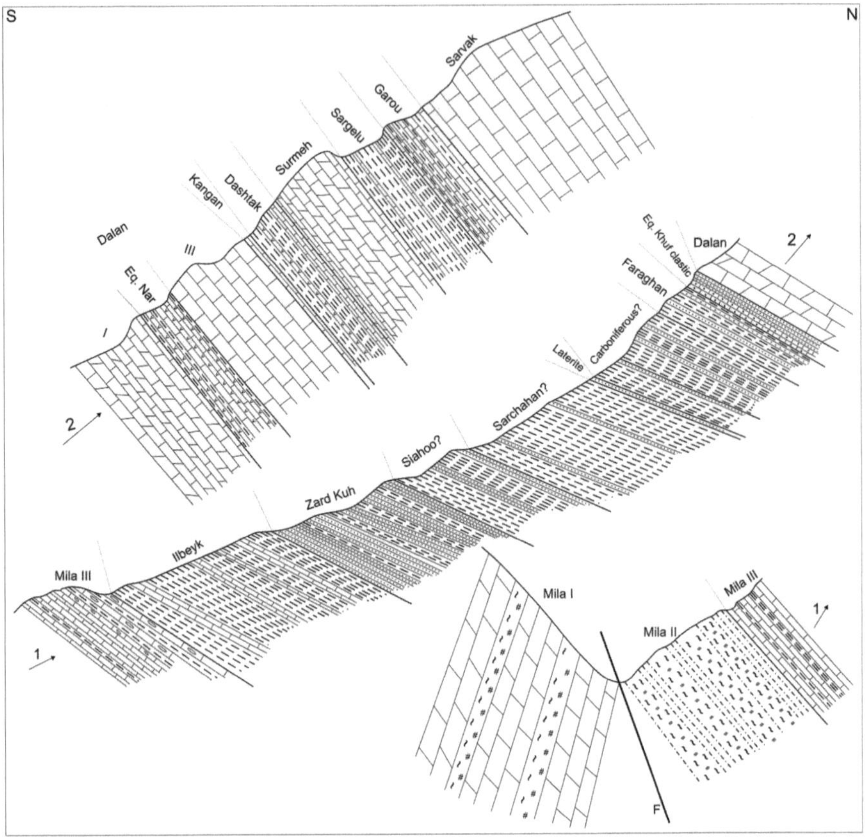

Fig. 3.13 Geological section of Zard Kuh area

The formation rests unconformably on the Surmeh Formation and in turn is overlain by the Fahlian Formation in the localities where the Hith Formation is exposed. The Fahlian Formation is conformably overlain by the Marl and limestone of Gadvan Formation. The base of the Gadvan Formation has been placed at the base of the lowest bed characterized by concave profiles resulting from the abrasion.

In general, the Fahlian Formation is a limey unit underlying the Higth Formation and overlying the Gadvan Formation. The formation is mostly preserved in the Fars area, but, it can be traced form Dezful Embayment and Lorestan where it gradually grades into the shale and clayey shale of the Garou Formation (Motiei 1993).

Gadvan Formation

The type section of this formation is located on the easternmost slope of Gadvan Mountain, 39 km north Shiraz.

This formation consists of 106.7 m of black grey limestone with shell fragments and grey to brownish grey marl. It is generally characterized by a highly weathered succession of shale and limestone underlying and overlying by Fahlian and Darian formation, respectively. Both of these contact are gradual and conformable, with the upper contact defined as the uppermost shale bed in the limestone and shale succession.

In southern Lorestan, the northern Dezful Embayment and southern Fars, the Gadvan Formation cannot be distinguished and the Fahlian and Darian formations constitute a continuous carbonate succession (Motiei 1993).

Darian Formation

The name of this formation was derived from Darian Village which is located on the southern slope of Gadvan Mountain. The 286.5 m thick, calcareous Darian Formation is composed of brown to grey, massive-thick-bedded, shallow water limestone containing abundant *Orbitulina* spp.

This formation overlies the shale, marl and limestone of Gadvan Formation with a gradational contact. Highly eroded shaly beds of the Kazhdomi Formation rest conformably on the Darian Formation with a gradual boundary placed at the uppermost limestone bed of the latter formation.

The Darian Formation can be traced in all parts of the Zagros Zone, except in southern and west southern Lorestan, where its facies changes to into the deep shaly sediments of the Garou Formation. In most of the Fars area, it is composed of various types of packstone and grainstone (Motiei 1993).

Garou Formation

The name of this formation was derived from the Kabir-Kuh area of Lorestan. Its type section has been chosen 10 km north east of Ghaleh-Darreh on the south western slope of the Kabir-Kuh Mountain.

The Garou Formation is lithologically composed of the following units (Motiei 1993):

Member 1 (387 m thick): a succession of black, radiolarian-bearing clayey limestones and black bitumen bearing pyritic shales.
Member 2 (140 m thick): black grey to brown highly clayey limestone.
Member 3 (113 m thick): brown silty and pyritic shale followed by highly clayey limestone with chert nodules.
Member 4 (58 m thick): fine-grained limestone with chert nodules
Member 5 (122 m thick): black to brown bitumen shale and highly clayey bitumen limestone bearing silt grain, radiolaria and *Globigerina* foraminifers.
Member 6 (74 m thick): black grey glauconitic limestone bearing pelagic foraminifers, echinoid fragments and larger benthic foraminifers. The uppermost 3 m of the formation includes sandy and highly glauconitic limestone.

Kazhdomi Formation

The name was obtained from the Kazhdomi Castle in the Dezful Embayment. Its type section is located in the Gargada or Gorgdan Pass in the southwestern slope of

Mish Mountain. The pass can be also accessed from the northeastern Do-Gonbadan area. The formation was estimated at 210 m in thickness; though this is not well established because the upper 120–150 m is covered.

The Kazhdomi Formation is characterized by the presence of black grey to black bituminous shale and is composed of highly clayey limestone with a regularly abraded profile, black shale and marl. It is underlain by the red iron oxide-bearing rocks of Darian Formation probably accompanied by a hiatus. The Kazhdomi Formation laterally changes to the Darian Formation in some areas. It is overlain by the Bangestan Limestone with an occasionally disconformable contact.

The Kazhdomi Formation can be observed in the Fars area and the Dezful Embayment but it gradually grades laterally into calcareous rocks in the northern Dezful Embayment-northeastern Lorestan. The formation is replaced by the Garo Formation in the central areas of southwestern Lorestan (Motiei 1993).

Sarvak Formation

The name of this formation was obtained from Bangestan Mountain, located in northwestern Behbahan, Khuzestan Province. Its type section is located in the Sarvak Pass on the southern flank of the Bangestan Anticline of northwestern Behbahan and the northeastern part of the Pars and Karanj oilfields.

The formation is mainly composed of 821.5 m of limestone divided into the following parts:

Lower part (245.5 m-thick): dark grey fine-grained clayey limestone with nodular bedding

Middle part (524 m-thick): massive limestone

Upper part (43 m-thick): thick to very thick-bedded limestone

The formation rests conformably on Kazhdomi Formation and is overlain by shaly and marly beds of the Gurpi Formation with a sharp contact.

The Sarvak Formation is widely distributed in the Zagros and especially in central Lorestan, Fars and the Dezful Embayment (Motiei 1993).

Surgah Formation

The name of the formation was obtained from the Surgah Mountain of Lorestan. Its type section is located in the Garab Pass on the southwestern flank of Surgah Mountain. In this area, the formation is 175.5 m-thick and is mainly composed of light to black grey pyritic shale alternating with fine-grained limestone with yellow weathering surfaces.

The Surgah Formation is mainly found in Lorestan, gradually disappearing to the southeast (Motiei 1993).

Ilam Formation

The name of this formation was obtained from Ilam City, with its type section chosen and measured in the Garab Pass on the southwestern Surgah Mountain on the northwestern flank of the Kabir-Kuh, 12 km south of Ilam City.

This formation, in type section, is composed of 190 m of fine-grained, light to black gray clayey limestone, sometimes showing white weathering surface and well- and thin-bedded shale intercalations. The same pelagic limestone beds grade from Lorestan southeastward into shallow limestone.

In its type section, the Ilam Formation is conformably underlain and overlain by the Surgah and Gurpi formations, respectively. The formation is characterized by two distinct facies in southwestern Iran with deep water facies in Lorestan and shallow water facies in the Dezful Embayment and Fars. However, both facies have recently been reported from these areas. In eastern Lorestan, Central Fars and the High Zagros, the Ilam limestone is replaced by a shaly succession. However, it is not known if this shaly succession belongs to the shaly Surgah Formation or to the Gurpi Formation (Motiei 1993).

Gurpi Formation

The name of this formation was obtained from the Gurpi Mountain of the northern Masjed-Soleyman City. Its type section was described in the Pabdeh Pass, located on the southwestern flank of Gurpi Mountain, north of the cities of Masjed-Soleyman and Lali. In this area, the formation is lithologically composed of 320 m of marl and bluish grey shale with thin-bedded clayey limestone intercalations. The Gurpi Formation forms the gentle topography of this area due to its softy nature.

In its type section, the formation rests disconformably on limestones of the Ilam Formation.

The shaly Gurpi Formation is widely distributed in southwestern Iran. In the Fars, the southern Dezful Embayment and adjacent areas, it is unconformably overlain by the Pabdeh Formation (Motiei 1993).

The lithostratigraphic features of Cretaceous rocks of Iran are summarized in Table 3.4 with selected diagnostic fossils from the Cretaceous of Iran illustrated in Plates 3.5, 3.6 and 3.7.

3.4 Mesozoic Sections

Some schematic geological sections of Mesozoic rock units of Iran are illustrated in Figs. 3.11, 3.12 and 3.13.

References

Abarghani, A. (2000). *Geochemistry and depositional environments of the Middle a and Upper Jurassic deposits (Lar and Dalichai Formation) of the Northeastern Maraghea, Azarbaijan* (Unpublished M.Sc. thesis), Azarbaijan: Shahid Beheshti University. (All Iranian universties thesis are written in Persian with English abstract.).

Abbasi, M. (2009). *Sequence stratigraphy of the Upper Dalan and Kangan Formations in some boreholes of the South Pars Oil Field, an explanation for the reservoir quality variation.*

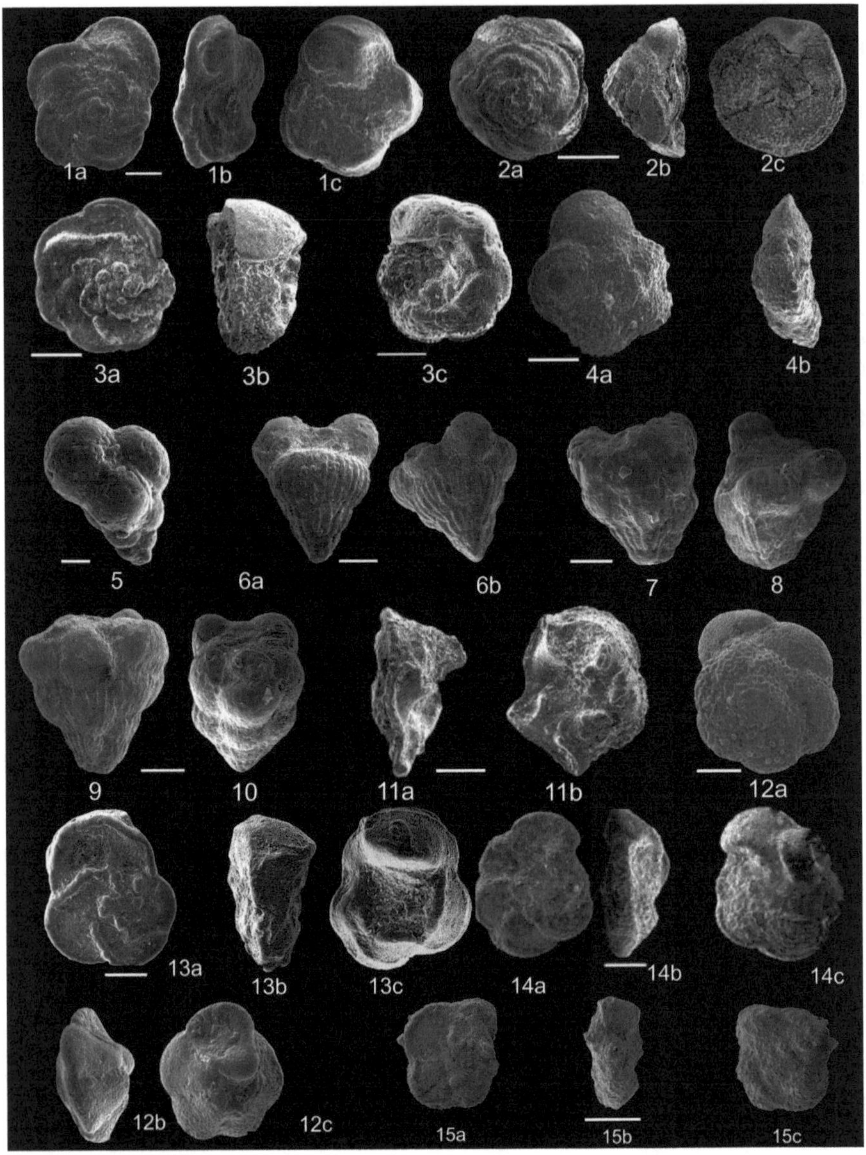

Plate 3.5 Selected diagnostic Cretaceous foraminifers of the Zagros and Central Iran, Figs. 1–12 were identified from Cretaceous of Zagros (Beiranvand and Ghasemi Nejad 2013) and Figs. 13–15 Central Iran (Ardestani et al. 2011, 2012). 1a.c. *Abathomphalus mayaroensis*; 2a.c. *Contusotruncana contuse*; 3a.c. *Gansserina gansseri*; 4a.b. *Globotruncanella havanensis*; *Guembelitria cretacea*; 6a.b. *Pseudotextularia intermedia*; 7.10. *Racemiguembelina fructicosa*; 11a.b. *Radotruncana calcarata*; 12a.c. I. *Gurina pusilla*; 13a.c. *Globotruncana aegyptiaca*; 14a.c. *Dicarinella concavata*, 15 a, b, c. *Dicarinella asymmetrica*

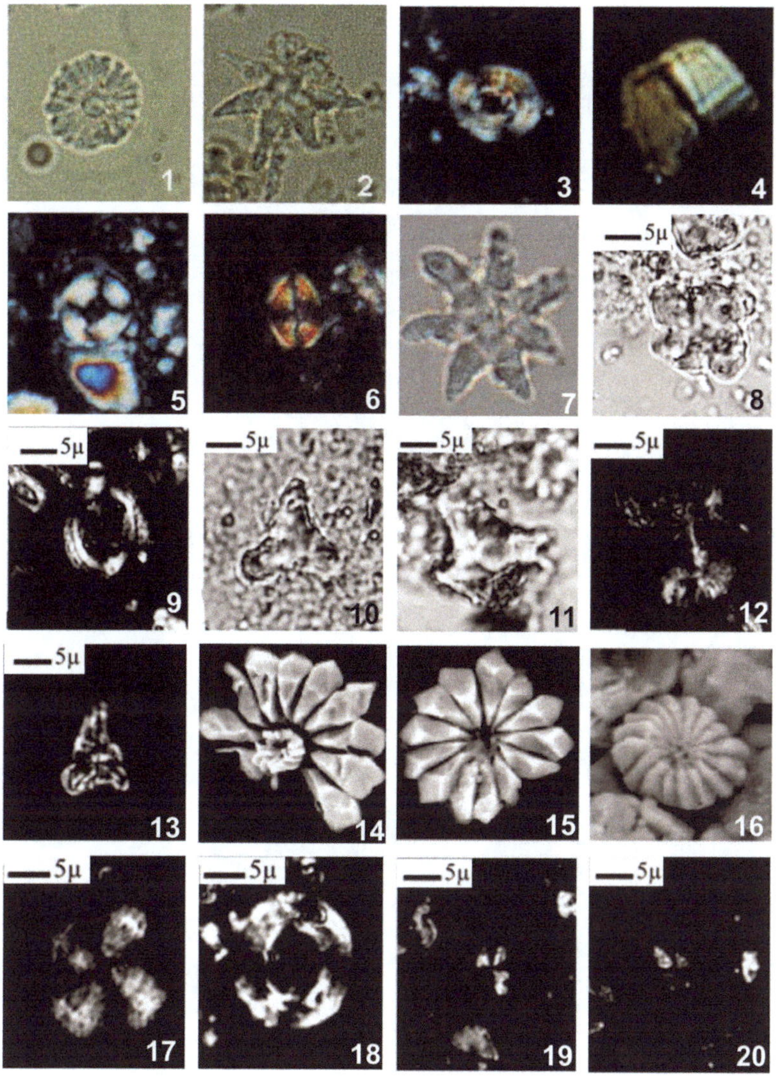

Plate 3.6 Selected diagnostic nannofossils of the Zagros and Kopeh Dagh (Scale bar: 5 μ). Figures 1–16 from are identified from Cretaceous rocks of Kopeh-Dagh (Foroughi et al. 2016) and Figs. 17–35 from Cretaceous of Zagros (Razmjooei et al. 2014). 1. *Monomarginatus quaternarius*; 2. *Monomarginatus quaternarius*; 3. *Staurolithites mielnicensis*; 4. *Staurolithites mielnicensis*; 5. *Reinhardtites anthophorus*; 6. *Reinhardtites* sp. cf. *R. anthophorus*; 7. *Reinhardtites anthophorus*; 8. *Reinhardtites anthophorus*; 9. *Eiffellithus parallelus*; 10. *Eiffellithus parallelus*; 11. *Uniplanarius sissinghii*; 12. *Uniplanarius sissinghii*; 13. *Uniplanarius trifidus*; 14. *Uniplanarius trifidus*; 5. *Uniplanarius trifidus*; 16. *Ceratolithoides verbeekii*; 17. *Eiffellithus eximius*; 18. *Tranolithus orionatus*; 19. *Micula murus*; 20. *Broinsonia parca constricta*; 21. *Lucianorhabdus cayeuxii*; 22. *Ceratolithoides kamptneri*; 23. *Reinhardtites levis*; 24. *Uniplanarius sissinghii*; 25. *Uniplanarius trifidus*; 26. *Cruciplacolithus tenuis*; 27. *Micula prinsii*; 28. *Discoaster multiradiatus*; 29. *Discoaster mohleri*; 30. *Fasciculithus tympaniformis*; 31. *Biantholithus sparsus*; 32. *Marthasterites furcates*; 33. *Bukryaster hayi*; 34. *Ceratolithoides verbeekii*; 35. *Lithraphidites quadratus*

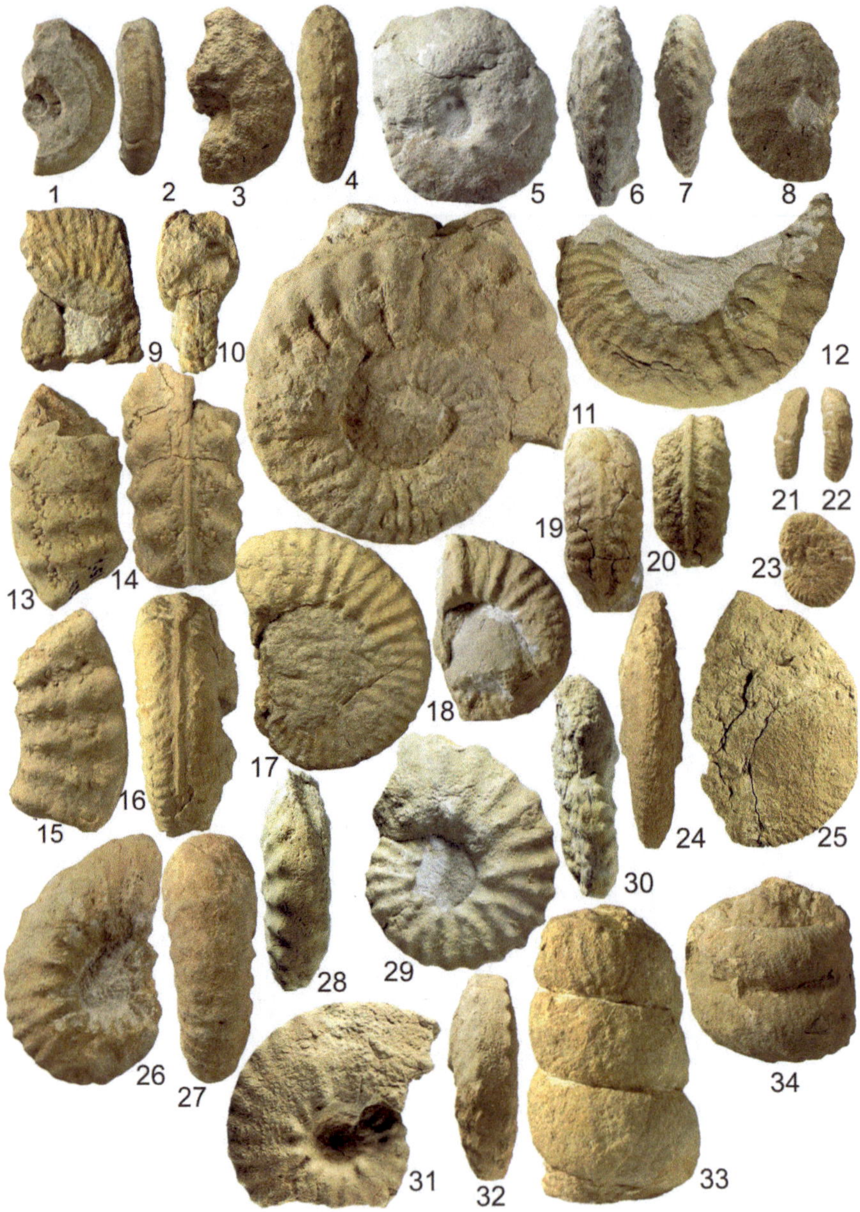

Plate 3.7 Selected ammonites of the Debarsou Formatin, Central Iran (Wilmsen et al. 2013b). 1, 2. *Puzosia (Puzosia) mayoriana*; 3, 4. *Schloenbachia* sp. cf. *S. varians*; 5–10. *Placenticeras grossouvrei*; 9, 10. *Mortoniceras* (*Mortoniceras*) sp. cf. *M. fissicostatum*; 11–12. *Mortoniceras* (*Mortoniceras*) sp.; 13–17. *Mortoniceras* (Subschloenbachia) rostratum; 18–19. *Mortoniceras* (*Subschloenbachia*) sp. cf. *M. perinflatum*; 20. *Mortoniceras* sp.; 21–23. *Mantelliceras saxbii*; 24–25. *Forbesiceras* sp.; 26–27. *Mantelliceras* sp. cf. M. *mantelli*; 28–32. *Acompsoceras renevieri*; 33. *Turrilites* sp. cf. *T. scheuchzerianus*; 34. *Pseudhelicoceras*? *robertianum*?

Abdolshahi, M. (2009). *Foraminiferal biostratigraphy of the Abderaz Formation in the Shourab Section* (Unpublished M.Sc. thesis). Faculty of Sciences, Ferdowsi Mashhad University.

Afshar-Harb, A. (1994). *The geology of Kopeh-Dagh, Geological book codification project*, Geological Survey of Iran, 274.

Afshin, S. (2007). *Microfacies, Geochemistry and depositional environment of the Sarvak Formation, Kuh-e Sefid Anticline, Kermanshah, and its correlation with Tang-e-Magin Borehole* (Unpublished M.Sc. thesis). Faculty of Earth Sciences, Shahid Beheshti University, Ilam Province.

Aghadadashi-Abhari, F. (2005) Biofaceis and lithofacies of tirgan formation, eastern Kopeh-Dagh. In *The 20th Symposium of Geology*, Geological Survey of Iran.

Aghanabati, A. (1975). *Etude geologique de la region de Kalmard These Doctorate thesis*, No AD 11/623 Gronoble, France.

Aghanabati, S. A. (1977). *Etude ge´ologique de la re´gion de Kalmard* (*W Tabas*). Stratigraphie et tectonique, Geological Survey of Iran, Report 35.

Aghanabati, A. (1998). Jurassic stratigraphy of Iran. *Geological Survey of Iran, 2*(65), 733.

Aghanabati, A. (2004). *Geology of Iran*. Geological Survey of Iran, Report (in Persian).

Ahmadi, W. (2011). Depositional environments of the Kangan Formation in the A and B boreholes of South Pars Field based on the lithofacies. In *The 1st Symposium of Geology*, Geological Survey of Iran.

Ahsani, A. (2009). *Biostratigraphy and depositional environment of Abderaz Formation of the Abderaz Section* (Unpublished M.Sc. thesis). Faculty of Earth Science, Shahid Beheshti University.

Aistov, L., Melnikov, B., Krivyakin, B., Morozov, L. (1984). *Geology of the Khur area (Central Iran), explanatory text of the Khur quadrangle Map 1:250,000.*

Akbar-Bass Kalateh, N. (2001). *Biostratigraphy and lithostratigraphy of the Sarvak and Ilam Formations of the Dezful Embayment, Kharak and Kuh-e-Mish sections* (Unpublished M.Sc. thesis). Faculty of Science, University of Tehran.

Akbari, T. (2006). *Lithostratigraphy and nannostratigraphy of Surgah Formation*, Ilam *area* (Unpublished M.Sc. thesis). Faculty of Earth Science, Shahid Beheshti University.

Akhtari, M. (2006). *Dinoflagellate biostratigraphy of the Garu formation in the Kabirkuh Section, type section* (Unpublished M.Sc. Thesis). Ilam: Faculty of Earth Science, Shahid Beheshti University.

Alavi, M. (2004). Regional stratigraphy of the Zagros fold-thrust belt of Iran and its proforeland evolution. *American Journal of Science, 304*(1), 1–20.

Alavi-Naeini, M. (1972). *Etude géologique de la région de Djam*, Geological Survey of Iran, Report 23, 293.

Alavi-Naeini, M. (2009). *An introduction on the stratigraphy of Iran*, Geological Survey of Iran, 507.

Alavi-Naini, M., Hajian, J., Amidi, A., & Bolurchi, H. (1982). Y Geology of takab Y Saein Qalel: Explanatory note of 1: 250,000 map of takab quadrangl, *Geol. Surv, Iran Rep.* (No. 50). Report.

Alidadi, N. (2006). *Biostratigraphy of the Darian Formation based on the calcareous algae in the northern Shiraz* (Unpublished M.Sc. thesis). Faculty of Sciences, Tabriz University.

Allenbach, P. (1966). *Geologie und Petrographie des Damavand und seiner Umgebung* (*Zentral-Elburz*), Iran: Doctoral dissertation.

Amini, A (2008) *Sedimentological and geological factors effective on the sediment load of Polroud River* (Unpublished M.Sc. thesis). Faculty of Sciences, University of Tehran, Guilan province.

Amiri-Bakhtiar, H. (1991). *Lithostratigraphy and biostratigraphy of tarbur formation, Fars area* (Unpublished M.Sc. thesis). Faculty of Earth Sciences, Shahid Beheshti University.

Amirzadeh, M. (2000). *Sedimentological and petrographical studies of the Sarvak Formation, Mokhtar Oil Field* (Unpublished M.Sc. thesis), Faculty of Earth Sciences, Shahid Beheshti University.

Asghari, M. (2005). *Geochemistry, depositional environment and diagenesis of Sarvak formation of the Tang-e-Rashid section, Izeh area, and its correlation with the Ahwaz Borehole* (M.Sc. thesis). Faculty of Earth Science, Shahid Beheshti University.

Ashouri, M. (2010). *Sequence stratigraphy and depositional environment of the Aitamir Formation in the southeastern Daregaz* (Unpublished M.Sc. thesis). Faculty of Sciences, Ferdowsi Mashhad University.

Assereto, R. (1966). *The Jurassic Shemshak Formation in central Elburz* (*Iran*).

Avarjani, Sh. (2005). *Sequence stratigraphy of Fahlian formation the type section and Dashtak Borehole No 1* (Unpublished M.Sc. thesis), Bu-Ali Sina Univsersity.

Bahrami-Samani, P. (1997). The *study of fish fossils of Pabdeh Formation, Babaheydar Section, Shahr-e-Kurd*, (Unpublished M.Sc. thesis). Faculty of Sciences, Ferdowsi Mashhad University.

Balazadeh, A. (1983). *Evaluation of the organic-rich beds of Kangan and Dalan formations, Salman Oil Field*, Unpublished M.Sc. thesis, Faculty of Science, University of Tehran.

Baud, A., Stampfli, G., & Steen, D. (1991). The Triassic Aghdarband Group: volcanism and geological evolution. *Abhandlungen der Geologischen Bundesanstalt Wien, 38*, 125–137.

Boluri-Moghadam, M. (1996). *An introduction on the biostratigraphy, microfaceis and palaeogegraphy condition of Abderzaz Formation* (Unpublished M.Sc. thesis). Faculty of Sciences, Ferdowsi Mashhad University.

Bozorgnia, F. (1964). *Microfacies and microorganisms of Paleozoic through Tertiary sediments of some parts of Iran*. Tehran, Iran: National Iranian Oil Company, 158.

Bozorgnia, F. (1973). *Paleozoic foraminiferal biostratigraphy of central and east Alborz Mountains*. Tehran, Iran: National Iranian Oil Company, Geological Laboratories Publication, 4, 185.

Bragin, Y., Jahanbakhsh, F, Golubev, S., Sadovnikov, G. (1976). *Stratigraphy of the Triassic–Jurassic coal-bearing deposits of Alborz* (Unpublished Report, National Iranian Steel Corporation v/o "Technoexport"). Tehran.

Brönnimann, P., Zaninetti, L., Bozorgnia, F., Dashti, G., & Moshtaghian, R. A. (1971). Lithostratigraphy and foraminifera of the upper triassic nayband formation. *Iran, Revue de Micropaléontologie, 14*(5), 7–16.

Cartier, E. G. (1971) *Die Geologie des unteren Chalus Tals Zentral-Alborz/Iran* (*No 164*), Geologisches Institut der Eidg, Technischen Hochschule und der Universität Zürich.

Dabagh, A. (2003). *The stratigraphy of Aptian-Albian sediments of the Darian Formation in the areas surrounding the Kazeroun Fault and southern Dezful Embayment* (Unpublished M.Sc. thesis). Faculty of Sciences, Kharazmi University.

Dalvand, F. (2006). *Petrography and depositional environment of the Sarvak Formation of the Tang-e-Bijar Gas Field, Ilam-Naftshahr* (Unpublished M.Sc. thesis). Faculty of Sciences, University of Tehran.

Davtalab-Zarghi, A. (2009). *Palynostratigraphy and palynofacies of the Sarcheshmeh Formation of the Anjirblagh Section* (Unpublished M.Sc. thesis). Faculty of Sciences, Ferdowsi Mashhad University.

Dehbozorgi, A. (2004). *Palynology of the Kashafroud Formation of the Senjadak Section, southeastern Mashhad* (Unpublished M.Sc. thesis). Faculty of Sciences, University of Tehran.

Dehghan, H. (2001). *Palynostratigraphy and palynofacies of the Upper Chamanbid Formation of the type section* (Unpublished M.Sc. thesis). Faculty of Sciences, University of Tehran.

Dehghan, F. (2002). *Biostratigraphy of Sarcheshmeh Formation based on the calcareous nannofossils of the type section* (Unpublished M.Sc. thesis). Faculty of Earth Sciences, Shahid Beheshti University.

Dellenbach, J. (1964). *Contribution et l'etude géologique de la region situee a l'est du Tehran* (*Iran*). Faculte Science Universite de Strasbourg (France), 117.

Eatemad Kermani, S. (2010). *Biostratigraphy and lithostratigraphy of Sarvak and Ilam Formation, southern limb of Kuh-e-Mish and Gachsaran Oil Field; implication on the Ilam-Sarvak boundary*.

Fahimi-Hanzaei, M. A. (2006). *Depositional environment and diagenesis of the Upper Sarvak Formation, Mishrif member, in the Reshadat and Siri-Esfand Fields, Persian Gulf* (Unpublished M.Sc. thesis). Faculty of Sciences, University of Tehran.

Fakhr-Rahimian, M. (2008). *Microstratigraphy of Darian Formation in the type section, northeastern Shiraz and Kuh-e-Aneh Anticline, Northeastern Gachsaran* (Unpublished M.Sc. thesis). Faculty of Earth Sciences, Shahid Beheshti University.

Fallahpour, K. (2002). *Depositional environment and diagenesis of the Dalichai Formation in the Firouzkuh and Soltanieh Sections*. (Unpublished M.Sc. thesis). Faculty of Sciences, University of Tehran.

Fanaei, Sh. (2010). *Calcareous nannofossils of the Gurpi-Pabdeh boundary in the Kangan area*. (Unpublished M.Sc. thesis). Paiam Nour university of Bojnourd.

Farhad, F. (2008). *Nannostratigraphy of the Surgah Formation in the Tang-e-Bijar Oil Field* (Unpublished M.Sc. thesis), Faculty of Earth Sciences, Shahid Beheshti University, Ilam province.

Farhoodi, Gh. (1975). *The study of Gurpi Formation*, Research Project, Faculty of Sciences, Shiraz University.

Fateh Bahari, L. (2010). The *study of depositional condition and post-depositional condition of the Abderaz Formation in the Hamam-Gahleh and Padeh sections* (Unpublished M.Sc. thesis). Faculty of Sciences, Ferdowsi Mashhad University.

Foroughi, F. (2004). *Microstratigraphy of Abderaz Formation in the eastern Kopeh-Dagh, north-eastern and eastern Mashhad* (Unpublished M.Sc. thesis), Faculty of Earth Sciences, Shahid Beheshti University.

Fürsich, F. T., Wilmsen, M., Seyed-Emami, K., & Majidifard, M. R. (2009). Lithostratigraphy of the Upper Triassic-Middle Jurassic Shemshak Group of Northern Iran. *Geological Society, London, Special Publications, 312*(1), 129–160.

Gaetani, M., Angiolini, L., Ueno, K., Nicora, A., Stephenson, M. H., Sciunnach, D., Rettori, R., Price, G. D., Sabouri, J. (2009). *Pennsylvanian–Early Triassic stratigraphy in the Alborz Mountains (Iran)*. Geological Society, London, Special Publications, *312*(1), 79–128.

Garouti, Z. (2005). *Lithostratigraphy and biostratigraphy of Gurpi Formation in the Mansouri Oil Field and its adjacent oil fields* (Unpublished M.Sc. thesis), Faculty of Sciences, University of Tabriz.

Ghalavand, H. (1996). *Lithostratigraphy and Biostratigraphy of the Darian and Kazhdomi in the southwestern Iran, Fars and Dezful Embayment* (Unpublished M.Sc. thesis). Faculty of Earth Sciences, Shahid Beheshti University.

Ghasemi-Nejad, A. et al. (2008). Paleodepth of the Gurpi Formation based on the benthic and planktic foraminifera, Scientific Journal of Tehran University, *34*(4).

Ghasemi-Nejad, A., Vaez-Javadi, F., Ghavidel-Syooki, M. (2001). Biostratigraphy of the Shemshak Formation in the Ozon Section, based on the dinoflagelates, Scientific Journal of University of Tehran.

Ghazanfari, P. (1991). *Microfacies, depositional environment and diagenesis of the carbonate-siliceous rocks if the Oxfordian-Kimmeridgian Mozduran Formation in the eastern Kopeh-Dagh basin* (Unpublished M.Sc. thesis). Faculty of Sciences, Kharazmi University.

Ghorbani, M. (2010). *Palobiogeography and compressional sequence stratigraphy of the Asmari and Qom Formations*.

Ghorbani, M. (2013). *The economic geology of Iran: Mineral deposits and natural resources*. Springer Science & Business Media.

Ghorichai, Sh. (2006). *Biostratigraphy of the Gurpi Formation in the Northern Kabirkuh, based on the foraminifers* (Unpublished M.Sc. thesis). Faculty of Sciences, University of Tehran.

Glaus, M. (1964). *Trias und Oberperm im zentralen Elburs (Persien). Geologisches Institut der Eidg*, Technischen Hochschule und der Universität Zürich.

Hadad-Tavakoli, A. (2005). *Palynostratigraphy of Kashafroud Formation of the Ghaleh-Sangi Section* (Unpublished M.Sc. thesis). Faculty of Sciences, University of Tehran.

Hadavi, F. (2005). Nannofossils of Abtalkh Formation in the Western Chehcheheh Village, Kalat-Mashhad Road, In *The 20th Symposium of Geology*, Geological Survey of Iran.

Hashemi-Hoseyni, P. (2006). *Biostratigraphy of the Fahlian Formation in the type section and borehole No 1 of Dara Oilfield* (Unpublished M.Sc. thesis), Faculty of Sciences, University of Tehran.

Hashemi-Yazdi, F. (2008). *Palynostratigraphy of Dalichi Formation in the Belou Section, Northern Semnan* (Unpublished M.Sc. thesis), Faculty of Sciences, University of Tehran.

Hekmatinia, S. (2004). *Nannostratigraphy of Garu Formation of the type section, Kabirkuh* (Unpublished M.Sc. thesis), Ilam: Faculty of Earth Sciences, Shahid Beheshti University.

Hemmat, M. (2009). *Geochemistry, diagenesis and depositional environment of the Ilam Formation in the Gurpi Mountain and its correlation with that of boreholes No 2, 5 of Kazeroun Oilfield* (Unpublished M.Sc. thesis). Faculty of Earth Sciences, Shahid Beheshti University.

Hoseini, A. (2001). *Plant fossils, macro and micro of Shemshak Formation of the Tazareh area, Shahroud* (Unpublished M.Sc. thesis). Faculty of Sciences, University of Tehran.

Hoseini-e Falehi, S. B. (2006). *Lithostratigraphy and nannostratigraphy of the Gurpi Formation of the Kuh-e-Mongasht and Kamestan Sections, Izeh area* (Unpublished M.Sc. thesis), Faculty of Earth Sciences, Shahid Beheshti University.

Hoseini-Jalilian, A. (1996). *Microfacies and depositional environments of Sarvak Formation of Khuzestan and Lorestan areas* (Unpublished M.Sc. thesis). Faculty of Sciences, Kharazmi University.

Houshiar, M. (2008). *Sedimentary rocks and depositional environments of the Gurpi Formation in the type section, Lali Oilfield, 140 km northeastern Ahwaz* (Unpublished M.Sc. thesis), Faculty of Earth Sciences, Shahid Beheshti University.

Hovaeizavi, A. (1998). *Depositional environment of the Gurpi and Pabdeh Formations, based on the geophysical analysis of the surface and undersurface samples, and determination of their boundary, Dezful Embayment* (Unpublished M.Sc. thesis). Faculty of Earth Sciences, Shahid Beheshti University.

Huckriede, R., Kürsten, M., Venzlaff, H. (1962) *Geotektonische Kartenskizze des Gebietes zwischen Kerman und Sagand*, (*Iran*), Bundesanstalt für Bodenforschung.

ICS (2015). http://www.stratigraphy.org/index.php/ics-chart-timescale.

Jahani, D. (2000). Microfacies, depositional environment and sequence stratigraphy of Elika Formation, In *The 4th Symposium of Geology*, Geological Society of Iran.

Jamshidi, Kh. (2010). *Magmatic studies of mafic rocks of the basal Shemshak Formation, Eastern Alborz Zone* (Unpublished M.Sc. thesis), Faculty of Sciences, Industrial University of Shahroud.

Kalantari, M. (2006). *Microbiostratigraphy, microfacies and depositional environment of the Carboniferous deposits, Mobarak Formation, in the northern and northwestern Damghan* (Unpublished M.Sc. thesis). Faculty of Sciences, Ferdowsi Mashhad University.

Kamiabi-Shadan, H. (2005). *Microstratigraphy of the Ilam and Sarvak Formations with special implication on their contact, Bangestan Mountain, Mansouri Oil Field* (Unpublished M.Sc. thesis), Faculty of Earth Sciences, Shahid Beheshti University.

Karimian, A. (2009). *Sequence stratigraphy and diagenesis history of Kalat and Pastaligh formations in the Shams Anticline, Daregaz* (Unpublished M.Sc. thesis). Faculty of Sciences, Ferdowsi Mashhad University.

Kazemi, A. (2010). *Lithostratigraphy and biostratigraphy of the Sarvak and Surgah Formations in the Siah-Kuh Anticline, southwestern Ilam, with special implication on the Subherceynian Orogenic phase* (Unpublished M.Sc. thesis), Faculty of Earth Sciences, Shahid Beheshti University.

Khasi, A. M. (2009). *Depositional environment, diagenesis of the Abhaji Formation deposits of the Sartakht and Cheshmeh-e-Bakhsi Section, Tabas area* (Unpublished M.Sc. thesis). Faculty of Earth Sciences, Shahid Beheshti University.

Khazaei, M. (1999). *Microfacies and depositional environment of the Tirgan Formation, Southwestern Shirvan-Southwestern Bojnourd, Western Kopeh-Dagh* (Unpublished M.Sc. thesis). Faculty of Sciences, Ferdowsi Mashhad University.

Khodaei, N. (1991). *Sedimentary facies, environment and depositional model of the Tirgan Formation, northeastern Kopeh-Dagh Basin, Iran*.

Khodaei, M. (2007). *Geochemistry and diagenesis of Kangan Formation of South Pars Field* (Unpublished M.Sc. thesis), Faculty of Earth Sciences, Shahid Beheshti University.

Khosravi Alghar, R. (2007). *Depositional environment of the Sanganeh Formation of the Ghalehno Section, Eastern Kopeh-Dagh, based on the organic matter* (Unpublished M.Sc. thesis). Faculty of Sciences, University of Tehran.

Krystyn, L., & Tatzreiter, F. (1991). Middle Triassic ammonoids from Aghdarband (NE-Iran) and their paleobiogeographical significance. *Abhandlungen der Geologischen Bundesanstalt, 38,* 139–165.

Kushesh, N. (2008). *Petrography, depositional environment and diagenesis of the Boeib, Zibar and Shoaiba Formations (equivalents of Gadvan and Darian Formations) in the Salman Oil Field* (Unpublished M.Sc. thesis). Faculty of Sciences, Ferdowsi Mashhad University.

Lasemi, Y. (1994). Petrographical analysis of the platform carbonate of Mozduran Formation, In *The 2nd Symposium of Geology of Eastern Iran.*

Lasemi, Y. (2000). *depositional environment and sequence stratigraphy of the Precambrian and Paleozoic rocks of Iran*, Geological Survey of Iran, 1–14.

Lotfi, S. (2006). *Depositional environment and reservoir quality of the Upper Surmeh Formation, equivalent of Arab Formation, in the Balal Oil Field* (Unpublished M.Sc. thesis). Faculty of Sciences, University of Tehran.

Mahanipour, A. (2010). *Integrated biostratigraphy of Sarcheshmeh and Sanganeh Formation, implications on the calcareous nannofossils* (Unpublished Ph.D. thesis). Faculty of Earth Sciences, Shahid Beheshti University.

Mahdavian-Rad, B. A. (2009). *Lithostratigraphy and biostratigraphy of Gurpi Formation in the Kamestan Anticline, northwestern Izeh, and its correlation with the type section located in the Northern Lali* (Unpublished M.Sc. thesis). Faculty of Earth Sciences, Shahid Beheshti University.

Mahmoudi, L. (2006). *Diagenesis and depostional environments of the Upper Surmeh Formation, equivalent of the Arab Formation, in the Salman Oil Field, Central Persian Gulf* (Unpublished M.Sc. thesis). Faculty of Earth Sciences, Shahid Beheshti University.

Mandanizadeh, A. (2005). *Biostratigraphy of Gurpi Formation in the Lar area, Kuh-e-Kurde section* (Unpublished M.Sc. thesis), Faculty of Sciences, University of Tabriz.

Minaei, M. (2002). *Geological reservoir characterizations of the Mirsharif and Ilam Formations in the Siri Oil Field, Persian Gulf* (Unpublished M.Sc. thesis), Faculty of Sciences, University of Tehran.

Mirshahani, M. (2011). *Depositional environment, diagenesis and geochemistry of the Surmeh Formation in the High Zagros and its correlation with those of subsurface sections in the Fars area* (Unpublished Ph.D. thesis), Faculty of Earth Sciences, Shahid Beheshti University.

Mirzaei-Ataabadi, M. (2003). An insight on the correlation of the plant and plant-feeders in the Hijadk deposits, northern Kerman. In *The 22nd Symposium of Geology*, Geological Survey of Iran.

Mohamadi Ostad-Kalateh, R. (2003). *Sequence stratigraphy of Darian Formation in the Aghajari borehole no 140 and Tango borehole no 1, southern Dezful Embayment* (Unpublished M.Sc. thesis). Faculty of Sciences, Bu-Ali Sina University.

Mohammad-Khani, H. (2003). *Depositional environment and sequence stratigraphy of Fahlian Formation in the Ragsefid and Khoiz oil Fields, Southern Dezful Embayment* (Unpublished M.Sc. thesis). Faculty of Sciences, Kharazmi University.

Moradi, M. (2002). *Reservoir characterization and depositional environment of the Darian Formation in the Persian Gulf, Qatar Tip through the northwestern Persian Gulf* (Unpublished M.Sc. thesis). Faculty of Sciences, University of Tehran.

Motiei, H. (1993). *Stratigraphy of Zagros*. Geological Book Codification Project, Geological Survey of Iran.

Mousavian, M. (1997). *Microbiostratigraphy of the Eocene rocks in the Neyzar area, southwestern Qom* (Unpublished M.Sc. thesis). Faculty of Earth Sciences, Shahid Beheshti University.

Mousavinia, A. (2008). *Stratigraphy and paleontology of the middle cretaceous strata in the Eastern Kopeh-Dagh, based on the ammonite fauna* (Unpublished Ph.D. thesis). Faculty of Earth Sciences, Shahid Beheshti University.

Movahed, Z. (2001). Introduction of reference section of Calcareous Khalik Member of Gadvan Formation in the Northern Gachsaran Area, In *The 20th Symposium of Geology*, Geological Survey of Iran.

Nabavi, M. (1972). Lower Cretaceous deposits in the Taft-Yazd and Khur area. *Geological Survey of Iran, Report, 106,* 1–127.
Nabavi, M. H. (1976). *An introduction to geology of Iran.* Tehran: Geological Survey of Iran.
Nabavi, M. H., & Seyed-Emami, K. (1977). Sinemurian ammonites from the Shemshak Formation of North Iran (Semnan area, Alborz). *Neues Jahrbuch für Geologie und Palaontologie Abhandlungen, 153*(1), 125–136.
Naddaf Kohneh-Quchan, R. (1998). *Lithostratigraphy and petrology of Shurijeh Formation in the Radkan-Bavar Section located in the southeastern Quchan and its correlation with the Cheshmeh-Gilas, Darkhor and Gardaneh Mozdouran sections* (Unpublished M.Sc. thesis). Faculty of Sciences, Ferdowsi Mashhad University.
Naderi-Roshanavandi, A. (2002). *Microfaceis, depositional environment and sequence stratigraphy of the Dalichai and Lar Formations in the Polour area* (Unpublished M.Sc. thesis). Faculty of Sciences, Kharazmi University.
Najafi, M., et al. (2005). New lithostratigraphic findings of the Fahlian Formation in the Dezful embayment, southwestern Iran, based on the borehole logs. In *The 20th Symposium of Geology*, Geological Survey of Iran.
Naseri, Z. (2002). *Reservoir characterization and depositional environment of the Darian Formation in the Hormuz neck to the Qatar Mountain, Persian Gulf* (Unpublished M.Sc. thesis), Faculty of Sciences, University of Tehran.
Naseri, N. (2005). *Geochemistry, depositional environment and diagenesis of the Saravak Formation in the type section located in the Bangestan Mountain, and its correlation with Parsi Borehole* (Unpublished M.Sc. thesis). Faculty of Earth Sciences, Shahid Beheshti University.
Nawrouzi-Chegini, T. (2005). *Biostratigraphy of Kazhdomi Formation, based on the calcareous nannofossils* (Unpublished M.Sc. thesis). Faculty of Earth Sciences, Shahid Beheshti University.
Negarestani, A. (1998). *Sedimentology and depositional environment of the equivalent unit of the Shemshak Formation in the northwestern Kerman* (Unpublished M.Sc. thesis). Faculty of Earth Sciences, Shahid Beheshti University.
Nekoeian, A. (2002). *Petrophyscial and petrographical studies of the Sarvak Formation in the Northeastern Persian Gulf; implication on the hydrocarbonic accumulation* (Unpublished M.Sc. thesis). Faculty of Sciences, University of Tehran.
NIOC. (2016). Paleontology and Biozonation of Paleozoic Sediments of Central Iran and Zagros Basins. National Iranian Oil Company, Tehran (no 89230).
NICS. (1962). Mesozoic rocks of Iran, National Iranian Committee on Stratigraphy. *Geol. Surv. Iran Rep.* (unpublished data).
Pashaei, H. (2009). *Biostratigraphy of the Permo-Triassic boundary of the Ruteh Section, Northern Tehran.* (Unpublished M.Sc. thesis). Geol. Surv. Iran Rep.
Pourbagheri, M. (2007). *Geochemistry, diagenesis and depositional environments of the Darian Formation in the Aneh Anticline and its correlation with the Chilinger Borehole 3* (Unpublished M.Sc. thesis). Faculty of Earth Sciences, Shahid Beheshti University.
Pourmoatamed, F., Amiribakhtiar, H., Akbarbas-Kalaleh, N. (2002). *Biostratigraphy of Sarvak Formation of the Khark and Kuh-e-Mish (southern part of Dezful Embayment and its surroundings).*
Raeesossadat, S N (1991) *Stratigraphy and depositional environment of the Sarcheshmeh and Sanganeh Formations of the Eastern Kopeh-Dagh Basin, Northeastern Iran* (Unpublished M.Sc. thesis), Faculty of Sciences, Kharazmi University.
Raeesossadat, S. N. (1995). Stratigraphy and depositional environment of the Sarcheshmeh Sanganeh Formation of the Eastern Kopeh-Dagh Basin, Northeastern Iran. In *The 2nd symposium of Northern Iran Geology*, Bojnourd University.
Rahaghi, A. (1976). *Contribution a l'etude de quelques grands foraminiferes de 1' Iran.* NIOC Publication No 6.
Rahimi, S. (2010). *Lithostratigraphy and biostratigraphy of the Gurpi Formation, Kuh-e-Sefid Section, Eastern Ramhormoz* (Unpublished M.Sc. thesis). Faculty of Earth Sciences, Shahid Beheshti University.

Raziani, M. (2010). *Lithostratigraphy and biostratigraphy of Surgah and Ilam Formation of the Kuh-e-Shahnakhjir Section and its correlation with the type section* (Unpublished M.Sc. thesis) Faculty of Earth Sciences, Shahid Beheshti University.

Rezaei, Z. (2009). *Palynology of the Abtalkh Formation of the type section* (Unpublished M.Sc. thesis). Faculty of Sciences, University of Tehran.

Rostami, A. (2002). *Reservoir quantity and depositional environments of the Ilam and Sarvak Formation of Persian Gulf* (Unpublished M.Sc. thesis). Faculty of Sciences, University of Tehran.

Ruttner, A. (1991). Geology of the Aghdarband Area (Kopeh-Dagh). *NE-Iran, Abhandlungen der Geologischen Bundesanstalt, 38,* 44–47.

Ruttner, A., Nabavi, M. H., Hajian, J. (1968) *Geology of the Shirgesht Area (Tabas area, East Iran).* pp 5–133, Geological Survey of Iran. (Report 4).

Sadeghi, R. (2009). *Biostratigraphy, Microfacies and sequence stratigraphy of Asmari Formation in the Fars area* (Unpublished Ph.D. thesis), Faculty of Sciences, Esfahan University.

Sadeghi-Saadatlou, Gh. (2003). *Petrography and depositional environment of the Asmari Formation in the southern flank of the Bangestan Anticline and its correlation with those of adjacent fields located in the Dezful Embayment* (Unpublished M.Sc. thesis). Faculty of Sciences, Bu-Ali Sina University.

Safarikhanlou, Z. (2000). *Diagenesis study of sandstone beds of the Shemshak Formation in the Eastern Alborz* (Unpublished M.Sc. thesis). Faculty of Sciences, University of Tehran.

Salahshour, Kh. (2004). *Palynostratigraphy, palynofacies and paleoenvironment of the Chamanbid Formation, Ghorban Section, Northeastern Mashhad* (Unpublished M.Sc. thesis), Faculty of Sciences, University of Tehran.

Salehi, F. (2001) *Nannofossils of type section of Gurpi Formation* (Unpublished M.Sc. thesis). Faculty of Earth Sciences, Shahid Beheshti University.

Schlagintweit, F., & Wilmsen, M. (2014). Orbitolinid biostratigraphy of the top taft formation (lower cretaceous of the yazd block, Central Iran). *Cretaceous Research, 49,* 125–133.

Seyed-Emami, K. (2003). Triassic in Iran. *Facies, 48*(1), 91.

Seyed-Emami, K., Fuersich, F. T., Wilmsen, M., Majidifard, M. R., Shekarifard, A. (2009). Upper Triassic (Norian) cephalopods from the Ekrasar Formation (Shemshak Group) of Northern Alborz, Iran. Rivista Italiana di Paleontologia e Stratigrafia (Research in Paleontology and Stratigraphy), *115*(2).

Seyed-Emami, K., Fürsich, F. T., Wilmsen, M. (2004). Documentation and significance of tectonic events in the northern Tabas Block (east-central Iran) during the Middle and Late Jurassic, Rivista Italiana di Paleontologia e Stratigrafia (Research In Paleontology and Stratigraphy), *110*(1).

Seyed-Emami, K., & Immel, H. (1995). Ammoniten aus dem Alb (Kreide) von Shir-Kuh (N Yazd, Zentraliran). *Paläontologische Zeitschrift, 69,* 377–399.

Seyed-Emami, K., Schairer, G., Aghanabati, S. A., & Fazl, M. (1991). Ammoniten aus dem Bathon der Gegend von Tabas-Nayband (Zentraliran). *Müncher Geowissenschaftliche Abhandlungen A, 19,* 65–100.

Shahsavani, D. (2003) *Palynostratigraphy, palynofacies and paleoenvironment of the Chamanbid Formation of the Jajarm section* (Unpublished M.Sc. thesis). Faculty of Sciences, University of Tehran.

Sharafi, M. (2009). Sequence stratigraphy and depositional history interpretation in the Sheikh Syncline and Bibahreh sections, Bojnourd area (Unpublished M.Sc. thesis). Faculty of Sciences, Ferdowsi Mashhad University.

Sheikhpour, H. (2005). *Diagenesis of the Upper Dalan and Lower Kangan Formations of the South Pars Gas Field* (Unpublished M.Sc. thesis). Faculty of Sciences, Tehran University.

Shemirani et al. (2001). The introduction of reference section of Khalij Limestone of the Gadvan Formation in the Gachsaran Area, In *The 20th Symposium of Geology*, Geological Survey of Iran.

Shemirani, A. (2002b). *The Handbook of Mesozoic of Iran.* Faculty of Earth Sciences, Shahid Beheshti University.

Shirvani, H. (2006). *Petrography and depositional environment of the Sarvak Formation in the Kabirkuh Section* (Unpublished Ph.D. thesis). Faculty of Earth Sciences, Shahid Beheshti University, Ilam.

Shoushtarian, F (2011) *Geochemistry, depositional environment, digenesis and sequence stratigraphy of the Ilam Formation in the type section and Kuh-e-Pion Section and its correlation with the subsurface sections of Danan, Ghaleh Nar and Lali, Northern Khuzestan* (Unpublished Ph.D. thesis), Faculty of Earth Sciences, Shahid Beheshti University.

Soheyli, S (2002) *Paleoecology of the Hojedk Formation of the Pabdana area, based on the palynological studies* (Unpublished M.Sc. thesis), Faculty of Sciences, University of Tehran.

Steiger, R. (1966) Die Geologie der West-Firuzkuh-Area (Zentralelburz/Iran), Mitteilungen aus dem Geologischen Institut der Eidgen6ssischen Technischen Hochschule und der Universitiit Zurich, *68*, 145 (in German).

Stöcklin, J., Eftekharnezhad, J., Hushmandzadeh, A. (1965). *Geology of the Shotori Range (Tabas area, East Iran)*, Geological Survey of Iran, Report 3, 1–69.

Stöcklin, J., & Setudehnia, A. (1991). *Stratigraphic Lexicon of Iran. Tehran, Iran*: Geological Survey of Iran, Report no. 18.

Szabo, F., & Kheradpir, A. (1978). Permian and Triassic stratigraphy, Zagros Basin, South-West Iran. *Journal of Petroleum Geology, 1*(2), 57–82.

Tahmasbi, A. (1997). *Microfacies, depositional environemet and sequence stratigraphy of the lower and middle Elika Formation in the eastern Central Alborz* (Unpublished M.Sc. thesis). Faculty of Sciences, Kharazmi University.

Taraz, H. (1972). *Geology de la Surmaq- Deh bid*. Central Iran Geological Survey of Iran report 37.

Taraz, H., Golshani, F., Nakazawa, K., Shimizu, D., Bando, Y., Ishii, KI., Murata, M., Okimura, Y., Sakagami, S., Nakamura, K., Tokuoka, T. (1981). *The Permian and the Lower Triassic systems in Abadeh region, central Iran.*

Tavakolinezhad, A. (2007). *Microstratigraphy and depositional environments of the Abderaz Formation of the Ghalehzou (Northern Shirvan) and Zanglanlou (southern Daregaz) Sections* (Unpublished M.Sc. thesis). Faculty of Sciences, Ferdowsi Mashhad University.

Teichert, C., Kummel, B., Sweet, W. C. (1973). *Permian-Triassic strata, Kuh-e Ali Bashi, northwestern Iran*, Harvard University.

Vakli-Baghmisheh, F. (1997). *Stratigraphy and paleontology of Upper Cretaceous-Cenozoic units of Eastern Kopeh-Dagh, and revision on the Abtalkh, Neyzar, Kalat, Pesteligh, Chehelkaman and Khangiran Formations* (Unpublished M.Sc. thesis). Faculty of Sciences, University of Tehran.

Valipour, Kh. (2010). *Microbiostratigraphy and facies analysis of the Fahliyan Formation in the borehole No 18 of Khark Oilfield* (Unpublished M.Sc. thesis), Faculty of Sciences, University of Damghan.

Vaziri-Moghadam, H., Safari, A., Shahriari, S., Taheri, A., Khazaei, A. (2010). The introduaction of red bedded, clastic-calcareous sediments of the High Zagros, Samirom-Ardel area, Scientific Journal of Tehran University, *36*(1).

Wilmsen, M., Fürsich, F. T., & Majidifard, M. R. (2015). An overview of the cretaceous stratigraphy and facies development of the yazd block, western Central Iran. *Journal of Asian Earth Sciences, 102*, 73–91.

Wilmsen, M., Fürsich, F. T., & Seyed-Emami, K. (2003). Revised lithostratigraphy of the middle and upper jurassic magu group of the northern tabas block, east central Iran. *Newsletters on Stratigraphy, 39*, 143–156.

Wilmsen, M., Fürsich, F. T., & Taheri, J. (2009). The Shemshak Group (Lower–Middle Jurassic) of the Binalud Mountains, NE Iran: stratigraphy, depositional environments and geodynamic implications. *Geological Society, London, 312*(1), 175–188. Special Publications.

Wilmsen, M., Storm, M., Fürsich, F. T., & Majidifard, M. R. (2013a). Upper Albian and Cenomanian (Cretaceous) ammonites from the Debarsu Formation (Yazd Block, Central Iran). *Acta Geologica Polonica, 63*(4), 489–513.

Wilmsen, M., Storm, M., Fürsich, F. T., & Majidifard, M. R. (2013b). Upper Albian and Cenomanian (Cretaceous) ammonites from the Debarsu Formation (Yazd Block, Central Iran). *Acta Geologica Polonica, 63*(4), 489–513.

Yahiaei, A. (2003). *Palynostratigraphy, palynofacies and depositional environment of the Chaman-bid Formation in the Navia Section* (Unpublished M.Sc. thesis). Faculty of Earth Sciences, Shahid Beheshti University.

Yazdani, M. (2006). *Microfacies and diagenesis of the Upper Sarvak Formation; implication on the reservoir quality, southwestern Iran* (Unpublished M.Sc. thesis). Faculty of Sciences, University of Tehran.

Yousef-Pour, M. (2004). *Depositional environment and reservoir characters of the Upper Surmeh Formation, equivalent of Arab Formation* (Unpublished M.Sc. thesis). Faculty of Sciences, University of Tehran.

Zabihi-ZaVarm, F. (2009). *Biostratigraphy and depositional environments of the Abderaz Formation of the Chehcheheh Section, based on the foraminifera* (Unpublished M.Sc. thesis). Faculty of Sciences, Ferdowsi Mashhad University.

Zandkarimi, K., Najafian, B., Bahrammanesh, M., & Vachard, D. (2014). Permian foraminiferal biozonation in the Alborz Mountains at Valiabad section (Iran). *Permophiles, 60*, 10–16.

Zandkarimi, K., Vachard, D., Najafian, B., Mosaddegh, H., & Ehteshami-Moinabadi, M. (2018). Mississippian lithofacies and foraminiferal biozonation of the Alborz Mountains, Iran: Implications for regional geology. *Geological Journal*. Article first published online. doi: https://doi.org/10.1002/gj.3242.

Zarea, M. (2003). *Depositional environment and diagenesis of the Fahlian Formation in Agha-jari borehole 140 and Mansouri Borehole 6* (Unpublished M.Sc. thesis). Faculty of Sciences, University of Tehran.

Zarea, A. (2005). *Palynology, biostratigraphy and paleoecology of the Gurpi Formation of the type section* (Unpublished M.Sc. thesis). Faculty of Earth, University of Tehran.

Chapter 4
Cenozoic

The Cenozoic Era begins at 65 Ma and continues to the present day (Fig. 4.1). At the beginning of this era, the earth was experiencing some tectonically-important evolutionary events that affected the geological and stratigraphical history of Iran. These results will be discussed below.

A stratigraphical gap and angular unconformity separate Cenozoic rocks from the underlying Cretaceous strata. This relationship is the consequence of the marine regression and shrinking of the Tethyan Ocean in Late Cretaceous time and also the Laramide structural event and its associated folding, the effects of which are seen from Europe to southern India. Because of its paleogeographical position in the middle of this fold belt, the effects of this event were especially severe in Iran, and also resulted in major facies changes at the K-P boundary.

After the short pause in sedimentation at the Cretaceous-P boundary, the margins of the Tethyan Ocean were once again submerged, leading to renewed marine deposition, with widely distributed Lower Paleogene sediments preserved in parts of Iran, Iraq, Pakistan and Afghanistan (Metallogenic map of the Middle East, 2011).

Paleocene-Eocene successions are either continuously or discontinuously exposed in the Africa, Europe and Middle East. The general pattern of sedimentation is not continuous throughout. As a consequence, it is accompanied by stratigraphical gap but in the Middle East by strata showing high facies change (this facies change is associated by increasing depth). In the Eocene, the facies are locally characterized by interfingering contact (e.g. Motiei 2003).

As is the case with the Paleocene-Eocene boundary, successions spanning the Eocene-Oligocene boundary represent either continuous or discontinuous deposition. The Eocene sediments are of more facies diversity than those of Oligocene. The Eocene-Oligocene boundary is associated with gaps and an angular unconformity. In Europe, a gradual transition across this boundary has been reported, while in the Middle East, major facies changes, stratigraphical gaps and angular unconformities are observed (e.g. Motiei 1993; Aghanabati 2004).

The Oligocene-Miocene interval is regarded as the time of "major marine and continental facies change". Sediments representing this chronostratigraphic interval are shallow water deposits changing to deeper marine facies in the latest Oligocene.

M. Ghorbani, *Lithostratigraphy of Iran*, Springer Geology,
https://doi.org/10.1007/978-3-030-04963-8_4

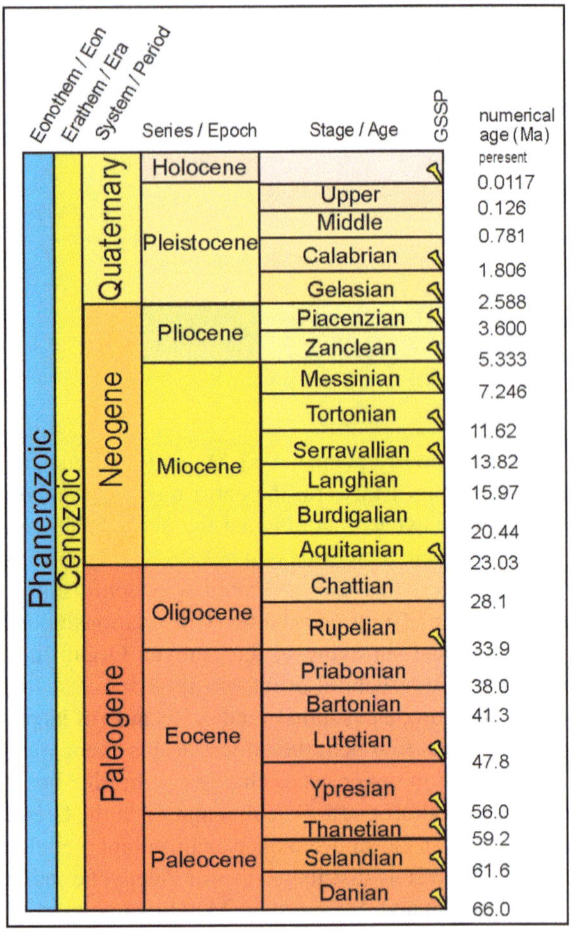

Fig. 4.1 International Chronostratigraphical Chart of the Cenozoic (ICS 2015)

The Miocene sediments are mainly composed of marine facies that are replaced by continental facies in the late middle Miocene and Pliocene (Aghanabati 2004). Alpine orogenic phase activities mainly started in the Eocene and continued until the latest Quaternary. This phase has created the extensive Alpine-Himalayan belt, and has been accompanied by volcanic activity (for more discussion, readers are referred to Ghorbani 2018). Most of the rock outcrops of the Alpian-Himalayan belt as well as Eocene facies diversity are directly related to Alpine-Himalayan orogenesis.

The late Eocene structural movements resulted in the angular unconformity at the base of the Oligocene and are also believed to have triggered volcano-magmatic activities in the Iranian Volcanic belt. This Eocene tectono-magmatic event, formed the recent basement of Iran and resulted in the uplift and exposure of most of Iran except for Zagros, Kopeh-Dagh and Makran, and the Eocene basin remnants, represented by shallow continental lakes (Aghanabati 2004).

The facies changed to a continental one, especially molas sedimentation, resulting in deposition of the Lower Red Formation in Central Iran, Azarbaijan and some pasts of northeastern Iran, including the Sabzevar Desert.

The Oligocene rocks of western Central Iran are red continental strata of the Lower Red Formation that are widely distributed in the Qom, Tafresh, southeastern Tehran, northern Garmsar and northern Semnan. The Lower Red Formation is referred to a succession of sedimentary rocks stratigraphically characterized by its position between the Eocene volcano-clastic rocks and Oligo-Miocene marine beds, i.e. Qom Formation (NIOC 1959). Its type section has not yet defined, but, in most areas, it is lithologically composed of conglomerate, sandstone, gypsum, salt rock, and locally of lava, siltstone and claystone. The ratio of mentioned rock types can be changed from place to place.

The remnant basins of the Eocene sea became continental in the Oligocene with deposition of evaporite in the Qom to Semnan areas. In southern Azarbaijan, Takab and Zanjan, there is absence of gypsum and salt in the Lower Red Formation in these areas (NIOC 2011).

The lithological and sedimentological evidence show that during the late Eocene (Chattian) and early Oligocene (Aquitanian), remnant basins of the Eocene sea were interconnected, a trend that gradually extended through northeastern and northwestern areas, until the marine basin with continental basement referred to as the Qom Basin evolved. This basin existed during Chattian through to latest Burdigalian time, though its evolution resulted in alternating regressions and transgressions in some areas. During Aquitanian time, it extended from northwestern Iran through Turkey and Armenia.

The recent Qom and Sahand Lakes are remnants of the former "Qom Basin". In the eastern part of Iran, the Qom Formation is found from Semnan to Torbat-e Heydarieh and in southeastern Iran to Makran. Due to the less extensive distribution of this basin in some areas during Oligo-Miocene time, the high mountain peaks were not covered by sediments. Consequently, the Upper Red Formation immediately overlies the Lower Red Formation and the Qom Formation is mostly represented by tongue-shaped sedimentary bodies. This has resulted in development of the shallow water, coral-bearing bioclastic limestones, while, carbonates and marls were deposited in deeper parts. Sediment slides and movements as well as basin floor movements have produced alternations of marls and limestones in the Aquitanian (NIOC 2011).

During the same interval, the distribution of the Qom Formation from Central Iran to northwestern Iran (in the direction of Qom, Tafresh, Saveh, Takab and Urmia) was accompanied by major volcanic activity, with the ratio of volcanoclastic to sedimentary rocks increasing and consequently forming the tuffaceous marls of this formation (Ghorbani 1999). The geographical extent of the Qom basin decreased into the Burdigalian, resulting in its exposure of some areas, a feature that is more obvious in central and eastern Iran. In northwestern Iran, no tangible changes have been recorded, and the same Aquitanian facies persisted.

The extensive volcanic activity in the areas of Qom, Tafresh, Saveh and Urmia and throughout Azarbaijan provided sufficient materials for the formation of evaporatic rocks. Due to the confining and closing of these basins after the Burdigalian, a favorable environment for deposition of evaporates was created that consequently formed the Salt Diapirs of Azarbaijan. It is worth mentioning that after closure of these basins, some 'ponds' covering several square kilometers still remained till the Pliocene, the time of deposition of the Upper Red Formation. The Mahneshan area contains some of these local ponds in which are now the Boron mines were formed (Ghorbani 1999).

4.1 Alborz

The Cenozoic lithostratigraphic units of northern Alborz completely from those of southern Alborz. Lithologically, the southern Alborz closely resembles to Central Iran and does not share any affinities to northern Alborz. Accordingly, none of the Cenozoic rock units, exposed in the southern Alborz, are observed in the northern Alborz. Indeed, the northern Alborz generally exposed in the Cenozoic and the Cenozoic marine rocks were not formed.

Also, the northern Alborz was generally exposed in the Cenozoic and, therefore, Cenozoic marine rocks were not deposited.

The Cenozoic succession of southern Alborz consists of the following units:

The Paleocene Fajan Formation; the Paleocene-Lower Eocene Ziarat Limestone; the Eocene volcanoclastic Karaj Formation and the Miocene Kond Formation.

Interestingly, equivalents of these rock units are exposed in Central Iran but with some differences. These include the Kerman Conglomerate (equivalent to the Fajan Conglomerate); the shaly-calcareous formation of the Kerman area (equivalent to the Ziarat Formation); and the volcanic and volcanoclastic Razak and Hazar Formations (equivalent to the Karaj Formation). Nevertheless, the rock units of these two areas show many facies and lithological difference.

4.2 Central Iran

The Paleogene rocks of Central Iran can be lithologically correlated with those of Alborz but the Neogene exhibits many differences. In general terms, the lithostratigraphy of Central Iran is as follows:

The shaly-calcareous Delfard Formation and its equivalents; the volcanoclastic Razak and Hazar formations of the Kerman area; the Lower Red Formation; the Qom Formation; and the Upper Red Formation. Of these, the Qom and the Upper Red Formation are the most widely distributed in Azarbaijan and Central Iran.

The Lower Red Formation which is attributed to the late Eocene-Early Oligocene, shows less facies and lithological variation than the Qom and Upper Red formations. Due to the economic significance of the Qom Formation, a more detailed description is provided in this chapter.

Lower Red Formation

The Lower Red Formation is composed of clastic rocks including conglomerate, microconglomerate and dark red sandstone, all of which are barren in terms of fossils and whose supposed age is based solely their stratigraphic context (NIOC 1959). Rahimzadeh (1994) attributed the same formation to the early Oligocene or even Late Eocene in some basins. However, the age of the overlying beds, the Qom Formation, depending on the sea level rise and the direction of transgression, can vary even to Late Oligocene in some places.

The lithostratigraphic column for the Lower Red Formation including some sections located in Central Iran can is shown in the Fig. 4.2.

Based on field observations of the Lower Red Formation and the overlying Qom Formation in Azarbaijan and areas of Central Iran such as the Qom area, Saveh, Tafresh and Kabudarahang by the author, deposition of this formation started in the Early Oligocene and ended in the Early Chattian. In Azarbaijan this was accompanied by the onset of volcanism and the formation is mainly of Rupelian age.

Qom Formation

In the western parts of Central Iran, an Oligocene-Lower Miocene, mainly-calcareous lithostratigraphic unit is extensively exposed in the Maku, Khoy, Takab, Qom, Kashan, Rafsanjan areas and on the margin of Jazmourian Embayment (Furrer and Soder 1955; Stöcklin and Setudehnia 1991; NIOC 2011).

Based on regional information, it is believed that, after the Pyrenean orogenic phase and the following Late Eocene continental deposition, these areas were progressivey inundated by the sea (Reuter et al. 2009). Apparently, the related regression and transgression did not occur simultaneously. So, the transgression begins in the Early Oligocene (Rupelian) of the areas including Sabzevaran and Kashan, and in the Aquitanian- Burdigalian of the some areas such as Azarbaijan (NIOC 2011) and Deh Namak (Daneshian and Dana 2007).

The strata of the Qom Formation are mainly composed of calcareous sediments that were mostly deposited in inner ramp to mid-ramp settings and to some extent on the outer ramp (Mohammadi et al. 2011). However, the presence of marl and evaporitic rocks show that the sea floor had not a uniform topography and depositional environments ranged from terrestrial to open marine including mangrove, restricted inner shelf lagoon, seagrass meadow, reefal, and deeper offshore environments (Reuter et al. 2009).

Due to the high temporal and spatial facies variation, the formal type section was not designated, until 1955, when it was established in the southern regions of the Qom plain, encompassing Mill, Do-Chah, Do-Baradar.

Furrer and Soder (1955) studied the informal type section of the eastern Shourab, located 30 km south of Qom. Here, they divided it into 6 members viz. “a, b, c, d, e, f”. Later studies by Bozorgnia (1963) further subdivided the C Member into C1, C2, C3 and C4 producing the 9 constituent members of the Qom Formation (Table 4.1).

However, in addition to these 9 members, the uppermost evaporite rocks, overlain by the Upper Red Formation, are termed “the Evaporite Member”. Also, according

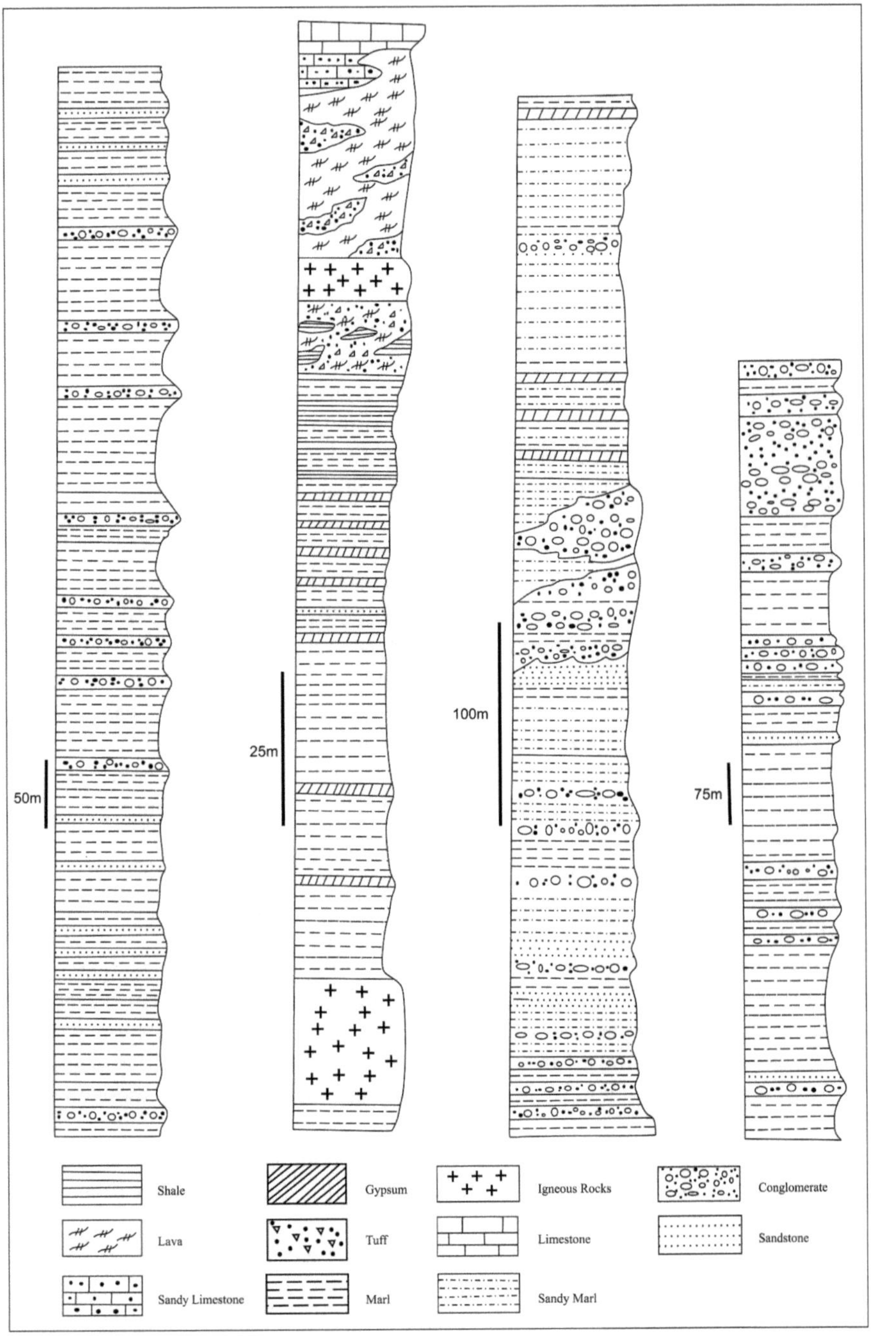

Fig. 4.2 Lithostratigraphical column of Lower red Formation of some parts of Central Iran and Azarbaijan

Table 4.1 The members of the Qom Fm. Furrer and Soder (1955)

Upper Red Formation	
Qom Formation	Member f
	Member e
	Member d
	Member c
	Member b
	Member a
Lower Red Formation	

to Bozorgnia (1966), there is an "Unnamed Member" in the lowermost Qom Formation of the Kashan area that contains planktonic fossils, presumably indicating transgression of the Qom Sea (Fig. 4.3).

Upper Red Formation

This formation is composed of mudstone, shale, siltstone, sandstone, microconglomerate and sometimes conglomerate (with red color and occasionally with colored pattern) (NIOC 1959). It rests on the limestone of the Qom Formation or on older rocks where the Qom Formation is absent. In some areas, the contact between the Qom Formation and the Upper Red Formation cannot be identified. The upper boundary, also, cannot be distinguished where the Pliocene conglomerate is not present.

The Upper Red Formation is characterized by the following features:

1. Lithological and stratigraphical evidence showing that the lower parts of the formation represent marine environments.
2. Volcanic rocks such as andesite, andesi-basalts and sometimes acidic tuffs are accompany sediments of the formation in the Azarbaijan area.
3. The presence of evaporatives in the basal beds of the middle part is such that they form small-scale salt diapirs in Azarbaijan (Figs. 4.2, 4.3 and 4.4).

4.3 Sanandaj-Sirjan

The most important difference between Sanandaj-Sirjan and Central Iran is the lack of Cenozoic lithostratigraphic units (especially the volcano-sedimentary unit found in the Urmia-Dokhtar volcanic belt) in the former area. Nevertheless, some sporadically-distributed outcrops of the Eocene pyroclastics and of the Qom Formation can be observed in northern parts of the Sanandaj-Sirjan area (Aghanabati 2004).

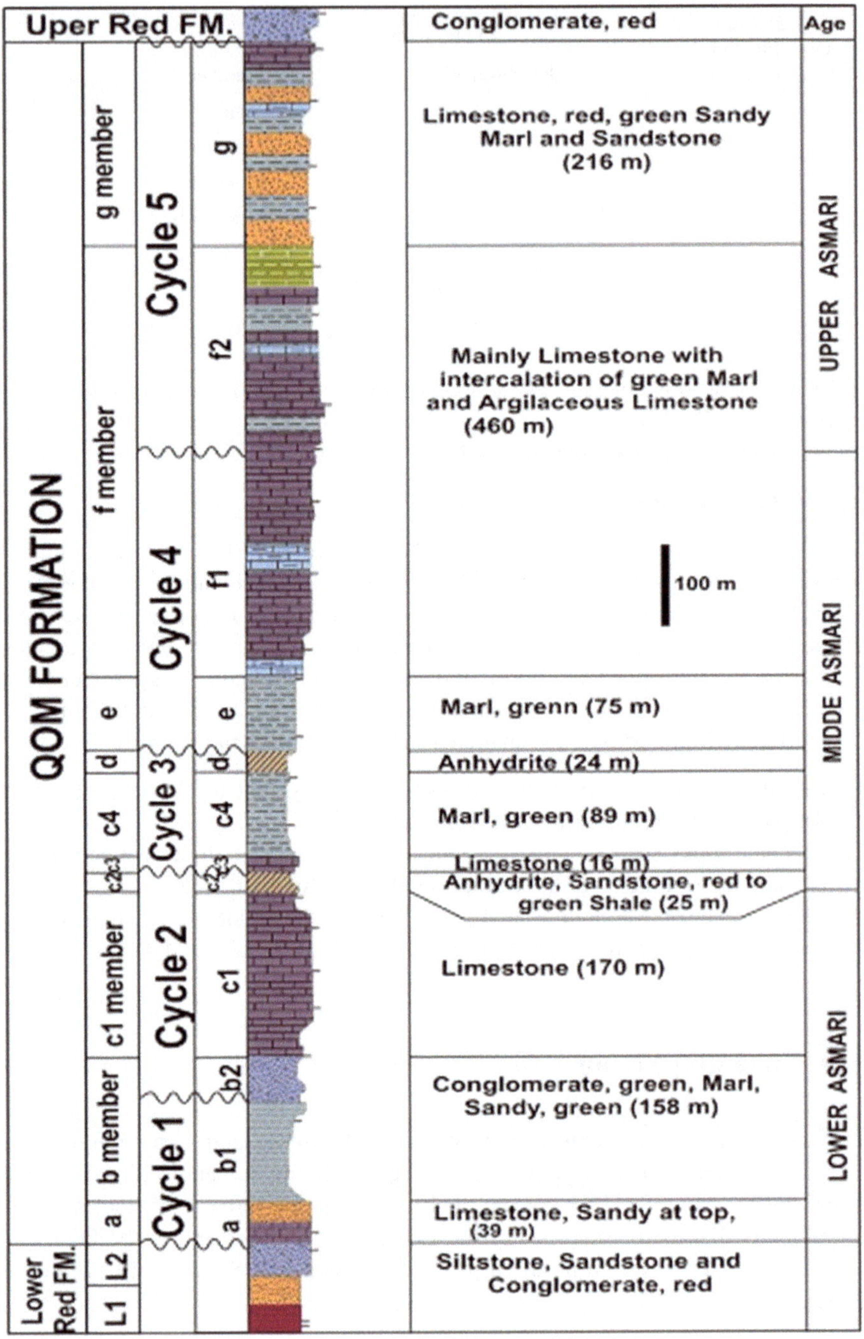

Fig. 4.3 Lithostratigraphical column of the Qom Fm. of the Do-Baradaran section, southern Qom, Modified after Baghbani (1996)

Fig. 4.4 A view of Upper Red Formation at Moshampa, Mahneshan area, westward view

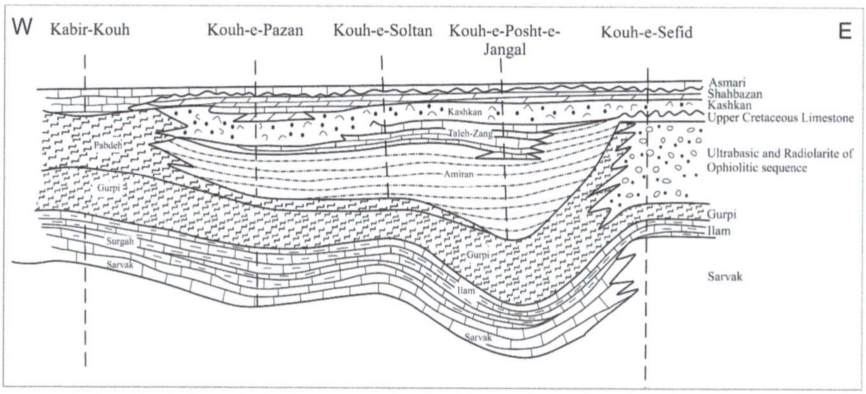

Fig. 4.5 Cretaceous and Cenozoic rock units of Zagros, modified after Shemirani (1991)

4.4 Zagros

The Cenozoic rocks of the Zagros are different to those from other parts of Iran due to marine condition having prevailed during almost all of the Cenozoic with lithologically varied formations representing foreland basin conditions in most areas (Alavi 2004). Furthermore, deposition continued across the Cretaceous-Paleogene boundary in most parts of the Zagros (Figs. 4.5 and 4.6). In general, the Cenozoic units of the Zagros are as follows: the Amiran, Sachon, Shahbazan, Talehzang, Kashkan, Jahrom, Pabdeh and Asmari formations, the Fars Group (comprising the Gachsaran, Mishan and Aghajari Formations), the Razak Formation and the Bakhtiari Formation.

As already mentioned, the Cenozoic of the Zagros shows many facies variations in Lorestan, Fars, the Dezful Embayment and the Hinterland of Bandar-Abbas. This variation is more obvious in the Pabdeh Formation and its lateral equivalents (Motiei

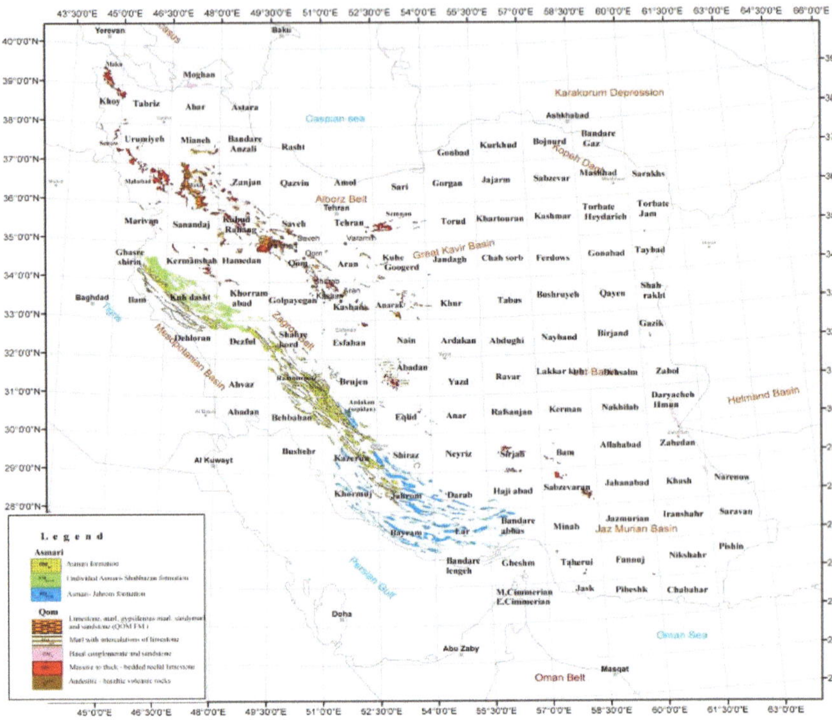

Fig. 4.6 Outcrops of the Qom and Asmari formations in Iran

1993) (Fig. 4.7). The Oligocene-Lower Miocene Asmari Formation is one of the more widely distributed units of the area, both lithologically and chronostratigraphically, compared to the Qom Formation of Central Iran (Aghanabati 2004). Due to economic significance of the Asmari Formation, a more detailed description is included here. The distribution of the Asmari Formation of Zagros and its equivalent in the Central Iran is shown in Fig. 4.6.

Asmari Formation

The Asmari Formation is the youngest reservoir rock unit of the Zagros area, and the first recognized reservoir of the Middle East, having a worldwide reputation. The formation is mostly composed of shallow water carbonates (Busk and Mayo 1918) deposited on a shallow carbonate ramp (e.g. Vaziri-Moghaddam et al. 2006) that developed during the extensional phase of the Zagros Basin (Motiei 1993). The Jarib Limestone, and Kalhor and Khamir formations were previously considered to be the lithostratigraphic equivalents of the Asmari Formation. Nevertheless, these rock units cannot be lithologically compared to the entire lithostratigraphic column of the Asmari Formation in the type section and are only comparable with some parts of the formation (Aghanabati 2004).

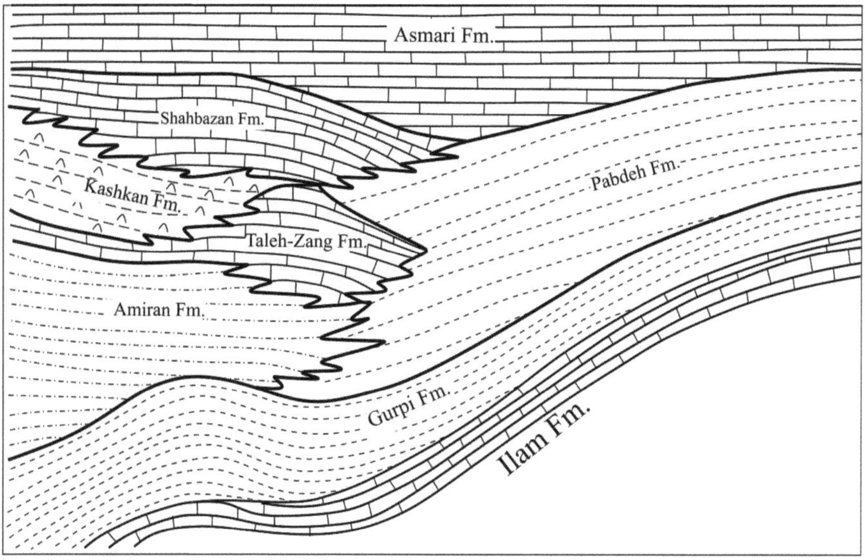

Fig. 4.7 Cenozoic rocks of the Lorestan, Zagros area, modified after Shemirani (1991)

The Asmari Formation is extensively exposed in the Zagros area and is most widely distributed in the Dezful Embayment. It extends from northwestern Iran, through Eastern Iraq and into the southwestern part of Oman with minor lithological and paleontological variations (Motiei 1993).

Numerous biostratigraphical and reservoir characterization studies that have been carried out on the formation since 1918, because of its economic significance. However, the accepted formal type section of the formation was not designated until 1924, when Richardson (1924) introduced and measured the type section in southeastern Masjed-Soleyman. Later revision of the description of the section by Thomas (1948) shows that the formation is composed of 314 m-thick strata of crème to brown resistant limestone having abundant fractures and shale intercalations. The same author subdivided the formation into three members, the lower, middle and upper Asmari, based on fossil contents and age. Interestingly, Adams (1969) added lithostratigraphic description for the three members and considered them to be lithostratigraphical subdivisions. It should be noted that, in the type section, the lowermost Oligocene part of the formation is not exposed.

Lower Asmari

The so called "Lower Asmari" is representative of the lower Oligocene part of the Asmari Formation. It consists of wackstone and occasionally dolomitized packstone. Its facies varies from place to place, as does the underlying formation (Motiei 1993). In most of the Lorestan outcrops, the Lower Asmari rests on the Eocene Shahbazan Formation. Toward the northern area, the Lower Asmari cannot be observed, seemingly, due to the exposure of the late Eocene to late Oligocene; whereas, in the

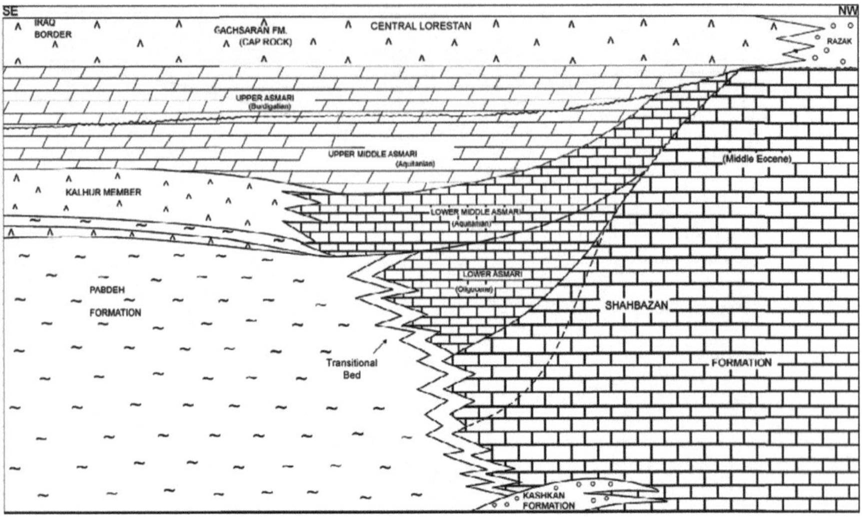

Fig. 4.8 The members of Asmari Fm, modified after Motiei (1993)

Central Lorestan, the lower Asmari grades downwards into shale of the underlying Pabdeh Formation. In the Dezful Embayment, the Lower Asmari overlies the Pabdeh Formation (Motiei 1993) (Fig. 4.8).

Middle Asmari

This member can be further subdivided into two distinct units, the "Lower Unit" and the "Upper Unit".

Lower unit. This includes a homogenous but repetitive succession of wackstone and highly dolomitized packstone indicating alternation of inner shelf and outer shelf environments.

Upper unit. This is composed of mudstone, wackstone and fine-grained dolomitized mudstone (Motiei 1993).

Upper Asmari

This unit consists of a succession of wackstone and packstone, containing anhydrite and dolomitized limestone to the south (Motiei 1993).

4.5 Kopeh-Dagh

The Kopeh-Dagh basin subsided during Jurassic to latest Cretaceous time, when it was connected, especially in its western part, to the Caspian Basin. However these basins were subsequently separated by the Cenozoic uplift of the Kopeh-Dagh and Alborz, while the Caspian Basin continued to subside (Afshar-Harb 1994).

The general pattern of sedimentation in the Cenozoic of Kopeh-Dagh is believed to indicate regression succession, resulting from the Cenozoic uplift described above.

As a consequence, the Cenozoic lithostratigraphic succession of the area contrasts with the Cretaceous which is dominated by clastics and evaporites rather than marine sediments (Afshar-Harb 1994).

According to Gharabeygi et al. (2008), Cenozoic rocks of western Kopeh-Dagh are less distributed than in eastern parts of the region.

4.5.1 *Paleogene*

Pesteligh Formation

This formation is mainly composed of clastic sediments, and is mostly exposed in southern areas. In its type section, it consists of a 280 m-thick succession of shale and sandstone (Afshar-Harb 1994).

In the Jajarm area, it consists of 1700 m of grey to brown conglomerate. The measured thickness of the Pesteligh Formation in the Jozak Section is 154 m. In northern Takal-Kuh, this formation is 126 m thick and is typically tripartite, consisting of lower sandstone, a middle sandy limestone and an upper shale and sandstone. This formation has been assigned a Paleocene age (Afshar-Harb 1994).

Chehel Kaman Formation

This name was applied to a succession of carbonate rocks with subordinate marl and limey shale which imply increases in depth in the Kopeh-Dagh Basin. The formation in its type section at Chehel Kaman consists of 229 m of limestone, dolomite, marl, sandstone and shale (Afshar-Harb 1994).

The Abtalkh Formation is overlain by the Chehel Kaman Formation (281 m-thick) in western Kopeh-Dagh is only exposed along the northern part of the Maraveh Tapeh Fault in the Atamir Syncline. In this area, the Chehel Kaman Formation consists of a 281 m-thick succession of limestone, and silty and sandy marls (Afshar-Harb 1994).

Khangiran Formation

This formation is one of the youngest marine units of the Kopeh-Dagh Basin (Afshar-Harb 1994). It is composed of sandstone and olive to grey limey shale in northern Takal-Kuh. Its thickness varies from 87 m in northeastern Bojnurd to 470 m in the Sugha section, where it is mainly composed of olive grey shales.

The Khangiran Formation has been assigned to the Late Paleocene to Eocene and probably Early Oligocene. Its basal beds in the eastern Kopeh-Dagh contain Late Paleocene foraminifers and in the western part, Early Eocene taxa, indicating that it becomes increasingly younger from east to west (Afshar-Harb 1994).

It seems that the upper beds of the Khangiran Formation were removed by erosion in the eastern Kopeh-Dagh because of its limited geographical distribution and age restricted to the middle Eocene (Gharabeygi et al. 2008).

4.5.2 Neogene

Compared to the Upper Red Formation and Neogene Conglomerate of Central Iran, the continental Neogene rocks of Kopeh-Dagh, deposited in the continental environments, are of less widely distributed. However, in some areas, such as Daregaz-Kenar in the southern slope of Zarrineh-Kuh, it can reach a thickness of 900 (Afshar-Harb 1994).

The Kopeh-Dagh Basin generally became a highland and was eroded in most parts. However, deposition of marine sediments similar to those of the southern Caspian in western parts of the Kopeh-Dagh Basin suggests that it was affected by a marine transgression from the Caspian Basin and that it became a marginal part of the latter basin (Afshar-Harb 1994).

4.6 Makran

Paleogene

Overlying the Cretaceous, the Paleogene-Pliocene rocks of Makran are mainly composed of sandstone, siltstone and shale and in some places marl with low carbonate content. These rocks are mostly of turbiditic facies. Resembling the Eastern Iran Flysch, the Paleogene rocks of the Makran are widely distributed from Pakistan in the East to the Minab area in the West (Hajian 1996).

In northern Makran (Mokhtar-Abad, Fahnouj and Tarmashk) and southern Jazmourian, Paleogene sediments are mixed in 'color mélanges'. The normal Paleocene parts of these areas are composed of limestone, shale and sandstone. However, near the color mélange complexes, e.g. the Mokhtar-Abad Complex, these tend to change into pelagic facies. In general, the Paleogene rocks of Makran consist of two complexes, the Bon-Roud Complex of the Fahnouj area and the Dorkan Complex of Taheroye that extends into northern Zabol (Hajian 1996).

The Eocene rocks of Makran are mostly clastic and turbiditic rocks including the following units:

1. The Bidak Unit which includes an early to late succession of coarse to fine-grained clastic sediments, including sandy conglomerate and clayey siltstone.
2. The White Sandstone Unit that is dated as early Eocene to late Oligocene.
3. The Marij Unit which is a fine-grained clastic unit overlying the Warakan Complex. This unit is more widely distributed in southern Makran, and is assumed to be Middle Eocene to Late Oligocene in age.
4. The Undivided Sandstone Unit. Consisting of a succession of sandstone and limestone, this unit is exposed in the southern part of Minab (Geological Map, Scale: 1/250000), where it rests on the Colored Mélange with angular unconformity (Hajian 1996).

Miocene Flysch units

Dahirdan Unit

The name of this unit is derived from the Dahirdan River of the Taheroei area, and was proposed for a succession of limey shale and sandstone. The sandstone facies is composed of a rhythmic, upward-coarsening sequence of sandstone and shale. This undivided sequence includes a succession of sandstone, shale and thick-bedded, fine-grained sandstone. The Dahiran Unit (600–1000 m thick) rests conformably on the Anguran Unit and is overlain by the Vaziri Unit. The former is found in the Minab, Taheroei, Fanhouj, Nikshahr and northern Pishin areas, and extends into Pakistan. In areas where the overlying Vaziri Unit is not exposed, the Dahirdan Unit is covered by one of the harzburgite conglomeratic units of Band-e Chaker and Darkhonish-Abshahr or Shahrpoum. Aquitanian-Burdigalian fossils have been recorded from the Dahirdan Unit (Rahimzadeh 1994).

Sabz Unit

The name of this unit was derived from the Sabz Mountain of western Taheroei. In the Minab quadrangle, it is mostly composed of grey to green mudstone, massive limey sandstone, brown conglomerate and crème limestone. The measured thickness of 3200 m solely represents the upper part of unit, while the total thickness is about 6000 m. Overlain unconformably by the Palami Conglomerate, the Sabz Unit consists of a flysch succession tat mainly formed in deep ramp depositional environments rather than in troughs. Its shallowing pattern is comparable to that of the Oligo-Miocene flysch (Rahimzadeh 1994).

Shahrpoum Unit

The name of this unit was obtained from Shahrpoum Village in the Taheroei area. The Shahrpoum Unit conformably overlies the Dahirdan unit and is composed of three subunits as follows:

Lower Sub-unit (5000 m thick). This includes a succession of sandstone, shale, some conglomerate, siltstone, mudstone and limestone. The sub-unit is characterized by some shaly successions (50–200 m).

Middle Sub-unit (2500 m thick). This consists of thin-bedded, fine-grained, mica-bearing sandstone and shale.

Upper Sub-unit (2000 m thick). This sub-unit includes coarse to fine-grained sandstone and silty limey shale. An early to middle Miocene age was proposed for the Upper Sub-unit but the fossils that indicate are probably reworked. This unit is the equivalent of the Band-e Chakr and Dar-Khonish shale. Towards the south and east of Shaprom, this unit is the age-equivalent of the neritic and evaporative Sabz Unit (Rahimzadeh 1994).

Hichan Unit

The name of this unit was derived from Hichan village in southwestern Nikshahr, where the Hichan Unit is approximately 3000 m thick and consists of light grey, fissile, brittle limey shale and flysch. The shaly beds are accompanied by some sandstone, mudstone and siltstone intercalations. The sandstone parts are thin-bedded and buff colored and are classified as subarkosic sandstone. The lower part also includes some conglomerates with pebble-sized clasts, gastropod-bearing micritic limestone and bivalve fragments. No unit is seen overlying the Hichan Unit. An Early Miocene age was proposed for this unit but it may extend up into the Middle Miocene (Rahimzadeh 1994).

Vaziri Unit

The name of this unit is derived from the village of Vaziri in the Minab area, where the unit mainly consists of reefal limestone, algal wackstone, packstone and clastic limestone. Its thicknesses highly variable, ranging from 300 m in northeastern Taheroei, to 2348 m in northwestern Taheroei, 2550 m in Band-e-Chaker, to 5000 m in the south, where its thickness is greatly increased. The Vaziri Unit rests conformably on either the Jaghin Unit or the Darpahn Unit (in the Anviri Syncline) and, in turn is overlain by the Dahirdan Unit. As the latter is dated as Middle Miocene, the Vaziri unit must be this age or older.

Darpahn Unit

The name of this unit was derived from Darpahn Village in the Taheroei area, where this unit includes lithic sandstone, shale and the upper polymictic conglomerate, composed of cherty and igneous cobbles and boulders with a high matrix content. This conglomerate is characterized by its red color in the Darpahn section, while in the other areas its color changes to brown. A middle to late Miocene age was proposed for this unit (Rahimzadeh 1994).

Jaghin Unit

The name of this unit was derived from the Jaghin River in the Taheroei area, where the Jaghin Unit is composed of three units, while to the east, only the lower sub-unit can be observed.

The **Lower sub-unit** is composed of sandstone, conglomerate and some shaly beds.

The **Middle sub-unit** includes red colored sandstone and thin-bedded conglomerate. This sub-unit lithologically and sedimentologically resembles to the lower subunit.

The **Upper sub-unit** consists of shale, sandstone, siltstone and conglomerate associated with lumashel and sandstone intercalations.

The thickness of this unit is at least 2750 m, but its top is not seen (Rahimzadeh 1994).

Ghasrghand Unit

The name of this unit was obtained from Ghasrghand Village in the Nikshahr area, where it is composed of gypsiferous mudstone and thin-bedded sandstone. The unit compared to the Roksha, Jarout and Pishin units, all of which are composed of neritic sediments with some muddy rocks. The Ghasrghand Unit is the age equivalent of the Sabz Unit (Rahimzadeh 1994).

Sahntang Unit

The name of this informal unit was derived from a river in southern Pishin, where it is composed of mixed clastic and carbonate rocks. Being a lateral facies variant of the Ghasrghand Unit, the Sahntang Unit is only exposed in a small area of southern Pishin. Based on its fossil content, the unit has been dated to early Middle Miocene (Rahimzadeh 1994).

Pishin Unit

The name of this unit was obtained from Pishin City in southern Pishin, where the unit is composed of a 100 m thick Miocene succession of sandstone showing an upward increase in sand content. The unit is characterized by some muddy parts and sedimentary structures such as graded bedding, oblique current ripples, groove casts and load casts. The Pishin Unit is interpreted as having been deposited in continental shelf to deep basin environments, based on the presence of mud flows and sand transportation above fair-weather wave base. It is comparable to the Ghasrghand Unit; the lower Roksha and lower Jarout units, and parts of the Hijan Unit. The lower part of the Pishin Unit shows evidence of marine regression in the form of an increase of sand content and the presence of cross bedding. The upper part contains the muddy deposits (Rahimzadeh 1994).

Jarout Unit

The name of this informal unit was derived from Jarout Mountain in the Pishin area. So far, six sections in this area and three sections in the Talar range have been studied.

In the Nikkshahr quadrant, the unit is divided into three subunits. The two lower subunits are of more and less similar lithology and differ only by shale and sandstone content. The lower sub-unit is composed of several shaly cycles ca. 3 m thick with their thickness deceasing upward. The middle sub-unit consists of cliff forming, massive to well-bedded sandstone 50 m thick. This unit contains fine to medium-bedded calcareous-cemented sandstone characterized by being pebble-bearing and containing lumashell layers with plant remains. Attributed to the early to middle Miocene, the Jarout Unit is the same age as the Roksha and Ghasrghand units and also some parts of the Sahntang and Pishin units (Rahimzadeh 1994).

Makran Unit

The name of this unit is derived from the Makran Coastal Range that extends eastward from the Makran region of the southwestern Baluchistan region of Iran, into Pakistan where it forms the northern coast of the Indian Ocean. The Makran Unit consists of a succession of neritic and molasse sediments distributed throughout the Baluchistan region of Iran and Pakistan. It is subdivided into four informal sub-units including:

1. The Gioshi Marl.
2. The Khakou Sandstone.
3. The Tiab Sandstone and the Minab Conglomerate.

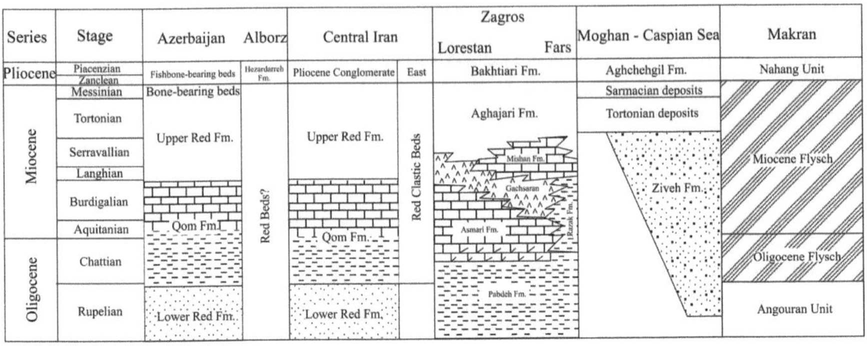

Fig. 4.9 The correlation of Oligocene, Miocene and Pliocene deposits of Iran

The Goshi Sub-unit has been assigned to the late Miocene based on the presence of the planktic foraminifer *Globorotalia acostaensis* and *Globorotalia humerosa*.

The Khakou subunit has been attributed to the late Miocene to early Pliocene on the basis of the presence of the foraminifer *Pulleniatina obliquilo culata*. A late Miocene to early Pliocene age has also been suggested for the Tiab Sandstone, because of the presence of *Globorotalia humerosa*. However, the Minab Conglomerate cannot be older than early Pliocene, because it rests on the Tiab sandstone, an age confirmed by the presence of vertebrate fossil *Planzenta* of (Rahimzadeh 1994).

Tahtoun Unit

The name of this informal unit was derived from the Tahtoun Range in the Taheroei area. This unit includes a transition from marine to fluvial facies, with marine beds overlain by fluvial conglomerate and sandstone, and then by conglomerate in the upper part. The unit has been divided into three sub-units as follows:

1. Marine-derived Sub-unit
2. Sandstone and Conglomerate Sub-unit
3. Conglomerate Sub-unit

The marine-derived subunit is composed of conglomerate and sandstone accompanied by shell fragments and foraminifer shells. Consisting of chert and quartz fragments, the clasts of the conglomerate are upward grading. This unit is generally characterized by the presence of sedimentary structures such as graded bedding, channels and mud cracks. Sandstone and conglomerate sub-units include the pebble-bearing sandstone with large clasts such as cobbles and boulders, and subordinate polymictic conglomerate with high matrix content.

The lithostratigraphic features of Cenozoic rock units of Iran are summarized in Table 4.2.

Figure 4.9 illustrates a correlation of the Oligo-Miocene and Pliocene of Iran but Kopeh-Dagh.

Plates 4.1, 4.2, 4.3 and 4.4 illustrate selected age-diagnostic fossils from Cenozoic deposits in Iran.

Table 4.2 Lithostratigraphic feature of Cenozoic rock units of Iran

Rock unit	Chronostratigraphic affinity	Lower/upper boundary	Lithology and thickness at type section	Note (references, comments)
Cenozoic (Alborz and Azarbaijan)				
Kond Fm.	Late Eocene	Lb. Karaj Fm. with disconformable contact and seemingly parallel Ub. Upper Oligocene Lower Red Formation with angular unconformity	Includes three informal units including lower unit: alternation of conglomerate and sandstone; middle gypsum; and upper limestone and marl Thickness: 250 m	Book: Hajian (1996, p. 173), Aghanabati (2004, p. 356), Alavi-Naeini (2009, p. 394)
Ziarat Fm.	Middle Eocene	Lb. Fajan Fm. Ub. Karaj Fm.	Includes two members viz. "Lower Member" consisting of gypsiferous yellowish marl (150 m-thick); and "Upper Member" including thick-bedded reefal limestone (330 m-thick). Thickness: 450 m	Book: Hajian (1996, p. 153), Aghanabati (2004, p. 353), Alavi-Naeini (2009, p. 391) Thesis: Alimi (2005), Ghorbani (2010)
Karaj Fm	Middle Eocene	Lb. Shemshak Fm. with unconformable contact Ub. Neogene Red Conglomerate	Includes 5 members of Lower Shale (limey and silty shale with intercalations of tuff and glassy tuff); Middle Tuff (white to light green thick-bedded, glassy tuff and upper limey shale); Asara Shale (limey shale); Upper Tuff (green tuff with tuffaceous shale intercalations, tuffaceous sandstone and limey shale); Kandovan Shale (limey shale and piceous limestone)	Book: Hajian (1996, p. 158); Aghanabati (2004, p. 353), Alavi-Naeini (2009, p. 392) Thesis: Babazadeh (1992), Safari (2010), Ghodsi (1995), Alimi (2005) Research Project: Shemirani (1991)

(continued)

Table 4.2 (continued)

Rock unit	Chronostratigraphic affinity	Lower/upper boundary	Lithology and thickness at type section	Note (references, comments)
Cheleken Fm.	Pliocene	Lb. Yellow Sarmasian deposits with angular unconformity or Middle Cretaceous rocks in the drilling cutting derived from the Farah-Abad area	Brown thick-bedded conglomerate associated by some layers of brownish red claystone and silty sandstone	Book: Aghanabati (2004, p. 362) Thesis: Jahanpanah (2003)
Ziveh Fm.	Late Eocene-Middle Miocene	Lb. Eocene deposits with disconformable contact and sometimes angular unconformity	The clayey fine- to coarse-grained conglomerate of Ziveh Fm. is divided into three distinct units "Lower Ziveh" is composed of conglomerate unit and sandstone-clayey siltstone unit (2300 m-thick) "Middle Ziveh' consist of sandstone with plant-bearing shale and thin-bedded, coal-bearing shale and gypsum intercalations (800 m-thick). "Upper Ziveh" consists of feldspar-bearing conglomeratic sandstone and yellow clayey siltstone with thin-bedded intercalations of limestone and dolomite	Book: Rahimzadeh (1994, p. 207), Aghanabati (2004, p. 359) Thesis: Faramarzian (2004) The most exposures of the formation are found in the Uromia in the around of Uromia dam The presence of a conglomerate unit is resulted to division of the Upper Ziveh into three subunit including "lower Siltstone", middle conglomerate" and "upper clayey siltstone"
Lower Red Fm.	Oligocene-Miocene	Lb. Karaj Fm. Ub. Qom Fm.	A red succession including marl, sandstone and conglomerate locally accompanied by gypsum and salt rocks and sometimes miliolid-bearing limestone in the lower part Thickness:	Book: Rahimzaadeh (1994, p. 51), Aghanabati (2004, p. 357) Thesis: Mozaffari (1999) This formation can be distinguished from the Upper Red Fm. by its dark red color and more coarse-grains

(continued)

Table 4.2 (continued)

Rock unit	Chronostratigraphic affinity	Lower/upper boundary	Lithology and thickness at type section	Note (references, comments)
Cenozoic (Central and Eastern Iran)				
Chupanan Fm.	Paleocene	Lb. Danian rocks with disconformable contact Ub. Not defined	A sandstone unit with different thickness and pelagic limestone	Book: Aghanabati (2004, p. 365)
Palang Fm.	Paleocene	Lb. Older rocks such as Karaj Fm. with conformable contact or angular unconformity Ub. Not defined	Alodepic limestone	Book: Aghanabati (2004, p. 365)
Kerman Con-glomerate	Cretaceous-Paleocene	Lb. older rocks everywhere with angular unconformity Ub. Note defined	Includes two distinct parts, viz. “Lower Part” consisting of thin-bedded, fine-grained silty and clastic beds, and “Upper Part” consisting of conglomerate accompanied by greywacke and coarse-grained sandstone intercalations with cross-bedding structure Thickness: 220–1450 m	Book: Aghanabati (2004, p. 365) Huckriede et al. (1962) introduced and measured the type section of the unit (300 m-thick) in the 7 km northern Kerman
Dokohaneh Fm.	Paleocene	Not recorded	Sandstone and fine-grained clastic beds (Flysch)	–
Lower Eocene sedimen-tary complex	Early Eocene	Lb. not reported Ub. Palangi conglomerate	Microconglomerate’, clacirodiate, feldspat-bearing greywacke, calcarenite, biocalcarnite, siltstone, sandy marl and micrite	Books: Hajian (1996, p. 195), Aghanabati (2004, p. 371)

(continued)

Table 4.2 (continued)

Rock unit	Chronostratigraphic affinity	Lower/upper boundary	Lithology and thickness at type section	Note (references, comments)
Bahr-Aseman Complex	Early Eocene	Lb. Jurassic Or Cretaceous rocks with angular unconformity Ub. Razak complex with angular unconformity	Igneous and acidic pyroclastics, and successions of pyroclastic, Andesite-Basalt Thickness: 7000 m	Book: Hajian (1996, p. 199), Aghanabati (2004, p. 371)
Middle Eocene sedimen-tary complex	Middle Eocene	Lb. Bahr-Aseman Complex Ub. Razak complex	Quartz sandstone, andesitic tuff and a turbiditic unit in the lower part, and the tuff succession of the upper part contains yellow limestone	Book: Hajian (1996, p. 202), Aghanabati (2004, p. 371)
Razak Complex	Uppermost Lutetian-Lowermost Priabonian	Lb. Middle Eocene sedimentary complex Ub. older rocks with angular unconformity	Basic and acidic rocks subdivided into three lithologically distinct members including "lower member" (mostly basic rocks); "middle member" (acidic rocks); "upper member" (basic rocks)	Book: Aghanabati (2004, p. 372), Hajian (1996, p. 208) Thesis: Morad-Pour (2009)
Hezar Igneous Complex	Lowermost Priabonian	Lb. Either Lower Red Fm. or Qom Fm Ub. Younger volcanic rocks	An inhomogeneous succession of tuffaceous sandstone, conglomerate, tuff, trachyandesite, trachybasalt, ignimbrite, thin-bedded limestone, agglomerates and so on	Book: Hajian (1996, p. 213), Aghanabati (2004, p. 372) Article: Ahmadi-Pour et al. (2003)
Lower Red For-mation	Oligocene-Miocene	Lb. Mostly Karaj Fm. Ub. Qom Fm.	Red colored deposits including marl, sandstone and conglomerate locally bearing gypsum and salt rocks, and occasionally contain miliolid-bearing limestone in the lower part	Book: Rahimzadeh (1994, p, 51), Aghanabati (2004, p. 357) Thesis: Mozaffari (1999) The formation can be distinguished from the Upper Red Fm. by its dark red color and coarse-grains

(continued)

Table 4.2 (continued)

Rock unit	Chronostratigraphic affinity	Lower/upper boundary	Lithology and thickness at type section	Note (references, comments)
Qom Fm.	Oligocene-Miocene	Lb. mostly Lower Red Fm. with sharp contact or weak angular unconformity Ub. Upper Red Fm. with conformable or in some areas disconformable contact	Consists of following members: "Binam"(a succession of red and green marl, sandstone, thin-bedded limestone); "a" (brown, thick-bedded to massive calcirudite and calcarenite with low sandy content (subarkose and sublitharenite); "b"(marl, sandy marl, glauconite bearing sandstone and a conglomerate horizon with volcanic fragments. "c1" (a succession of biosparodite, biomicrodite and marl); "c2" (red and grey shale, sandstone and gypsum); "c3"(oolite-bearing limestone, bryozoan-bearing limestone (biosparodite and biomicrite); "c4" (green to yellow marl); "d" (gypsum and some shale layers); "e" (green marl with some gypsum and clayey limestone intercalations); "f" (massive limestone with clacirudite composition); and "evaporative mb." Thickness: 1200 m	Book: Rahimzadeh (1994, p. 70), Aghanabati (2004, p. 377) Thesis: Mozaffari (1999), Ganji (1998), Karimi-Mossadegh (1999), Keshavarzi (1997), Rahimi-Matin (2007), Bakhshi (1998), Sartipzadeh (2002), Dadfar (2003), Bakhtiari (2002); Ramezani-Dana (2003), Naeiji (1997); Pahlavanpour (1997), Azimi (1997), Poursalehi (2004); Aftabi (2009) Article: Daneshian and Aftabi (2010) Research Project: Pourabrishami et al. (1993)
Pliocene continental facies	Pliocene	Not recorded	Mostly conglomerate deposits	Book: Aghanabati (2004, p. 391)

(continued)

Table 4.2 (continued)

Rock unit	Chronostratigraphic affinity	Lower/upper boundary	Lithology and thickness at type section	Note (references, comments)
Pliocene lacustrine facies	Pliocene	Not recorded	Silt and clay, lignite-bearing rocks; marly rocks the most parts of which are composed of yellow claystone with thin-bedded gypsum layers; lake clay and volcanic ash	Book: Aghanabati (2004, p. 391)
Cenozoic (Zagros)				
Amiran Fm.	Paleocene in Lorestan and Masstrichtian to Paleocene in Khorram-abad-Kermanshah	Lb. Gurpi Fm. with conformable and gradual contact Ub. Telehzang Fm.	Siltstone and dark olive to brown sandstone and some limestone and conglomerate with chert pebbles Thickness: 871.7 m	Book: Motiei (1993, p. 289), Alavi-Naeini (2009, p. 401) Thesis: Yazdi-Moghadam (1995)
Sachon Fm.	Paleocene to Early Eocene	Lb. Tarbour Fm. with conformable and gradual contact Ub. Jahrom Fm. with sharp and conformable contact	Includes some successions of marl-gypsum-dolomite; marl-dolomite-gypsum; marl-dolomite-limestone as well as marl-dolomite Thickness: 1414.25 m	Book: Motiei (1993, p. 289), Aghanabati (2004, p. 398), Alavi-Naeini (2009, p. 401)
Ghorban Calcare-ous mb.	Paleocene	Lb. Tarbour Fm. Ub. Pabdeh Fm.	Brown to yellow, highly recrystallized limestone bearing calcite veins and iron-oxide Thickness: 183 m	Book: Motiei (1993, p. 293)
Shahbazan Fm.	Middle to Late Eocene	Lb. Clastic Kashkan Fm. with conformable contact Ub. Asmari Fm.	Dolomite and limey dolomite. Thickness: 332.8 m	Book: Motiei (1993, p. 299), Aghanabati (2004, p. 400), Alavi-Naeini (2009, p. 403) Thesis: Najafi (1998), Abdi (1998)

(continued)

Table 4.2 (continued)

Rock unit	Chronostratigraphic affinity	Lower/upper boundary	Lithology and thickness at type section	Note (references, comments)
Taleh-Zang Fm.	Paleocene to Middle Eocene	Lb. Amiran Fm. with conformable and sharp contact Ub. Kashkan Fm. with apparently conformable contact	Grey to brown, massive to thick-bedded limestone Thickness: 176 m	Book: Motiei (1993, p. 294); Aghanabati (2004, p. 399), Alavi-Naeini (2009, p. 402) Thesis: Zohdi (2007), Najafi (1998), Yazdi-Moghadam (1995)
Kashkan Fm.	Paleocene to Middle Eocene	Lb. Taleh-Zang with sharp and apparent conformable contact Ub. Shahbazan Dolomite with disconformable contact	Red clastic sediments such as siltstone, sandstone and conglomerate Thickness: 370 m	Book: Motiei (1993, p. 297). Aghanabati (2004, p. 399). Alavi-Naeini (2009, p. 402) Thesis: Najafi (1998)
Jahrom Fm.	Paleocene to Middle Eocene	Lb. Sachon with conformable contact Ub. Asmari Fm.	Dolomite and limey dolomite Thickness: 467.5 m	Book: Motiei (1993, p. 301), Aghanabati (2004, p. 398), Alavi-Naeini (2009, p. 403) Thesis: Shaiesteh (2006), Karimpour (2009), Zakerzadeh (1998)
Pabdeh Fm.	Paleocene to Oligocene in Fars and Khouzistan area, Late Paleocene to Miocene in Lorestan	Lb. Shale and marl beds of Gurpi Fm. Ub. Asmari Fm. with conformable and gradual contact	Shale, blue to purple marl, thin-bedded, clayey limestone informally called "purple shale"; grey shale and thin-bedded, clayey limestone; thin-bedded limestone with cherty nodules informally called "chert limestone" grey shale and rarely with thin-bedded clayey limestone intercalations with shale and limestone. Thickness: 176 m	Book: Motiei (1993, p. 303), Aghanabati (2004, p. 397), Alavi-Naeini (2009, p. 404) Thesis: Hovaeizavi (1998), Hadavandkhani (2009), Heydari (2010), Gholamifard (2008), Mohseni (2003), Delavar-Najafi (1998), Bahrami-Samani (1997), Tabrizi (2001), Baiat (2006), fanaei (2010), Behbahani (2006) :

(continued)

Table 4.2 (continued)

Rock unit	Chronostratigraphic affinity	Lower/upper boundary	Lithology and thickness at type section	Note (references, comments)
Asmari Fm.	Oligocene-Early Miocene (Burdigalian)	Lb. in the most of areas Pabdeh Fm., but in the Central Lorestan and interior Fars is carbonatic Shahbazan Fm. and Jahrom Fm., respectively with paraconformity Ub. in the most of areas the Formation is overlain by the anhydrite Gachsaran Fm., nevertheless in the interior Fars where the latter formation is replaced by clastic Razak Fm., the overlying formation is Razak Fm.(Aghanabati, 2004)	Cliff-forming resistant limestone with some shaly intercalations Thickness: 314 m	Book: Motiei (1993, p. 327), Aghanabati (2004, p. 401), Alavi-Naeini (2009, p. 405). Article: (2010 Najafi (1998), Omidpour (2003), Aram et al. (2003), Ainolahi (2008) Article: Hashemi (2012) Thesis: Moeini (1998), Tahmasebi-Sarvestani (2010), Shaiesteh (2006), Naseh (2002), Sadeghi-Saadatlou (2003), Ghafrani (2001), Hazavehei (2002), Lotfi (2002), Ranjbaran (2009), Nezafat (2009), Soltanian (2012), Dayer (2010), Hosesini (2011), Sadeghi (2009), Saadat (2005), Tavakoli (1983) Research Project (Torabi (1994)

(continued)

Table 4.2 (continued)

Rock unit	Chronostratigraphic affinity	Lower/upper boundary	Lithology and thickness at type section	Note (references, comments)
Gachsaran Fm.	Miocene	Lb. Eocene rocks in the around of Geshm Island, however northward is Lower Miocene (Burdigalian) rocks	Consists of seven members including evaporates, salt rock, anhydrite, marl and thin-bedded limestone beds; anhydrite-marl alternation of salt rock, marl, anhydrite and some limestone beds; alternation of anhydrite, marl, salt rock and thin-bedded limestone; anhydrite, marl, limestone, salt rock; alternation of anhydrite, marl and limestone	Book: Motiei (1993, p. 414); Aghanabati (2004, p. 405); Alavi-Naeini (2009, p. 406). Thesis: Samavi (2009) Article: Hajeb (1985) First formation of Fars Gr.
Anhydritic Chehel mb.	Early Miocene	Lb. Asmari Fm. Ub. Champeh mb. Both contacts are sharp and conformable	Gypsum and anhydrite alternating with marl and thin-bedded limestone Thickness: 295.6 m	Book: Motiei (1993, p. 418), Aghanabati (2004, p. 406)
Carbonate Chempeh mb.	Early Miocene	Lb. gypsiferous Chehel mb. Ub. Mol mb. Both contacts are conformable	Limestone, white gypsiferous dolomite, red to grey gypsiferous marl and nodular massive gypsum. Thickness: 110 m	Book: Motiei (1993, p. 421), Aghanabati (2004, p. 406) Thesis: Samavi (2009)
Marly Mol mb.	Early Miocene	Lb. Chamopeh mb. with conformable contact Ub. Calcareous Guri mb. of Mishan Fm. with sharp and conformable contact	Red and greyish green gypsiferous marl alternating with gypsiferous limestone and gypsum Thickness: 53.3 m	Book: Motiei (1993, p. 422), Aghanabati (2004, p. 406) Thesis: Samavi (2009)

(continued)

Table 4.2 (continued)

Rock unit	Chronostratigraphic affinity	Lower/upper boundary	Lithology and thickness at type section	Note (references, comments)
Razak Fm.	Miocene	Lb. Jahrom-Asmari Fm. Ub. Limestones of Mishan Fm. with gradual and conformable contact	Red, green to grey silty marl with some silty limestone beds Thickness: 774.4 m	Book: Motiei (1993, p. 425), Aghanabati (2004, p. 406), Alavi-Naeini (2009, p. 407). Thesis: Moradpour (2009), Abbasi (2009) This formation is diachronous, because it was attributed to Burdigalian to Middle Miocene in the Lorestan, Oligocene to Early Miocene in the Bandar-Abbas and Aquitanian to Middle Miocene in the Sisakht around
Mishan Fm.	Early to Middle Miocene	Lb. Gachsaran Fm. with sharp and conformable contact Ub. Aghajari Fm. with gradual but sharp contact	Grey marl and clayey limestone Thickness: 710 m	Book: Motiei (1993, p. 429); Aghanabati (2004, p. 407), Alavi-Naeini (2009, p. 408) Thesis: Feshki (2003), Homayun-Zadeh (2002)
Clastic Lahbari mb.	Pliocene	Lb. Aghajari Fm. with conformable contact Ub. Bakhtiari Fm.	Siltstone, silty marl, limey sandstone and gypsum Thickness: 1575 m	Book: Motiei (1993, p. 436), Aghanabati (2004, p. 409)
Bakhtiari Fm.	Pliocene-Pleistocene	Lb. Aghajari Fm. is covered	Conglomerate, cherty limey sandstone. Thickness: 550 m	Book: Motiei (1993, p. 441), Aghanabati (2004, p. 410), Alavi-Naeini (2009, p. 408)
Carbonatic Guri Mb.	Early Miocene to Late Miocene	Lb. Gachsaran Fm. Ub. Mishan Fm. both contacts are sharp, gadual and conformable	Crème limestone alternating with grey thin-bedded marl Thickness: 112.5 m	Book: Motiei (1993, p. 431); Aghanabati (2004, p. 408).

(continued)

Table 4.2 (continued)

Rock unit	Chronostratigraphic affinity	Lower/upper boundary	Lithology and thickness at type section	Note (references, comments)
Aghajari Fm.	Middle Miocene to Pliocene	Lb. Mishan Fm. with sharp. Ub. Bakhtiari Conglomerate	Alternation of limey sandstone and red marl; gypsum and red siltstone Thickness: 2966 m	Book: Motiei (1993, p. 434), Aghanabati (2004, p. 408), Alavi-Naeini (2009, p. 408)
Cenozoic (Kopeh-Dagh)				
Aghchehgil Fm.	Pliocene	Lb. Cretaceous rocks specially Sanganeh Fm.	Yellow limey sandstone, bioclastic porous limestone with cross bedding Thickness: 53.3 m	Book: Afshar-Harb (1994, p. 200), Aghanabati (2004, p. 416); Alavi-Naeini (2009, p. 414)
Khangiran Fm.	Late Paleocene to Late Eocene and probably early Oligocene	Lb. Chehel Kaman Fm. with conformable contact	A homogenous succession of silty limey shale with olive to bluish grey sandstone intercalations Thickness: 1880 m	Book: Afshar-Harb (1994, p. 189), Aghanabati (2004, p. 414); Alavi-Naeini (2009, p. 413) Thesis: Senemari (1998) Article: Mafi (2003)
Chehel-Kaman Fm.	Late Paleocene	Lb. Pesteligh Fm. concordant but disconformable contact. Ub. Khangiran Fm. with gradual and continuous contact.	Limestone and dolomite with few intercalations of marl and limey shale, locally contained thin-bedded sandstone and gypsum. Thickness: 229 m	Book: Afshar-Harb (1994, p. 178), Aghanabati (2004, p. 414), Alavi-Naeini (2009, p. 411) Thesis: Mafi (1997), Mahbubi (2000)
Pesteligh Fm.	Early Paleocene	Lb. Kalat Fm. with concordant but disconformable contact. Ub. Eighter Chehel-Kaman or Khangiran Formations	An arrhythmic succession of reddish brown shale, mudstone, limey sandstone and conglomerate Thickness: 280 m	Book: Afshar-Harb (1994, p. 172), Aghanabati (2004, p. 411); Alavi-Naeini (2009, p. 411) Thesis: Zakerian-Moghaddam (2009), Karimian (2009); Ramezani (2009)

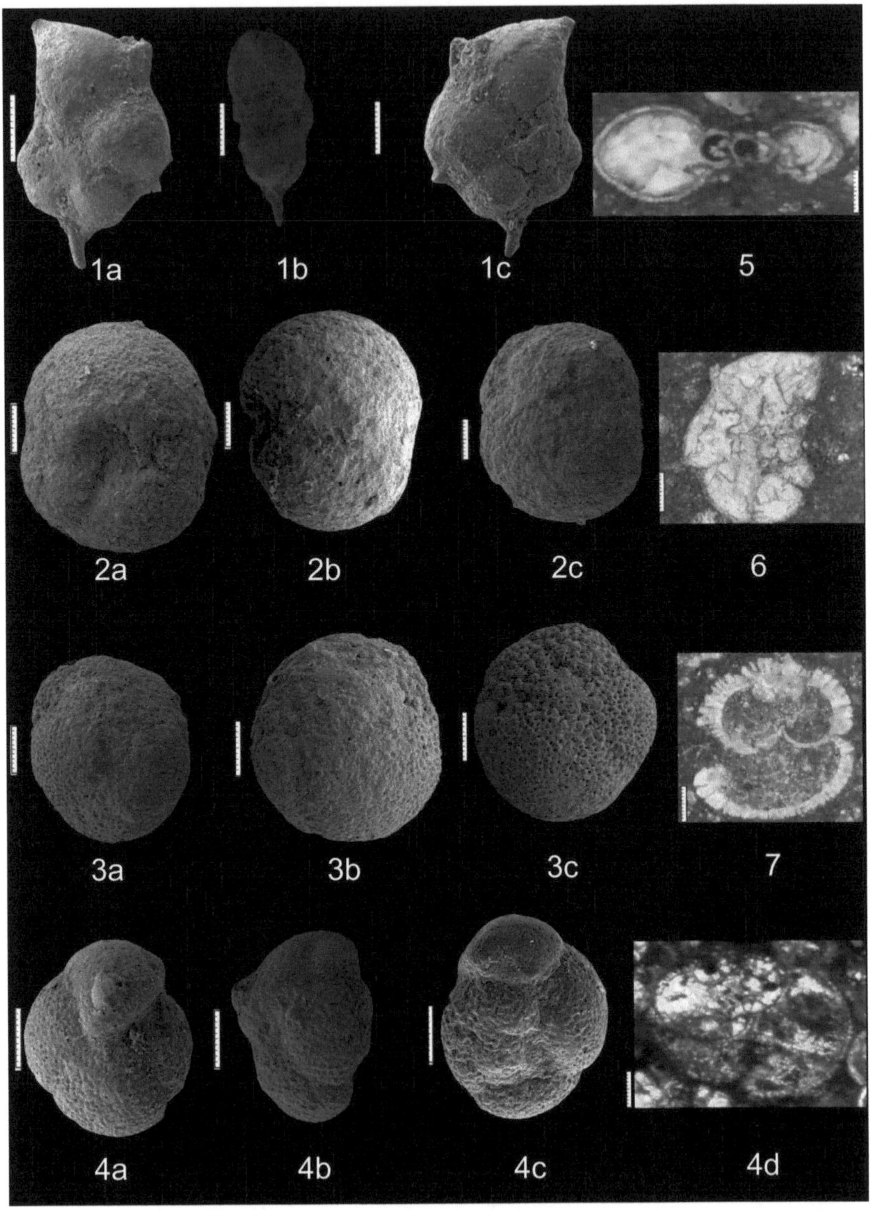

Plate. 4.1 Selected Paleogene foraminifers of the south Iran, Izeh Zone (Hadavandkhani et al. 2018). 1a–c: *Hantkenina alabamensis*; 2a–c: *Globigerinatheka index*; 3a–c: *Globigerinatheka seminvoluta*; 4a–d: *Turborotalia ampliapertura*; *Pseudohastigerina naguewichiensis*; *Hantkenina alabamensis*; *Globigerinatheka index*

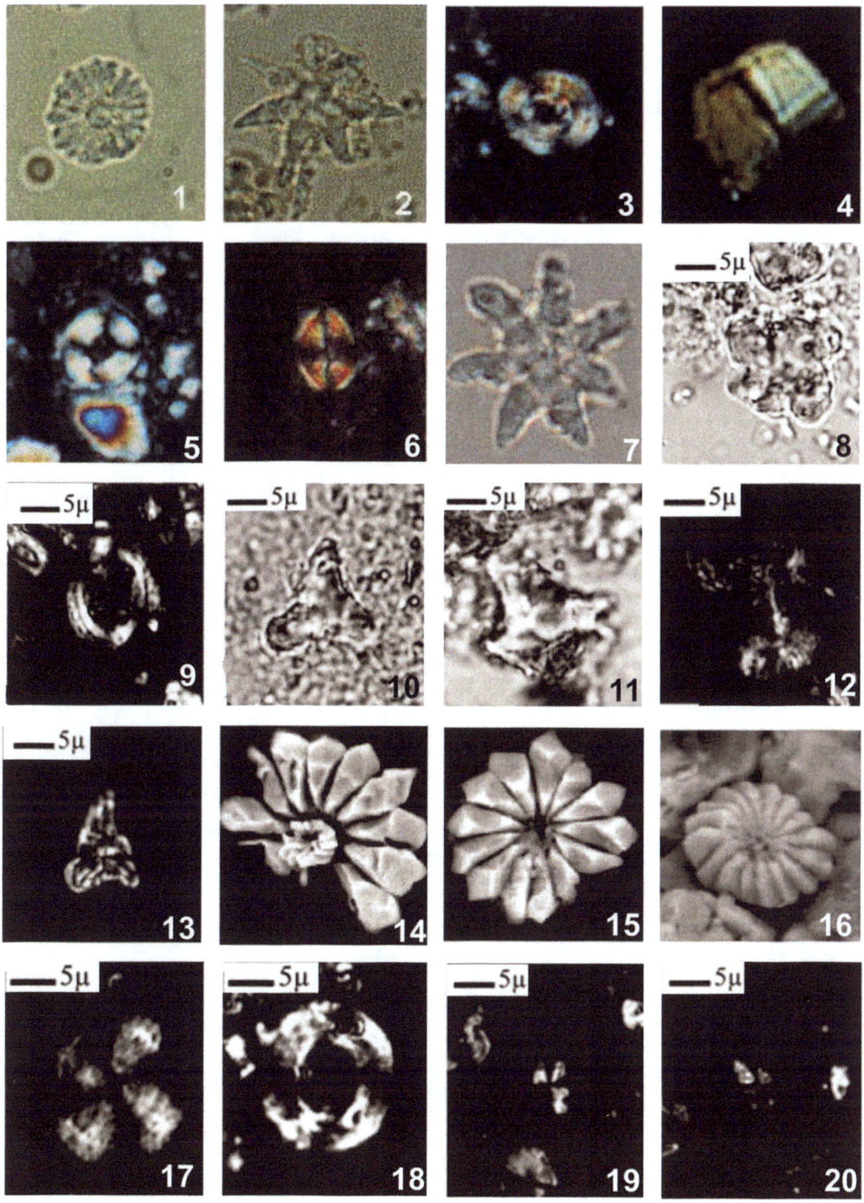

Plate. 4.2 Selected Paleogene nannofossils of the Zagros Zone (Figs. 1–7, 14, 16 adapted from Khavari Khorasani et al. 2013 and Figs. 8–13, 17, 20 from Ahifar et al. 2015). 1. *Discoaster multiradiatus*; 2. *Discoaster lodoensis*; 3. *Reticulofenestra hillae*; 4. *Fasciculithus tympaniformis*; 5. *Ericsonia Formosa*; 6. *Pontosphaera enormis*; 7. *Discoaster saipanensis*; 8. *Nannotetrina fulgens*; 9: *Chiasmolithus solitus*; 10. *Tribrachiatus orthostylus*; 11. *Tribrachiatus contortus*; 12. *Ellipsolithus macellus*; 13. *Sphenolithus pseudoradians*; 14–15. *Discoaster barbadiensis*; 16. *Discoaster wemmelensis*; 17: *Heliolithus Kleinpelli*; 18: *Reticulofenestra umbilicus*; 19–20: *Sphenolithus spiniger*

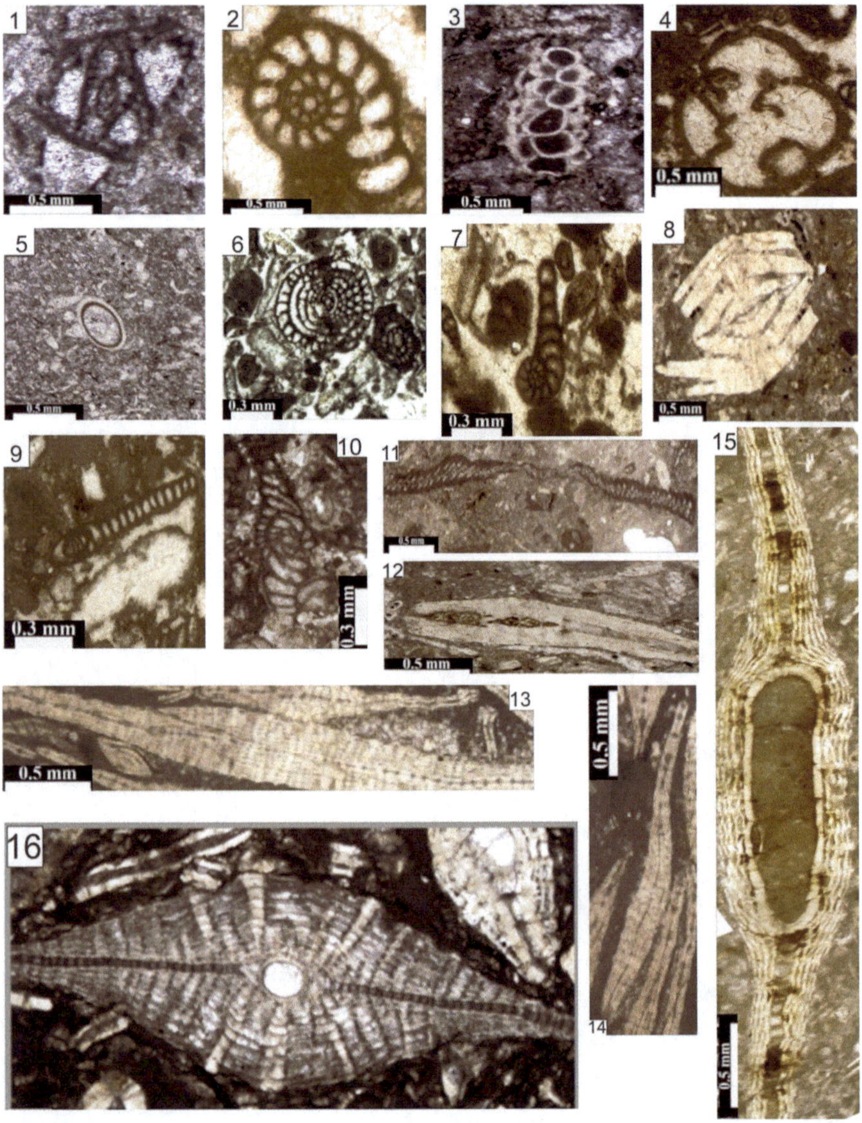

Plate. 4.3 Selected Oligo-Miocene foraminifers of the Zagros basin (Roozpeykar and Moghaddam 2016). 1. *Austrotrilina* sp.; 2. *Dendritina rangi*; 3. *Miogypsina* sp.; 4. *Valvulinid* sp.; 5. *Ditrupa* sp.; 6. *Borelis melo*; 7. *Spirulina* sp.; 8. *Nummulites* sp. cf. *N. viscus*; 9. *Peneroplis* sp.; 10. *Peneroplis thomasi*; 11. *Meandropsina iranica*; 12. *Operculina* sp. cf. *O. complanata*; 13. *Spiroclypeus blankenhorni*; 14. *Spiroclypeus blankenhorni*; 15. *Eulepidina* sp.

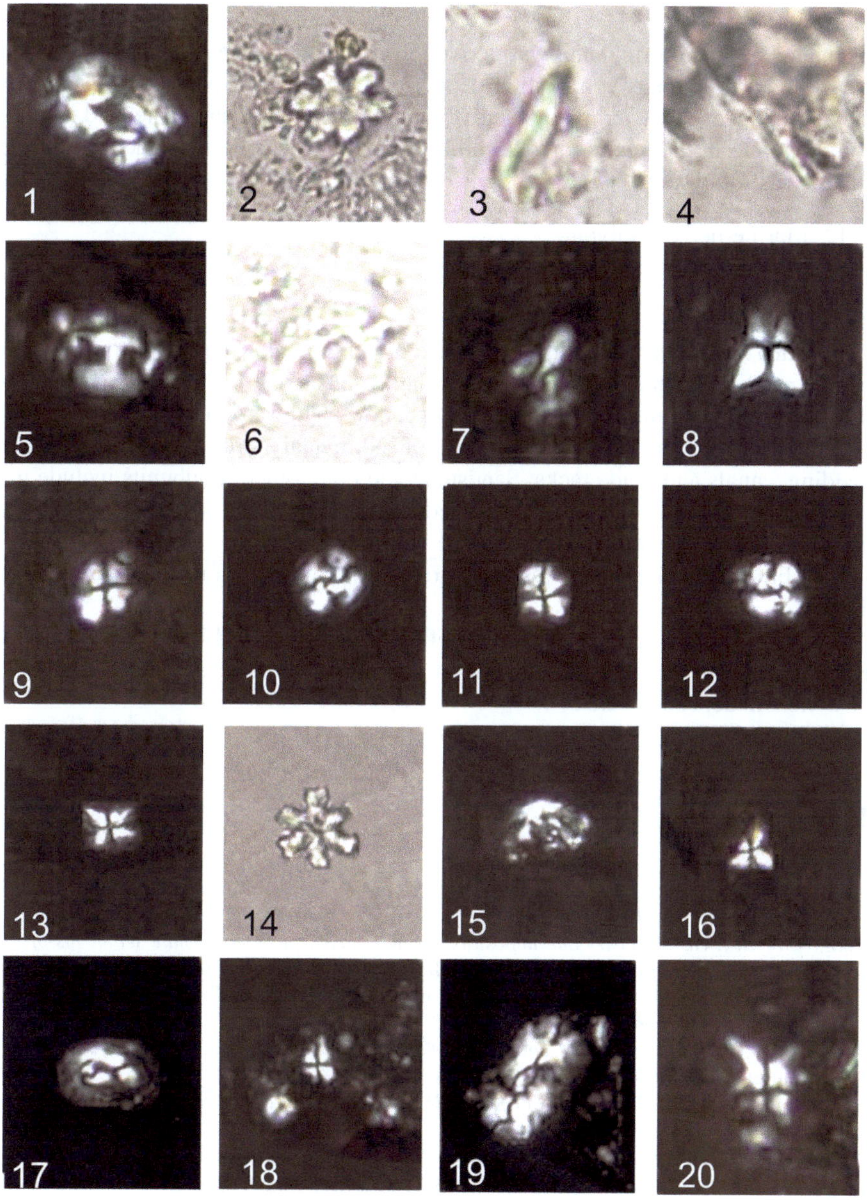

Plate. 4.4 Selected Neogene nannofossils of the Zagros Zone (Senemari 2016). All figures light microghraphs × 2000 to × 2500. 1. *Helicosphaera ampliaperta*; 2. *Discoaster druggii*; 3–4. *Triquetrorhabdulus carinatus*; 5–6. *Helicosphaera recta*; 7. *Sphenolithus ciperoensis*; 8. *Sphenolithus belemnos*; 9. *Sphenolithus compactus*; 10. *Cyclicargolithus floridanus*; 11. *Sphenolithus dissimilis*; 12. *Cyclicargolithus abisectus*; 13. *Sphenolithus calyculus*; 14. *Discoaster deflandrei*; 15. *Heliosphere granulate*; 16. *Sphenolithus delphix*; 17. *Helicosphaera scissura*; 18. *Sphenolithus conicus*; 19. *Helicospheara euphratis*; 20. *Sphenolithus capricornutus*

Tahtoun Unit
The name of this informal unit was derived from the Tahtoun Range in the Taheroei area. This unit includes a transition of marine facies to the river facies, such that the marine layers are overlain by the river conglomerate and sandstone, and then by conglomerate in the upper part. Accordingly, the unit has been divided into three subunits as follows:

1. Marine-derived subunit
2. Sandstone and conglomerate subunit
3. Conglomerate subunit

The marine-derived subunit is composed of conglomerate and sandstone accompanied by shell fragment and foraminifer shell. Consisting of chert and quartz fragments, the components of conglomerate part show an upward increase of grading. It is generally characterized by the presence of sedimentary structures such as graded bedding, canals and mud cracks. Sandstone and conglomerate subunits include the pebble-bearing sandstone with large-sized fragments such as cobble and boulders, and subordinate polymictic conglomerate with rich matrix.

The lithostratigraphic features of Cenozoic rock units of Iran can be availed of Table 4.2.

The Fig. 4.9 illustrates a correlation of Oligo-Miocene and Pliocene of Iran but Kopeh-Dagh and the Plates 4.1, 4.2, 4.3 and 4.4 include selected age diagnostic fauna of Cenozoic deposits of Iran.

References

Abbasi, M. (2009). *Sequence stratigraphy of the Upper Dalan and Kangan Formations in some boreholes of the South Pars Oil Field, an explanation for the reservoir quality variation.*

Abdi, A. (1998). *Geochemistry and depositional environment of the carbonate sediments of the Shahbazan Formation in the Shabikhoun Section, northern Khorram-Abad and its correlation with the Darabi Section, northern Kuh-dasht*, Unpublished MSc thesis, Faculty of Earth Sciences, Shahid Beheshti University.

Adams, T. D. (1969). *The Asmari Formation of Lurestan and Khuzestan providences: National Iranian Oil Company*, Report 1151, unpublished.

Afshar-Harb, A. (1994). *The geology of Kopeh-Dagh, Geological Book codification project*, Geological Survey of Iran, p. 274.

Aghanabati, A. (2004). Geology of Iran. *Geological Survey of Iran, Report* (in Persian).

Ahifar, A., Kani, A. & Amiri-Bakhtiar, H. (2015). Nanofossil stratigraphy of the Pabdeh Formation in the Gurpi Anticline. *Earth Science Journal, 24*(95), 107–120.

Ahmadi-Pour, H. et al. (2003). *Genetic classification ofpyroclastic rocks of the upper Hazar Volcanic (Kerman Province)*, In: 22th Iranian Symposium of Earth Science.

Ainolahi, A. (2008). Diagenesis and its effects on the reservoir characterizations of the Asmari Formation in the Lali Oil Field, northwestern Masjed Soleyman, J*ournal of applicable geology, 6*(2), 118–128.

Alavi, M. (2004). Regional stratigraphy of the Zagros fold-thrust belt of Iran and its proforeland evolution. *American Journal of Science, 304*(1), 1–20.

Alavi-Naeini, M. (2009). *An introduction on the stratigraphy of Iran*, Geological Survey of Iran, 507.

Alimi, N (2005) *Biostratigraphy and lithostratigraphy of the Ziarat and Karaj Formations of the northwestern and northern areas of Shahroud* (Unpublished M.Sc. thesis, Faculty of Sciences, University of Tehran).

Aram, et al. (2003). *An introduction on the sequence stratigraphy of Asmari Formation of the Ramin Oil Field*, In: The 22th Symposium of Geology, Geol. Surv. Iran Rep.

Azimi, R. (1997). *Microfacies and depositional environment of the Qom Formation in the Jalayer Anticline, southwestern Saveh county* (Unpublished M.Sc. thesis, Faculty of Sciences, Kharazmi University).

Babazadeh, A. (1992). *Microstratigraphic studies of the Karaj Formation of the Lavasanat-Kalun Basteh, Central Alborz* (Unpublished M.Sc. thesis, Faculty of Earth Sciences, Shahid Beheshti University).

Baghbani, D. (1996). The study of complete Permian Section of northeastern Shahreza. In *The 9th symposium of Earth Science*, Geological Survey of Iran.

Bahrami-Samani, P. (1997). *The study of fish fossils of Pabdeh Formation, Babaheydar Section, Shahr-e-Kurd* (Unpublished M.Sc. thesis, Faculty of Sciences, Ferdowsi Mashhad University).

Baiat, S. (2006). *Lithostratigraphy and nannostratigraphy of Pabdeh Formation of the type section, Tang-e-Pabdeh, and its correlation to the Tang-e-Maghar Section, Banestan Mountain* (Unpublished M.Sc. thesis, Faculty of Earth Sciences, Shahid Beheshti University).

Bakhshi, A. (1998). *Petrography and sedimentology of Qom Formation of the Southwestern Zanjan, Saeidabad and Kersef Sections* (Unpublished M.Sc. thesis, Faculty of Earth Sciences, Shahid Beheshti University).

Bakhtiari, F. (2002). *Biostratigraphical and Lithostratigraphical studies of the Qom Formation, Beryeh and Mesrghan Sections, Northwestern Saveh* (Unpublished M.Sc. thesis, Faculty of Sciences, Kharazmi University).

Behbahani, R. (2006). *Microfaceis, organic geochemistry and depositional environment of bitumen-bearing strata of Pabdeh Formation, Western Ivan* (Unpublished M.Sc. thesis, Faculty of Sciences, Boalisina University of Hamedan).

Busk, H. G., & Mayo, H. T. (1918). Parsian Oil fields. *Jour Ins Petr Tech*, S(17).

Bozorgnia, F. (1966). Qom formation stratigraphy of the Central Basin of Iran and its intercontinental position. *Bulletin of the Iranian Petroleum Institute, 24*, 69–75.

Dadfar, H. (2003). *Microfaceis and microstratigraphy of the member E and F of the Qom Formation, Southeastern Kashan, Navab Anticline* (Unpublished M.Sc. thesis, Faculty of Earth Sciences, Shahid Beheshti University).

Daneshian, J., & Aftabi, A. (2010). Foraminiferal biostratigraphy of the Qom Formation, based on the new findings in the Navab Anticline Section, Southeastern Kashan. *Tehran Journal of Science, 35*(4), 137–154.

Daneshian, J., & Dana, L. R. (2007). Early Miocene benthic foraminifera and biostratigraphy of the Qom Formation, Deh Namak, central Iran. *Journal of Asian Earth Sciences, 29*(5), 844–858.

Dayer, N. (2010). *The study of larger benthic foraminifer accumulation as the indicator of the paleoenvironment of the Asmari Formation in the Mardouk area, Northeastern Kazeroun*. Unpublished MSc thesis, Payam-e-Nour University.

Fanaei, S. (2010). Calcareous nannofossils of the Gurpi-Pabdeh boundary in the Kangan area (Unpublished M.Sc. thesis, Paiam Nour university of Bojnourd).

Faramarzian, A. (2004). *Petrofacies analysis of the Ziveh Formation, Moghan Plain and their application in the sequence stratigraphy* (Unpublished M.Sc. thesis, Faculty of Sciences, University of Tehran).

Feshki, A. A. (2003). *Biostratigraphical characters of the Mishan Formation in the Gachsaran area, based on the calcareous nannofossils* (Unpublished M.Sc. thesis, Faculty of Earth Sciences, Shahid Beheshti University).

Furrer, M. A, & Soder, P. A. (1955). The Oligo–Miocene marine formation in the Qom region (Central Iran). In *Proceedings of the 4th World Petroleum Congress*, Rome, Section I/A/5, 267–277.

Ganji, M. (1998). *Petrography and microfacies of the Qom Formation in the Northeastern Takab quadrate* (Unpublished M.Sc. thesis, Faculty of Earth Sciences, Shahid Beheshti University).

Ghafrani, A. (2001). *Reservoir quality of the Asmari Formation in the eastern part of Ahwaz Oil field* (Unpublished M.Sc. thesis, Faculty of Sciences, University of Tehran).

Gharabeygi, Gh. et. al. (2008). *Structural analysis and priority of the western Kopeh-Dagh, Subsurface Branch*. National Iranian Oil Company, No 2156 GR.

Gholamifard, A. (2008). *Nannostratigraphy of the Pabdeh Formation in the Gachsaran Oil Field* (Unpublished M.Sc. thesis, Faculty of Earth Sciences, Shahid Beheshti University).

Ghorbani, M. (1999). *Petrological analysis and metallogeny of the Tertiary-Quaternary magmatic rocks*. Geological Survey of Iran Report, p. 263.

Ghorbani, M. (2010). *Palobiogeography and compressional sequence stratigraphy of the Asmari and Qom Formations*.

Ghorbani, M. (2019). *The magmatism and metamorphism of Iran*, in preparation.

Hadavandkhani, N. (2009). *Lithostratigraphy and biostratigraphy of Pabdeh formation in the Kamestan Anticline, Northeastern Izeh, and its correlation with type section*, Unpublished MSc thesis, Faculty of Earth Sciences, Shahid Beheshti University.

Hadavandkhani, N., Sadeghi, A. & Tahmashbi-e Sarvestani, A. (2018). Lithostratigraphy and biostratigraphy of the Pabdeh Formation in the Chaardeh area, Izeh Zone.*Earth Science Journal of Shahid Beheshti University, 107*, 137–150.

Hajian, A. (1996). *Cenozoic rocks of Iran*. Geological Book Codification Program. Geological Survey of Iran Report, p. 395.

Hashemi, H. (2012). Biostratigraphy of the Asmari Formation in the borehole No 281 of the Maroon Oilfield. In *The 16th symposium of geology*, Geological Society of Iran.

Hazavehei, F. (2002). *Reservoir quality evaluation of the Asmari Formation in the Danan Oilfield*. Unpublished MSc thesis, Faculty of Sciences, University of Tehran.

Heydari, A. (2010). *Lithostratigraphy and nannostratigraphy of the Pabdeh Formation of the Ahwaz Oil Field* (Unpublished M.Sc. thesis, Faculty of Earth Sciences, Shahid Beheshti University).

Homayun-Zadeh, S. (2002). *Lithostratigraphy and biostratigraphy of the Mishan Formation in the central Dezful Embayment*. Unpublished MSc thesis, Faculty of Sciences, Kharazmi University.

Hosseini, S. R. (2011). *Microfacies and depositional environment of the Asmari Formation of the Masjed-Soleyman Oil Field*. Unpublished MSc thesis, Bu-Ali Sina University.

Hovaeizavi, A. (1998). *Depositional environment of the Gurpi and Pabdeh Formations, based on the geophysical analysis of the surface and undersurface samples, and determination of their boundary, Dezful Embayment*. Unpublished MSc thesis, Faculty of Earth Sciences, Shahid Beheshti University.

Huckriede, R., Kürsten, M., & Venzlaff, H. (1962). *Geotektonische Kartenskizze des Gebietes zwischen Kerman und Sagand, (Iran)*. Bundesanstalt für Bodenforschung.

ICS (2015). http://www.stratigraphy.org/index.php/ics-chart-timescale.

Jahanpanah, A. (2003). *The study of reservoir sediments of Cheleken 'Series' of the southern part of Caspian Basin* (Unpublished M.Sc. thesis, Faculty of Sciences, University of Tehran).

Karimian, A. (2009). *Sequence stratigraphy and diagenesis history of Kalat and Pastaligh formations in the Shams Anticline, Daregaz* (Unpublished M.Sc. thesis, Faculty of Sciences, Ferdowsi Mashhad University).

Karimi-Mossadegh, Z. (1999). *Biostratigraphy, lithostratigraphy and depositional environment of the Qom Formation in the Saveh area* (Unpublished M.Sc. thesis, Faculty of Earth Sciences, Shahid Beheshti University).

Karimpour, B. (2009). *Petrography and depositional environment of Jarhrom Formation in the Sarkhon Oil Field (Bandar Abbas) and type section of Jahrom Formation (Southern Jahrom)*. Unpublished MSc thesis, Faculty of Earth Sciences, Shahid Beheshti University.

Keshavarzi, S. (1997). *Biozonation of the Qom Formation based on the foraminiferal family Lepidocyclinidae and Alveolinidae* (Unpublished M.Sc. thesis, Faculty of Sciences, Ferdowsi Mashhad University).

Lotfi, M. (2002). *Zonation of Asmari Formation in the Dorud Oil Field* (Unpublished M.Sc. thesis, Faculty of Sciences, University of Tehran).

Mafi, A. (1997). *Nannostratigraphy of Chehlkaman Formation in the reference and type sections* (Unpublished M.Sc. thesis, Faculty of Sciences, Ferdowsi Mashhad University).

Mafi, A. (2003). Calcareous nannofossils and age determination of Neogene rocks of Arsang Mountain, northwestern Kashmar, In: *The 4th symposium of geology*, Geological Survey of Iran.

Mahbubi, A. (2000). *Petrology, depositional environment, geochemistry and diagenesis of the Upper Paleocene Chehelkaman Formation, Kopeh-Dagh basin* (Unpublished M.Sc. thesis, Faculty of Sciences, Ferdowsi Mashhad University).

Metallogenic map of the Middle East (2011). Geological Survey of Iran, Report.

Moeini, A. (1998). *Sedimentology, ecology and microfaunstic of the Persian Gulf's shoreline, from the Bandar-Magham though Bandar-Nakhl Taghi* (Unpublished M.Sc. thesis, Faculty of Earth Sciences, Shahid Beheshti University.

Mohammadi, E., Safari, A., Vaziri-Moghaddam, H., Vaziri, M. R., & Ghaedi, M. (2011). Microfacies analysis and paleoenviornmental interpretation of the Qom Formation, South of the Kashan, Central Iran. *Carbonates and Evaporites, 26*(3), 255.

Mohseni, H. (2003). *Depositional environment and diagenesis of the Pabdeh Formation in the Northern Dezful Embayment* (Unpublished Ph.D. thesis, Faculty of Sciences, University of Tehran).

Moradpour, P. (2009). *Microstratigraphy of the Eocene deposits and Razak Complex in the Northern Baft, Kerman*. Unpublished MSc thesis, Faculty of Earth Sciences, Shahid Beheshti University .

Motiei, H. (1993). *Stratigraphy of Zagros, Geological Book Codification Project*, Geological Survey of Iran.

Mozaffari, B. (1999). *Biostratigraphy and lithostratigraphy of Qom Formation and lithostratigraphy of Upper and Lower Red Formations in the Northwestern Takab, Sheikh-Jaber-Moshampa and Andabad* (Unpublished M.Sc. thesis, Faculty of Earth Sciences, Shahid Beheshti University).

Naeiji, M. (1997). *Depositional environment and diagenesis of the Qom Formation in the northern Iwanaki Section*. Unpublished MSc thesis, Faculty of Sciences, University of Tehran.

Najafi, D. (1998). *Microstratigraphy of the Pabdeh Formation in the Northeastern Dezful Embayment* (Unpublished M.Sc. thesis, Faculty of Earth Sciences, Shahid Beheshti University).

Nasehi, A. (1992). *Microstratigraphy and micropaleontology of Karaj Formation in the Kan-Soulghan areas, Imamzadeh Davud, Keloun Basteh County* (Unpublished M.Sc. thesis, Faculty of Earth Sciences, Shahid Beheshti University).

Nezafat, A. (2009). *Foraminiferal biostratigraphy of the Asmari Formation in the southern flank of Khoweiz Anticline, northeastern Behabad* (Unpublished M.Sc. thesis, Faculty of Sciences, Ferdowsi Mashhad University).

NIOC. (1959). *Explanatory notes to the Geological Map of Iran, scale 1: 2,500,000*, Tehran.

NIOC. (2011). A research on the paleogeography, biostratigraphy and sequence stratigraphy of the Qom (Central Iran) and Asmari Formations (Zagros), Tehran.

Omidpour, A. (2003). *Diagenesis history of Asmari Formation, Kupal Field, and its effects on the reservoir quality* (unpublished Ms.C thesis, Shahid Beheshti University.

Pahlavanpour, B. (1997). *Microbiostratigraphy and depositional environment of Qom Formation in the Khoy and Tasouj Sections* (Unpublished M.Sc. thesis, Kharazmi University).

Pourabrishami, Z. (1993). *Stratigraphical, sedimentological and structural studies of Qom Formation of the Azarbaijan, and its correlation to that of Central Iran*. Research Project, University of Tabriz.

Poursalehi, F. (2004). *Biostratigraphy of the Qom Formation in the northwestern Semnan (M.Sc. thesis)*. Kharazmi University.

Rahimzadeh, F. (1994). Oilgocene, Miocene and Pliocene of Iran, Geological Book Codification Project, Geol. Surv. Iran Rep. no 12.

Rahimi-Matin, L. (2007). *Sequence stratigraphy and biostratigraphy of the Qom Formation of southern Varamin, Davazdah Emam, Yorteh Shah* (Unpublished M.Sc. thesis, Faculty of Sciences, University of Tehran).

Ramazani, Sh. (2009). *Sedimentary facies and history of the Pesteligh Formation in the Chehel Kaman and Tang-e-Neyzar areas*, Unpublished MSc thesis, Faculty of Sciences, Ferdowsi Mashhad University.

Ramezani-Dana, L. (2003). *Biostratigraphy of Qom Formation of the Garmsar area, Northern Dehnamak* (Unpublished M.Sc. thesis, Faculty of Sciences, Kharazmi University).

Ranjbaran, M. (2009). *The diagenesis history of the Asmari Formation in the northern Dezful Embayment* (Unpublished Ph.D. thesis, Faculty of Sciences, Kharazmi University).

Reuter, M., Piller, W. E., Harzhauser, M., Mandic, O., Berning, B., Rögl, F., et al. (2009). The Oligo/Miocene Qom Formation (Iran): evidence for an early Burdigalian restriction of the Tethyan Seaway and closure of its Iranian gateways. *International Journal of Earth Sciences, 98*(3), 627–650.

Richardson, R. K. (1924). The geology and oil measures of south-west Persia. *Journal of the Institute of Petroleum Technology, 10*, 256–283.

Roozpeykar, A., & Moghaddam, I. M. (2016). Benthic foraminifera as biostratigraphical and paleoecological indicators: An example from Oligo-Miocene deposits in the SW of Zagros basin. *Iran, Geoscience Frontiers, 7*(1), 125–140.

Saadat, K. (2005). *Archie coefficient determination of the several petrofacies of Asmari Formation in the Ghachsaran Oil Field* (Unpublished M.Sc. thesis, Faculty of Sciences, University of Tehran).

Sadeghi, R. (2009). *Biostratigraphy, Microfacies and sequence stratigraphy of Asmari Formation in the Fars area* (Unpublished Ph.D. thesis, Faculty of Sciences, Esfahan University).

Sadeghi-Saadatlou, G. (2003). *Petrography and depositional environment of the Asmari Formation in the southern flank of the Bangestan Anticline and its correlation with those of adjacent fields located in the Dezful Embayment* (Unpublished M.Sc. thesis, Faculty of Sciences, Bu-Ali Sina University).

Safari, S. (2010). *Petrography and geochemical studies of the igneous rocks of the upper part of the Karaj Formation in the Zanjan area* (Unpublished M.Sc. thesis, Faculty of Sciences, Zanjan University).

Samavi, A. (2009). *Lithostratigraphy and biostratigraphy of the Chehel, Champeh and Mol Members of Gachsaran Formation of the type section, Gachsaran Mountain, northwestern Bandar-e-Lengeh* (Unpublished M.Sc. thesis, Faculty of Earth Sciences, Shahid Beheshti University).

Sartipzadeh, S. (2002). *Microstratigraphy and microfacies of Oligo-Miocene Qom Formation of Khoy Section, Northwestern Azarbaijan* (Unpublished M.Sc. thesis, Faculty of Earth Sciences, Shahid Beheshti University).

Senemari, S. (1998). *Biostratigraphy of the Khangiran Formation, based on the calcareous nannofossils, Eastern Chehel Kaman Syncline* (Unpublished M.Sc. thesis, Faculty of Sciences, Ferdowsi Mashhad University).

Senemari, S. (2016). Biostratigraphy of the Qom Formation deposits based on calcareous nannofossils in the south east Kashan. *Iranian Journal of Earth Science, 48*, 121–145.

Shaiesteh, M. (2006). *Biostratigraphy of Asmari-Jahrom Formations of the Nargesi Oil Field and its correlation with the Kuh-e-Gisan Section, Borazjan area* (Unpublished M.Sc. thesis, Faculty of Sciences, Kharazmi University).

Shemirani, M. (1991). Cenozoic rocks of Iran. Handbook of Shahid Beheshti University, p. 32 p

Soltanian, N. (2012). *Biostratigraphy and microfacies of Asmari Formation of Bavara Anticline, Southeastern Firozabad* (Unpublished M.Sc. thesis, Faculty of Sciences, University of Esfahan).

Stöcklin, J., & Setudehnia, A. (1991). Stratigraphic Lexicon of Iran. Tehran, Iran: Geological Survey of Iran, Report No. 18.

Tabrizi, G. (2001). *The study of oil potential of Cenozoic deposits, Pabdeh Formation, of the Kabirkuh area, Lorestan* (Unpublished M.Sc. thesis, Industrial University of Esfahan).

Tahmasebi-Sarvestani, A. (2010). *Bio-Lithostratigraphy, sequence stratigraphy and depositional environment of the Asmari Formation in the western Dezful Embayment* (Unpublished Ph.D. thesis, Faculty of Earth Sciences, Shahid Beheshti University).

Thomas, A. N. (1948). *The Asmari limestone of Southwest Iran AIOC*, Report 705 (Unpub.).

Torabi, H. (1994). *Petrography and depositional environment of the Asmari Formation of Gardbisheh-Emamzadeh Hamzehali, Broojen-Taghanak.* Research Project, University of Esfahan.

Vaziri-Moghaddam, H., Kimiagari, M., & Taheri, A. (2006). Depositional environment and sequence stratigraphy of the Oligo-Miocene Asmari Formation in SW Iran. *Facies, 52*(1), 41–51.

Yazdi-Moghadam, M. (1995). *Biostratigraphy of Amiran and Tale-Zang in the Lorestan basin, Zagros Mountain* (Unpublished M.Sc. thesis, Faculty of Sciences, Ferdowsi Mashhad University).

Zakerian-Moghaddam, M. (2009). *The interpretation of the depositional environment and provenance of the siliciclastic beds of Pastaligh Formation of the Paskamar and Shourijeh areas* (Unpublished M.Sc. thesis, Faculty of Sciences, Ferdowsi Mashhad University).

Zakerzadeh, M. (1998). *Lithofaceis determaining of the Gahrum Formation of the southwestern Dezful Embayment and its stratigraphical connection with those of Fars area* (Unpublished M.Sc. thesis, Faculty of Sciences, Kharazmi University).

Zohdi, A. (2007). *Geochemistry, diagenesis and depositional environment of the carbonate deposits of the Taleh-Zang Formation at the type section, Langar Anticline (Dezful) and its correlation with subsurface section of Kialou (Dehloran)* (Unpublished M.Sc. thesis, Faculty of Earth Sciences, Shahid Beheshti University).

Chapter 5
Plio-Quaternary

The upper Pliocene rocks in Iran cannot be distinguished from those of lower Quaternary owing to the same dominant environment conditions. In this context, some geological events started from the late Pliocene and continued through the early Quaternary such as sedimentation of Hezardarreh Formation of the Alborz and Bakhtiari Conglomerate of the Zagros whose lower parts has been assigned to the late Pliocene and the upper Part to the early Quaternary. Most of the Quaternary volcanism also have started since the late Pliocene. As a consequence, before the introduction of Quaternary rocks of Iran, that of Plio-Quaternary will be discussed in below.

During the late Pliocene till the early Quaternary, Iranian microcontinent experienced many changes, caused by Pasadanian orogenic phase, resulting in the recent geological features of Iran. Following this phase, some lakes dried up, and the lands and seas formed as they are recently observed. In the late Pliocene and early Quaternary four geologically important events occurred in the Iran:

1. Continuing of the tectonic movements and resulted formation of elevated range of northern and Iran (Alborz and Zagros mountain ranges, respectively) accompanied by the formation and uplifting of hillside and inter-mountain plains.
2. River floodings, due to the rough morphology and heavy rains, and its resulted formation of thick and wide fan, and hillside sediments.
3. Intense seismic activities, due to sever tectonic activities and their development in these times.
4. Relatively wide volcanic activities in the some parts of Iran contributing in the local morphology, formation of the volcanic rocks of Quaternary and its related development of travertine rocks.

The mentioned events had such the significant results that are not comparable to what occurred in the present time of Iran. It is assumed that in Plio-Quaternary, the Mountains of the Zagros, Sahand and Sabalan were much higher than the present time. The high topographic dip in one hand and the climate pattern of the Early Quaternary in the other hand resulted in the torrential rains and flooding the river, confirmed by the wide distribution and high thickness of deposits and alluvial, and

M. Ghorbani, *Lithostratigraphy of Iran*, Springer Geology,
https://doi.org/10.1007/978-3-030-04963-8_5

their low textural maturity (Hezardarreh Formation of Northern Iran, and Bakhtiari Conglomerate and detached alluvial deposits of the southern Iran).

The distributions of the Quaternary lands are relatively higher than those of other time units, so that they include in all of the piedmonts and plains of Iran. The study of Quaternary deposits is believed to plays significant roles in the geology of Iran, because the most of Iranian roads, ground water and agricultural lands are directly related to the Quaternary lands. Furthermore, most of the urban and rural buildings have been constructed on the Quaternary deposits as well the many ore deposits are covered by the same deposits.

The Quaternary lands can be divided into following classes:

The Quaternary formations whose their lower parts were deposited in the Late Pliocene are mostly folded, coarse-gained and are represented by foothills.
The Quaternary formations are weakly folded and are mostly characterized by the foothills and plain-shape.
The Quaternary lands being plain-like morphology and clay, sand and silt composition
The Fluvial terraces without specified grading and folding
Aeolian sediments and desert dunes
Travertine rocks deposited by the hot or cold water formed around of the Neogene-Quaternary magmatic areas.

5.1 Alborz

Southern Alborz

The Plio-Quaternary sediments of southern Alborz are mainly distributed in the Tehran surroundings; accordingly the Plio-Quaternary of Tehran will be discussed hereafter.

Tehran and environs have been constructed on the Quaternary alluvial sediments, firstly studied by Rieben (1953, 1955, 1960, 1966) and then other geologists (e.g. Knill and Jones 1968; Drecsh 1961; Huber 1960; Bassir 1971; Vita-Finzi 1969). Approximately, all of these geologists adapted the lithostratigraphical classification of Rieben (1955–1966) and so far no significant modifications were applied.

All of the alluvial deposits exposed in the southern Alborz to southern Tehran, Shahre-Rai, Eslam-Shahr, and in the Eastern Tehran to Jajrud (Babaei and Tehran-Pardis Highways) as well as those deposits cropped out in the Western Tehran to Karaj, have resulted from the activities of river and seasonal floods originated from the southern slopes of Alborz. Generally, the alluvial deposits of Tehran have been divided into five formations as follows (from old to young order):

1. Alluvial Hezardarreh Formation ("A" Formation)
2. Inhomogeneous alluvial formation of northern Tehran (northern Tehran "B" Formation) and clayey siltstone Kahrizak Formation (southern Tehran "B" Formation)
3. Alluvial Tehran Formation ("C" Formation)
4. Alluvial Khorram-Abad Formation in the western Ivanaki
5. Recent Alluviums ("D" Formation).

"A" Formation (Hezardarreh Formation)

The oldest alluvial deposits of the Tehran area is called by Rieben (1955) to Hezardarreh Formation or "A" Formation, exposed in the Kan, Evin, Gheytarieh, Abbas-Abad, Ghochak defile and Tehran-Pardis highway pass as the form of high hills or deep valleys.

The Hezardarreh Formation is an extensive flood deposit of a river originated from the northeastern Tehran Plain and from the northern Alborz mountains. It has flowed towards the southern and southeastern Tehran, simultaneous with the latest phase of Alborz uplifting. The thickness of the formation most probably tends to decrease southward.

The Hezardarreh Formation (1200 m-thick) is lithologically composed of the light grey, highly cemented, homogenous conglomerate. It locally contains clayey and sandy lens and generally appears as the form of hills with deep valleys. The layer dip may reaches to 90°.

The evident characters of the formation are as follows:

1. High thickness (approximately 1200 m) and homogeneity
2. Regular bedding locally bearing clayey and sandy lenses
3. Average size of grading (10–25 cm) and the advanced roundness
4. Light grey color
5. High layer dip and folding.

Semi-rounded pebbles, whose 90% of components were derived from the Karaj Formation and 10% from the rocks of other ages. This indicates that during the sedimentation of alluvial Hezarderreh Formation, the Alborz Mountains experienced the uplifting and subsequent erosional phase. In this context, one can compare the Hezarderreh Formation with Bakhtiari Formation, formed during the uplifting of the Zagros (Pilgrim 1908; James and Wynd 1965).

The Hezarderreh Formation, affected by folding, faulting as well as high erosional process, is overlain by the inhomogeneous alluvial formation of the northern Tehran, northern area of Shahid Beheshti University of the eastern Darakeh.

In the southeastern Tehran, the Hezarderreh Formation overlies the Upper Red Formation without any angular unconformity but with gradual contact, confirmed the attribution of the formation to Pliocene-Quaternary.

"B" Formation (Inhomogeneous Formation of the Northern Tehran)

This formation is an unsolidated conglomerate, consisting of pebble, gravel and cobble, and sometimes fragments with several of meters in thickness. The matrix is composed of clay, silt and clay grains. In the southern Tehran, there are fine-grained deposits, called Kahrizak Silt, and compared to "B" Formation of the northern Tehran (Rieben 1955).

The "B" Formation is of the less distribution than "A" Formation, and is underlain by the latter formation. It was measured 60 m in thickness which is decreasing southward. Having layer dip about 15°, the "B" Formation is exposed in the Bagh-Feyz, Shahrak-e-Ghods, Shahid Beheshti University, Shamiran Road and Shemiran old road, southward to Abbas-Abad.

"C" Formation (Alluvial Tehran Formation)
Consisting of pebble, gravel and cobble grains as well as sand and silt matrix, the "C" Formation (till 60 m-thick) is an alluvial fan, characterized by an unsorted, homogenous and unconsolidated grading. Most of Tehran buildings have been constructed on this formation (Rieben 1955).

"D" Formation (Modern Alluviums)
Consisting of the uncemented, rounded gravel and cabble grains in the northern Tehran, and of the silt and clay in the southern Terhan, the "D" formation is the youngest sedimentary deposits of the recent river. The formation, also, includes horizontal layers with less than 10 m thickness.

The distinguishing of the "B", "C" and "D" formations is found to be difficult. Accordingly, they can be considered as the unconsolidated and unfolded sediments of greater Tehran under the name of "Major Tehran Formation".

Northern Alborz
Although, the Paleogene and Neogene rocks are not exposed in the northern Alborz; however, the unconsolidated Plio-Quaternary rocks can be observed in the Mazandaran, Guilan and Moghan Plain. The mentioned rocks tend to show lesser thickness and more fine-grained rocks in comparison to those of southern Alborz. Represented the lacustrine and marine sediments of the Plio-Quaternary of northern Alborz, a continental and roughly marine succession in the northern Alborz is called "Apsheron Formation" (Aghanabati 2004).

Apsheron Formation
One of the Quaternary formations whose type section is located in the Caspian basin, i.e. Apsheron peninsula, is "Apsheron Formation", extending as the narrow belt in the western Mazandaran, Sisangan area of the eastern Noshahar, and Moghan Plain and probably Guilan. Consisting of the limestone, grey and blue marl accompanied by carbon-bearing lenses, bivalve-bearing mud, calcareous-cemented sandstone, and ostracod-bearing oolitic limestone and intrabasinal conglomerate, the Apsheron Formation tends to increase thickness towards the Royan area. The Apsheron Formation (724 m-thick) is underlain by the Aghchehgil Formation., and is overlain by lumashel limestone subsequently overlain by the Baku Layers with disturbed contact resulted from the activity of fault. The youngest unit of the formation in the Mazandaran is a succession of marly sandstone and fossiliferous limestone containing bivalve fossil such as *Dreissensia* sp., *Dessensia polimorfa*, *Dressensia rostriformis*; ostracod

species *Limnocyther*; and gastropods such as *Lipthoghyhus pylumida* and *Micromelania griimi*. The depositional facies of the Apsheron Formation of the area is assumed to be an alternation of the alluvial, meandering river, floodplain, deltaic and lacustrine facies. The facies analysis and its correlation with recent depositional environments indicate that the Apsheron formation was deposited in a lacustrine. The marine regression and transgression, caused by the tectonically active environment, resulted in the formation of alternation of deltaic, alluvial and fluvial-flood facies. The fluvial-deltaic facies show poor textural maturity owing to the low distance from the source rock to sedimentary basin. The formation was attributed to the early-late Pleistocene in age.

5.2 Zagros

A most lithologically important Plio-Quaternary formation of Zagros is "Bakhtiari Formation", which is observed approximately in all of the Zagros Zone especially its mountain high-slopes where the Asmari and calcareous Cretaceous formations are exposed.

Bakhtiari Formation

Its name is derived from the Bakhtiari Tribe. Its type section is located in the Kondar Defile in the northern Masjed-Soleyman where the Karoun River deeply cut of the Bakhtiari Formation and forms a narrow defile (Pligrim 1908). The Bakhtiari Formation is underlain by the Aghajari Formation with a sharp angular unconformity; however, in the some areas the lower contact is vague. It is affected by the erosional process, and in some places is covered by the modern sediments. Although the Bakhtiari Formation is more consolidated than Hezar darreh Formation, nevertheless, it is not yet considered as a true conglomerate (Semi-consolidated) (Aghanabati 2004). The lower part of the formation is an alternation of very hard, resistant and massive conglomerate with conglomerate sandstone, microconglomerate and grainstone. The sandstone and grainstone, forming a special morphology (depression) due to the weathering, are, also, included in the conglomerate as the lenses shape particles.

Fig. 5.1 A view of the Bakhtiari Conglomerate, Bazoft Valley (northward view)

Fig. 5.2 A view of the Bakhtiari Conglomerate, Bazoft Valley (southward view)

The components of the conglomerate are mostly composed of the limestones and locally some cherty nodules, derived from the Sarvak, Ilam and Asmari formations. The matrix is mostly composed of more fine-grained components including sand, gravel and cobble, and sometimes calcareous grains. The component of the Gachsaran, Mishan and Aghajari formations (Fars Group Elements) cannot be traced in the Bakhtiari conglomerate (Figs. 5.1 and 5.2).

In the coastal Fars, the formation grades into the thin-bedded, progressive limey conglomerate, overlying the Aghajari Formation with an angular and sharp unconformity. Based on the stratigraphic position, the Bakhtiari Formation was assigned to late Pliocene- Pleistocene in age by the petroleum geologists (Motiei 1993).

The other Quaternary deposits of the Zagros, reaching their maximum thickness in the Khuzestan Plain, can be found in the plains such as Khuzestan, Mahdasht, Zahab, Shiraz, Darab, Firzouzabad plains and those of Chaharmahle-Bakhtiari. In the other portions of Zagros, no modern formations can be observed but river flooding alluviums.

Although, the Zagros and Khuzestan plains are most drilled areas in the Iran, however, the Quaternary deposits of Zagros are rarely known, because they are not the targets of oil companies (Motiei 1993).

The geophysical studies of the Khuzestan plain indicate that the maximum thickness of the Quaternary deposits is concentrated in the Iraq-Iran boarders especially that of Hoveyzeh where it reaches 100 m in thickness. The minimum thickness (0–10 m), also, can be found in the marginal heights of plains.

5.3 Kopeh-Dagh

The Quaternary deposits of Kopeh-Dagh include water deposit, losses, terraces, alluvial fans and dues. The most distribution of the deposits, consisting mostly of unconsolidated and loss components including clay, silt and rubbles, can be observed in the Gorgan Plain, Sarakhs, Mashhad, Bojnourd, Darehe-Gaz and Ashkhaneh (Afshar-Harb 1994).

The losses of the Kopeh-Dagh are of the more thickness in the northeastern and northwestern areas, and cannot be observed in the central and southern Kopeh-Dagh. Its thickness may reaches to 50 m in the Atrak valley of the Korn Village vicinity. There is no bedding in the losses (Afshar-Harb 1994).

The terraces, mostly consisting of the relatively loose, unsorted, medium- to coarse-grained conglomerate with the matrix of clay, silts, sand and sometimes weak calcareous cement, are observed in the margin of the streams. The terrace surfaces are covered by the newly-formed, fine-grained sediments such as slit and clay

The dunes of the Kopeh-Dagh area are solely exposed in the northern Gorgan plain and Sarakhs Plain.

5.4 Central Iran and Sanandaj-Sirjan

The Plio-Quaternary deposits of the Central Iran and Sanandaj-Sirjan differ from those of Zagros and Alborz in terms of sedimentary and volcanic deposits, and specially their high diversification. Conglomerate deposits, resembling to the Bakhtiari Formation and even Hezardarreh Formation, form a part of Plio-Quaternary deposits of the Central Iran and Sanandaj-Sirjan; however, in the former area, they are rarely observed. The Plio-Quaternary deposits of the Central Iran are represented by the pediment and hillside alluvial fans, foothill and intermountain plain deposits.

Fig. 5.3 The Takht-e-Soleyman Travertine

The stream, lacustrine, aeolian and evaporate deposits, alluvial traces, Takht-e-Soleyman Travertine (Fig. 5.3) and pyroclastic and lava rocks are the components of Quaternary of the Central Iran and Sanandaj-Sirjan. There are many alluvial plains in these areas with unmeasured alluvium thickness, however, there is no doubt that the plains bounded by the mountains, if the boundary with mountains is faulted the thickness increases, and if not, it decreases. The lithology and morphotectonic of the plain surroundings control the type and size of the sediment components. In the following, some plains of Central Iran and Sanandaj-Sirjan are introduced:

Kabudarahang, Bahar-e Hamedan, Ghorveh, Asad-Abad, Kangavar, Malayer, Nahavand and Arak plains as well as those of around of Esfahan, Abadeh, Sirjan.

Generally, the plains located in the Sanandaj-Sirjan are characterized by the more fine-grained and low-dip topography, while those of Central Iran are accompanied by the variation of grading and rock type diversity.

The most important sediments of Central Iran plains, which have not previously well studied, are evaporite sedimentary deposits of the drying lakes (Hoz-Soltan of Qom and Gavkhooni) as well as the desert deposits of Dasht-e- Kavir and Lut Desert (Figs. 5.4 and 5.5).

Fig. 5.4 A view of the Pliocene conglmoerate along the Julfa-Poldasht road

Fig. 5.5 An angular uconformity resluted from Pasadanian tecctonic movement between Upper Red Formation and Pliocene conglomerate

5.5 Spatial and Temporal Correlation of Iranian Lithostratigraphic Units

Since 1960, many efforts have been done to reveal the facts of stratigraphy of Iran by geologists, such as Stöcklin (1971), who carried out many works in the all portions of Iran, the details of which published in the Stratigraphy Lexicon of Iran (Stöcklin

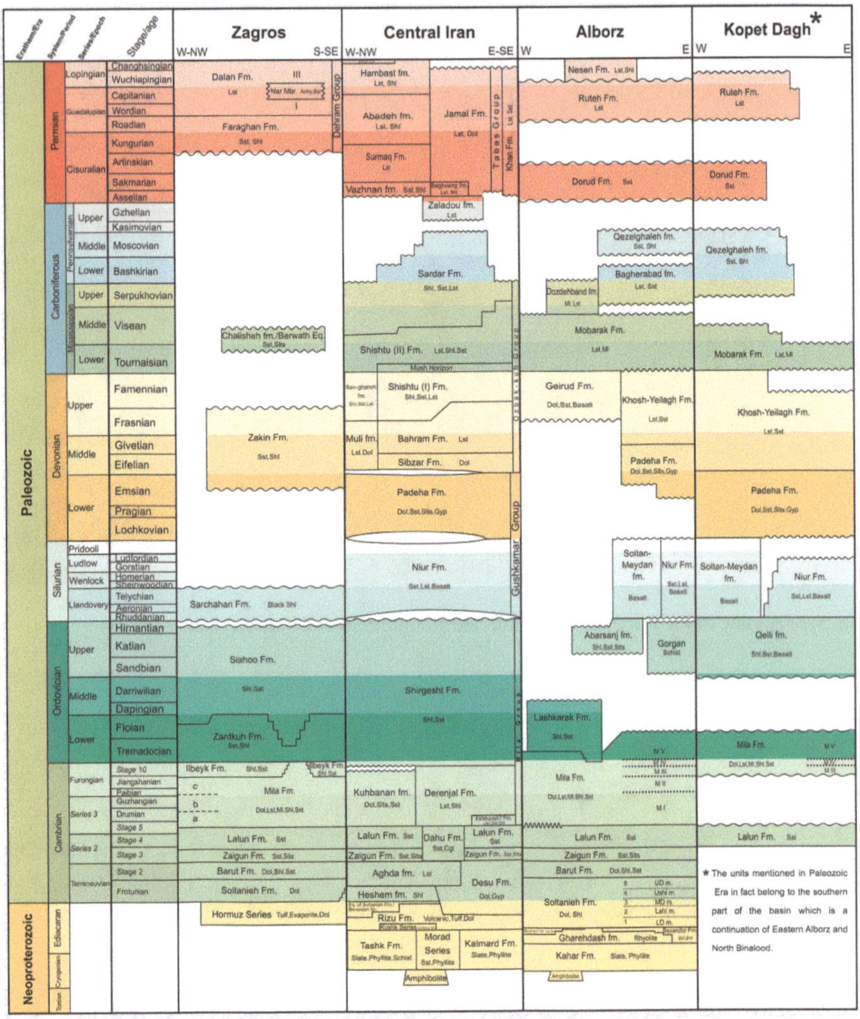

Fig. 5.6 Temporal- Spatial correlation of Late Proterozoic and Paleozoic of Iran (Ghorbani 2016)

1971). Subsequently, Aghanabati (2004) provided a relatively complete column of the lithostratigraphic units and their spatial and temporal relationship. Nevertheless, considering the new results and subsequent modifications of the Neoproterozoic to Paleozoic ages of lithostratigraphic units, the revised lithostratigraphic column is presented here. For comparison, the spatiotemporal stratigraphical columns of Cenozoic and Mesozoic, provided by Ghorbani (2016) which is the generalized compilation of stratigraphic results carried out by researchers, are presented in the Figs. 5.6, 5.7 and 5.8.

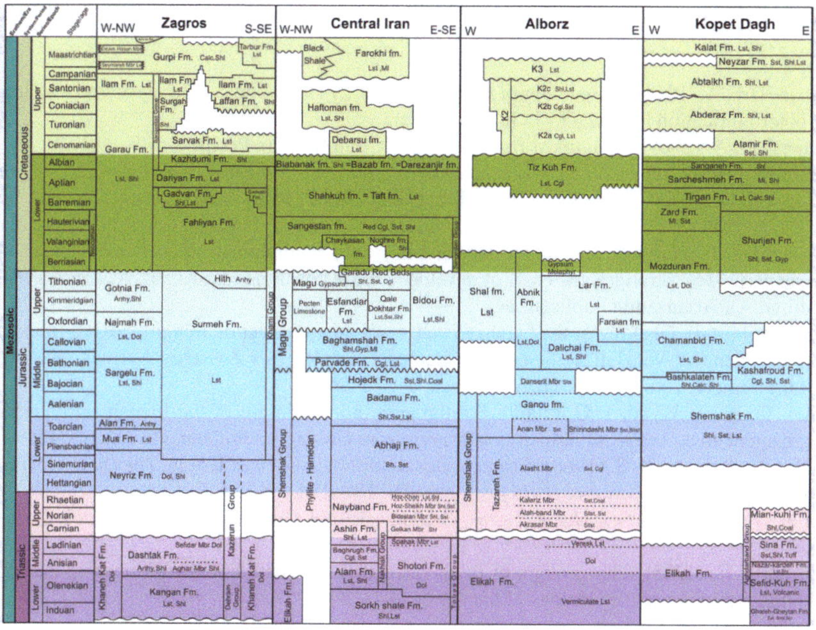

Fig. 5.7 Temporal-Spatial correlation of Mesozoic rock units of Iran (Ghorbani 2016)

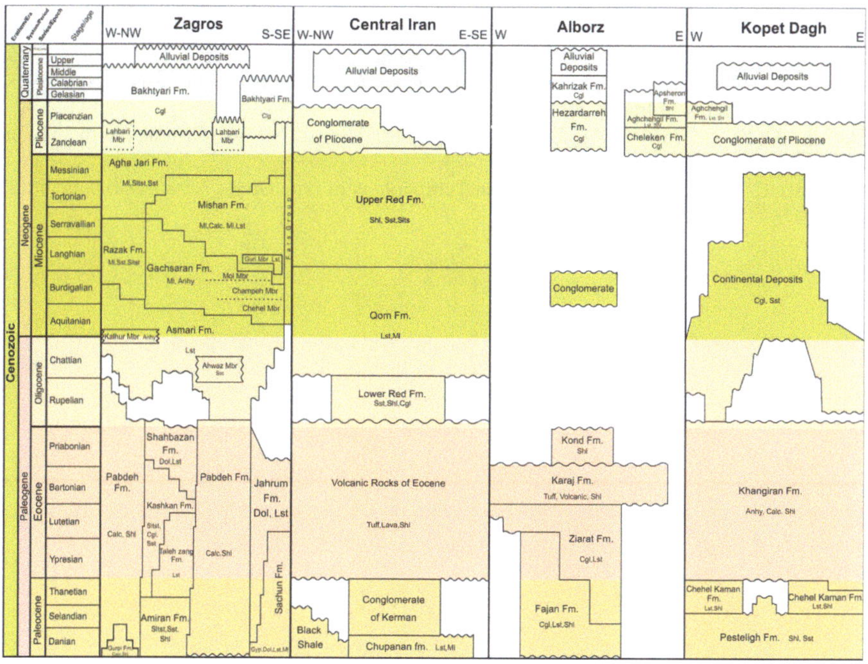

Fig. 5.8 Temporal-Spatial correlation of Cenozoic rock units of Iran (Ghorbani 2016)

References

Afshar-Harb, A. (1994). *The geology of Kopeh-Dagh* (p. 274). Geological Survey of Iran: Geological Book Codification Project. Geol. Surv. Iran Rep.

Aghanabati, A. (2004). *Geology of Iran.* Geol. Surv. Iran Rep. (in Persian).

Bassir, M. (1971). *Ingenieur geologische Baugrundunt ersuchungen in der Region Grofs-Tehran Iran*, (Ph.D. thesis). Rheinisch—Westfalische Technische Hochschule, Aachen.

Dresch, J. (1961). Le piedmont de Tehran, In observations de geographie physique en Iran septentrional: Centre Centre Documentation Cartographique et géographique, Mem et Docum, 8, 85–101.

Ghorbani. (2016). Stratigraphic chart of developed basins of Iran (eds: Aftabi, A, Zandkarimi, K, Ahifar, A), Arianzamin Publications.

Huber, H. (1960). Geology of the Farahzad new town area (district of Tehran), NIOC, Huckriede, R, Kuersten, M, Venzlaff, H (1962) Zur Geology des Gebietes Zwischen Kerman und Sagand (Iran) Beihefte zum Geologischen Jahrlmch, 51, 24–26.

James, G. A., & Wynd, J. G. (1965). Stratigraphic nomenclature of the Iranian oil consortium agreement area. *American Association of Petroleum Geologists Bulletin, 49,* 2182–2245.

Knill, J. L., & Jones, K. S. (1968). Groundwater condition in Greater Tehran. *The Quaterly Journal of Engineering Geology and Hydrogeology, 1*(3), 181–194.

Motiei, H. (1993). *Stratigraphy of Zagros.* Geological Book Codification Project, Geol. Surv. Iran Rep.

Pilgrim, G. E. (1908). The geology of the persian gulf and the adjoining portions of Persia and Arabia, Memoirs of the Geological Survey of India, 34, 177.

Rieben, E. H. (1953). Note preliminaire sur les terrains alluviaux de Tehran et particulierement du terriotoire de Shemran, Bulletin de la Laboratoire de Geologie Mineralogie et Geophysique, Musee Geologique de l'Universite de Lausanne, 105, 1–12.

Rieben, H. (1955). The geology of the tehran plain. *American Journal of Science, 253,* 617–639.

Rieben, E. H. (1960). *Les terrains alluviaux de la region de Tehran: Vol. 4.* (p. 41), Arid Zone, Research Center: Univiversity of Tehran, Publication.

Rieben, H. (1966). *Geological observation on alluvial deposits in Northern Iran* (Report No. 9). Geological Survey of Iran. p. 40.

Stöcklin, J. (1971). *Stratigraphic lexicon of Iran*, Part I, (Report No. 18). Central, North and East Iran: Geological Survey of Iran, p. 384.

Vita-Finzi, C. (1969). Late quaternary alluvial chronology of iran. *Geologische Rundschau, 58*(2), 951–973.

Annex: Geological Studies Conducted in Iran

A.1 Introduction

Most of the geological activities in Iran are carried out by governmental organizations particularly Geological Survey of Iran, and National Iranian Oil Company (Exploration Directorate), and the papers, M.S., and Ph.D. thesis published by individual researchers. Atomic Energy Organization deals with the uranium exploration, and most of water resources including surface and groundwater studies conducted by Ministry of Energy. Also, the topographical and geographical survey conducted by the Geographical Survey of Army and National Cartographic Center of Iran. The topographical produced maps in different scales are the basis for geological studies. The aerial photographs and satellite images are produced by Satellite Center of Iran.

The activities of these organizations related to the geology of Iran briefly are as follows.

Geological Studies by Geological Survey of Iran.

Geological studies to determine the events of Earth's crust are the main aims and objectives of the Geological Survey of Iran. So, the production of geological maps and related reports on different scales are the main task of this organization as follows:

Geological maps with a scale 1:250,000.

After the foundation of Geological Survey of Iran, in order to obtain comprehensive information about the general geological structure of Iran, rock and mineral potential, the main priority and efforts focused on geological mapping of scale 1:250,000 to comply with the strict international standards. The results are complete coverage of the country. Each geological quadrangle map (1:250,000) is published with a descriptive report. Each quadrangle area of approximately 15,000 square kilometers encompasses the range of 1.5° latitude and 1° longitude. 121 quadrangle maps (1:250,000) cover the entire surface of the country.

M. Ghorbani, *Lithostratigraphy of Iran*, Springer Geology,
https://doi.org/10.1007/978-3-030-04963-8

The National Iranian Oil Company conducts the geological maps (1:250,000) in oil-rich regions of southwestern Iran, so, the geological projects of the mentioned areas eliminated from the rest of country, and Geological Survey of Iran is responsible for the rest of areas of the country. Till now, all the field geological studies and maps preparation (1:250,000) of the entire country has been completed and published (Fig. A.1).

Each geological map (1:250,000) has a descriptive report which, is published separately. Regarding the regional and global application, most of these reports have been written in English. The report of some of these geological maps has not been finished yet.

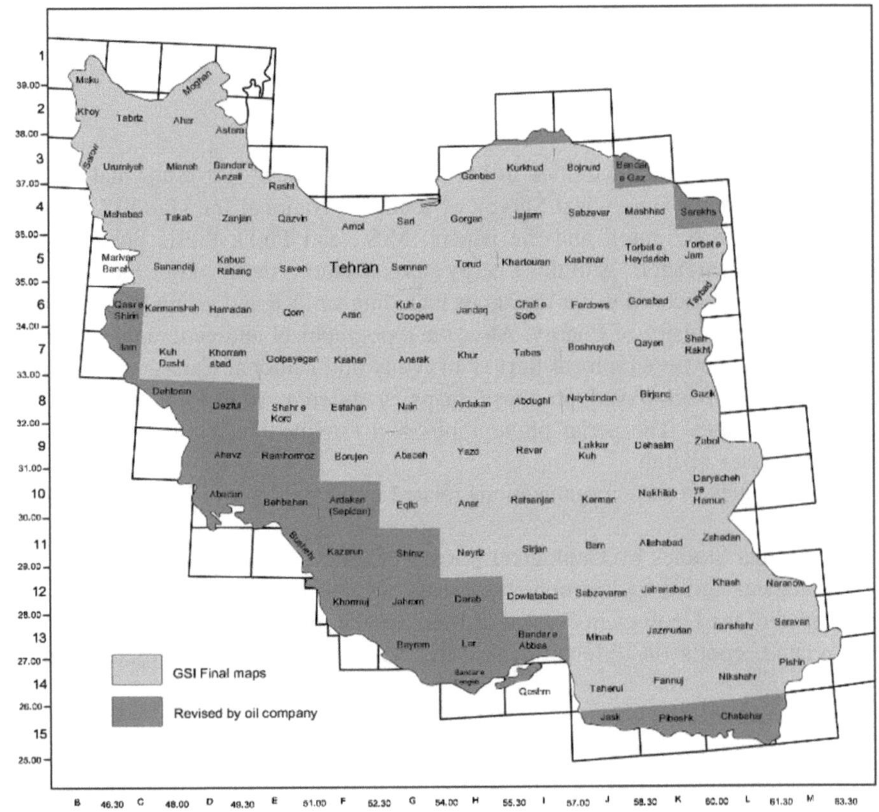

Fig. A.1 The final guideline of Geological Maps of Iran (1:250,000)

A.2 Geological Maps Scale 1:100,000

Along with the geological mapping (1:250,000), the other surveys and geological studies (1:100,000) have been conducted in the areas with high potential for the occurrence of mineral deposits to solve the key geological problems and development the national projects. The entire country is covered by 659 geological quadrangle maps (1:100,000). Each map is about 2500 km^2 located between two latitude and two longitudes encompasses half degree. A brief descriptive geological report printed on the marginal side of each geological map (1:100,000). The published and under preparation maps is shown in Fig. A.2. As Fig. A.2 shows, almost all 1:100,000 maps are prepared in priority sectors. Those quadrangle maps which are not in priority, have not been prepared include mostly desert zones.

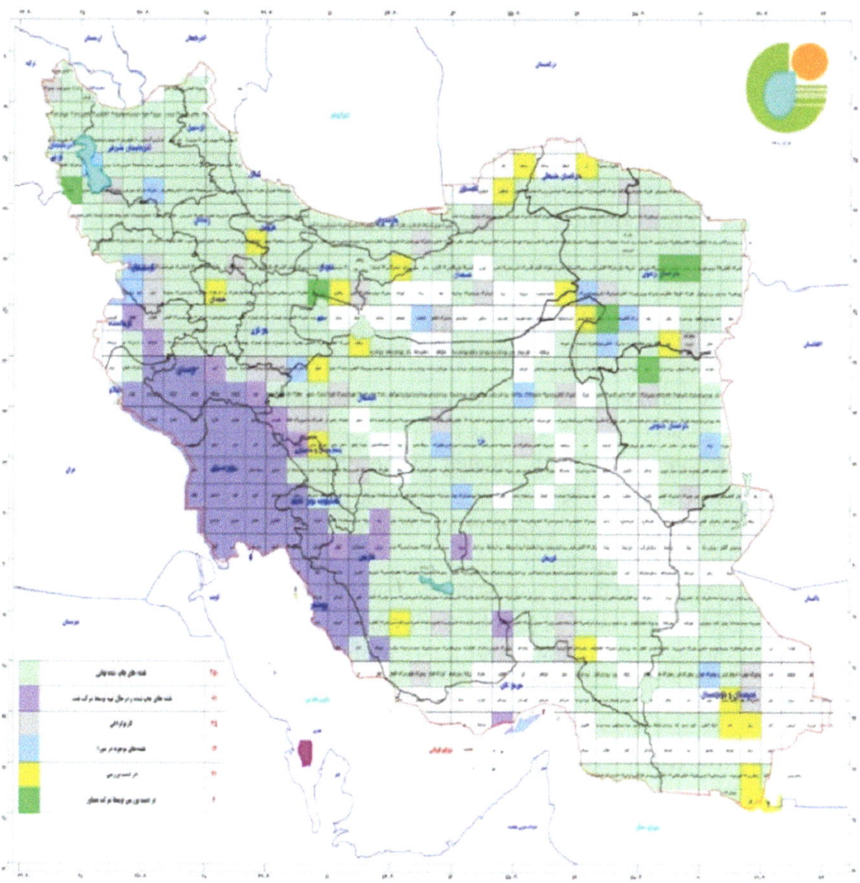

Fig. A.2 The latest status of maps 1:100,000

A.3 Geological Maps Scale 1:25,000

Geological Survey of Iran plans to cover the entire country by geological maps at scale 1:25000, to provide with information about the geological rock units, their age, structural elements, environmental information, and seismo-tectonic data in a GIS system. Conducting the program is based on the priority areas, where have the specific characteristic including mineral potential and environmental matters.

A.4 Airborne Geophysical Maps

In order to obtain more comprehensive information on regional geology and tectonics, as well as identifying areas for detailed exploration, especially hidden mineral deposits, Geological Survey of Iran plans to survey Airborne Geophysical Maps. To further ease the integration of geophysical and geological data, Airborne Geophysical Maps of 1:250,000 have been published in three colors. In addition to the above maps, the maps of the Magnetic Total Intensity 1:100,000 and Magnetic Anomaly maps 1:250,000; the Base Magnetic Map, and Magnetic Isopach Maps are prepared and published.

Also, Geological Survey of Iran plans to conduct Magnetic and Radiometric survey by intervals of 500 m and Electromagnetic survey by intervals of 250 m for priority zones of the country.

A.5 Geological Research Studies

In order to solve some of the key issues and developing the exploration geology, some parts of geological research has been allocated in this matter. Although some of these studies are not directly related to the objectives and tasks of Geological Survey of Iran, but could be helpful to provide increasing quantitative and qualitative studies, especially on Engineering Geology and have an important role for geological hazards investigation.

In this regard, a significant part of the researches and reports are the result of the geological exploration, and research studies in the framework of the current programs or in the form of regional and international cooperation.

A.6 Exploratory Studies

The most geological exploration program focuses on prospecting and exploration of Metallic and non-metallic mineral deposits. At the beginning of the activity, according to the needs of industrial markets and economic importance, most of the exploration programs were subjective exploration, but later exploration projects focused on industrial needs and economic necessities. In other words, it was necessary to concern, in addition to the thematic exploration, the systematic exploration program which, is of greater importance widely for the long-term development of the country. Therefore, the current aim of geological exploration tendency towards the regional systematic exploration is intended to conduct these studies particularly at 1:100,000 scale.

The geological exploration activities in Geological Survey of Iran are focused on three main factors:

Mining Exploration: Geological Survey of Iran has been successful to discover a number of valuable resources such as metallic and non-metallic deposits by implementation topical and systematic exploration projects, particularly, phosphate, bauxite, magnesite, alunite, perlite, nepheline syenite, diatomaceous earth, asbestos, feldspar, sand casting, refractory soil, celestite, and vermiculite as non-metallic deposits and Copper, Iron, Gold, Lead and Zinc, and other metallic ones that some of them have significant economic value.

It should be noted that since 1998 Geological Survey of Iran has been responsible for the exploration of the entire country. So the geological exploration and the systematic exploration increased widely.

Geochemical Explorations in Iran: Applying the geochemical exploration technique, along with the foundation of Heavy Minerals and Atomic Absorption Laboratories, goes back to the late 1962.

In early 1970s, in order to explore the mineral resources geochemical exploration have been done for copper, lead, zinc and molybdenum, particularly in forested areas in Northern Iran, west of Qazvin, Northeastern Khorasan province, central Iran, Azarbaijan and Zahedan. The final results are presented as prospecting, semi-detailed and detailed characteristics phases. During the same decade, the geochemical exploration program in central Iran in the range of 42,000 km^2 and eastern Iran (270,000 km^2) took place by cooperation of several foreign consulting engineers and companies.

In late 1980s, according to the economic policies of the Islamic Republic of Iran in understanding the mineral potential of the country, more emphasis was given to the Regional (1:100,000) Geochemical Exploration in, Tarom, Zanjan, Manjil, Ghareghosh, Ahar, Kharvana and other areas. In addition, the subjective studies of copper deposits in Sungun and Nojamyr were continued.

In 1990s the systematic geochemical studies in the Varcheh, Aran, Nofal-Lou Chateau, Khoy, shazand, Mahallat, and other areas were carried out, which only the results of Golpaiegan, Aligoodarz, Jebale-barez, Hana and other areas published during this period. At the present time, the entire geochemical maps of the entire country have been done and published.

Geological Survey of Iran plans to revise the previous geochemical maps, and conduct the hydro-geochemical maps at 1:100,000 scale.

The Geochemical Exploration Department of Geological Survey of Iran, with highly sophisticated facilities, equipment and advanced software, is active in the following areas:

1. Research and identify all valuable heavy minerals such as precious minerals (gold, silver), radioactive minerals (monazite), copper minerals (chalcopyrite, covellite, etc.), base metal minerals, lead and zinc, iron ore group, etc.
2. Qualitative and quantitative content of heavy minerals grades.
3. Determine the source of the minerals in relation to depositional environments.
4. Geochemical statistical calculations
5. Plotting the distribution elements anomalous single or multi-element maps and relevant reports
6. Design and implementation the technical specification of geochemical exploration projects at different scales
7. Determination the target area to follow up exploration including drilling and underground working.

A.7 Geophysical Explorations

Department of Geophysics of Geological Survey of Iran has a very constructive role in mineral Prospecting. The experts of the Geophysical Department apply the static methods (gravity, and magnetic radiometric) and dynamic (seismic, magnetic and electromagnetic) to investigate the physical properties of rocks for metallic and non-metallic mineral and ground water exploration and to solve the unknown geological matters. The Geophysical Department of Geological Survey of Iran has the most advanced geophysical equipment and software to provide not only services to conduct its own projects, but also provide services to the private clients.

In addition to geophysical maps preparation, the Geophysical Department of Geological Survey of Iran after managing to educate its scientific experts and provide the most technologically advanced software and hardware, start reprocessing the National Geophysical Data of the previous projects to provide Vertical Gradient, Residual, Horizontal Gradient, as shadowing or Image. Further analyzing and interpreting of rocks and depth of anomalies by applying 2D, 21/2D, and 3D methods will be conducted.

All geophysical applications such as gravity, magnetic survey, autogenic potential, radiation, induced polarization, and resistance VLF survey conducted by following the pre-compiled programs for the entire country, according to request of other client, governmental and private organizations. The entire above projects carry out by the geophysical department's equipment, and Implementation of international standards, practices in all field observations, calculations and mapping, run by the most sophisticated software programs.

A.8 The Geological, Exploration, and Research Projects

Some parts of the objectives and tasks of the Geological Survey are carried out in the framework of geological and exploration projects are implemented. So far, Geological Survey of Iran has been able to plan the exploration of evaporate salts deposits, geological studies Outside Makran Project, clay projects, Geochemical Exploration Projects, Seismic Atlas project, and the Systematic and Subjective Exploration Projects which achieve some parts of its goal in the field of geology and exploration.

A.9 Thematic Geological Maps

In addition to the geological maps 1:250,000, 1:100,000, and 1:25,000, Geological Survey attempts to publish some of the Subjective Geological Maps to show the structural geology, tectonics and exploration characteristic of Iran (Table A.1).

Based on the aims of "Compiling the Geological Book of Iran" project, a tremendous numbers of articles and books about the geology and mineral resources of Iran and thematic maps were developed and published, Tables A.2 and A.3.

In addition to focused and systematic studies of the Geological Survey, the Industries & Mines Organization carry out great numbers of exploration projects within their territories in different states (although it has not achieved satisfactory results).

In the framework of Masters and Ph.D. thesis, at the universities of the country many important research and studies on different subjects of geology of Iran are conducted each year.

On the other hand, AEOI (Atomic Energy Organization of Iran) has an exploration Department where particularly deals with geological exploration of radioactive materials.

A.10 National Iranian Oil Company

National Iranian Oil Company Exploration Management is a governmental organization performs particularly the geological studies and the surface and underground exploration for oil and gas in the entire country. Since the main exploration activities focused on oil and gas, the Zagros and Kopeh-Dagh zones are the main exploration target areas. But the rest of country is also considered for the second priority.

1. The major studies conducted by the National Iranian Oil Company are as following:
2. Geological mapping, in the Zagros region and the Kopeh-Dagh Basin, scale 1:250,000, 1:100,000 and in some specific areas 1:50,000.

Table A.1 Thematic maps published by the Geological Survey of Iran

Raw	Map name	Scale	Pablication date	Remark
1	Preliminary Metallogenic Map of Iran	1:2,500,000	1965	Finished
2	Central Lut Map	1:50,000	1972	Finished
3	Preliminary Seismic focuses Map	1:2,500,000	1973	Finished
4	Tectonic map of Iran	1:2,500,000	1973	Finished
5	Late Carboniferous geological map of Kerman	1:50,000	1973	Finished
6	Mining geological map of Kerman	1:50,000	1973	Finished
7	Geological Map of Biabanak–Bafgh	1:50,000	1973	Finished
8	Sismotectonic Map of Iran	1:500,000	1975	Finished
9	Mineral Distibution Map of Iran	1:2,500,000	1975	
10	Earthquake Risk Map	1:5,000,000	1977	
11	The Recored Intensity of Earthquakes Map of Iran	1:5,000,000	1977	
12	Intensity Earthquake zoning Map of Iran	1:5,000,000	1977	
13	Historical seismicity Map of Iran	1:5,000,000	1977	
14	Isograde Seismic Map of Iran	1:5,000,000	1977	
15	Structural and geological mapping of hard rock	1:500,000	1983	In two separate sheet
16	Metallogenic map of Anarak	1:50,000	1984	
17	Geological Map of Iran	1:2,500,000	1984	Second edition 1369
18	Tectonic map of Anarak	1:50,000	1984	
19	Postcards geology of Iran	1:10,000,000	1986	Second edition 1369
20	Postcards geology of Iran	1:27,000,000	1986	Second edition 1369
21	Geological map of the Middle East	1:5,000,000	1986	Second edition 1369
22	Metamorphism Map of Iran	1:2,500,000	1986	
23	The Total Magnetic Intensity Map of Iran	1:1,000,000	1989	In six separate sheets
24	Magnetic Anomaly Map	1:2,500,000	1989	
25	Postcards geological Middle East	1:27,000,000	1989	
26	Hot Mineral Waters Map of Iran	1:2,500,000	1990	
27	Postcards distribution of minerals	1:27,000,000	1990	
28	Seismotectonics map of the Middle East	1:2,500,000	1991	
29	Tectonics map of the Middle East	1:5,000,000	1991	
30	Metallogenic map of the Middle East	1:5,000,000	2010	

Table A.2 The books on geology, and mineral resources published by Compiling the Geological Book of Iran Project

Row	Book	Author
1	Coal Geology of Iran	Myr-Behzad Razavi Armghany , H. Moeinosadat
2	Sodium sulfate, talc and sulfur	Mohamad Dadkhah, Reza Jafari, Mojtaba Hijazi
3	Feldspar and Feldspathoid	H. Sheikhi Karyzky
4	Kaolin and refractory clays	M. Ghorbani, K. Arzani
5	Bentonite and zeolite	M. Hijazi, M. Ghorbani
6	Silica	Mansour Ghorbani
7	Iron ore	M. Ghorbani, Pezeshkpur, Jafar zadeh
8	Manganese Deposits of Iran	B. Aghaebrahim Samani
9	Phosphate	Hashem Halalat, M.H Bolorchi
10	Petroleum Geology	H. Motiei
11	Antimony, Arsenic, Mercury	Mansour Ghorbani
12	Copper deposits in Iran	N. Khoei, M. Ghorbani, P. Tajbakhsh
13	Barite deposits in Iran	A. Khoshjoy
14	Lead and Zinc in Iran	M. Ghorbani, P. Tajbakhsh, N. Khoei
15	Perlite and ball clay	Mansour Ghorbani
16	Evaporates	Farideh Helmi

Table A.3 Thematic maps developed by Compiling the Geological Book of Iran Project

Row	Thematic map	Scale
1	Geological Map of Iran	1:1,000,000
2	Seismotectonic Map of Iran	1:1,000,000
3	Mineral Distribution Map of Iran	1:1,000,000
4	Tectonics map of Iran	1:1,000,000
5	Mount hydrography Map of Iran	1:1,000,000
6	Magmatism Map of Iran	1:1,000,000
7	Geomorphological map of Iran	1:1,000,000
8	Magmatic rocks distribution map of Iran	1:2,500,000
9	Metamorphic rocks distribution map of Iran	1:1,000,000
10	climatology Map of Iran	1:2,500,000
11	Quaternary map of Iran	1:2,000,000

3. Preparation the first geological thematic map of Iran, scale 1:1,000,000.
4. Stratigraphic and Paleontological studies in many areas, especially in the Zagros, Kopeh-Dagh Basin, and more recently in central Iran, Makran and south of Caspian Sea.
5. The structural geology of the Zagros and the Kopeh-Dagh Basins.

Uncited References

Abaie, I., Ansari, H. J., Badakhshan, A., & Jaafari, A. (1964). History and development of the Alborz and Sarajeh fields of Central Iran. *Bulletin of Iranian Petroleum Institute, 15,* 561–574.

Abdi, A. (2008). *Geochemistry and depositional environment of the carbonate sediments of the Shahbazan Formation in the Shabikhoun Section, northern Khorram-Abad and its correlation with the Darabi Section, northern Kuh-dasht* (Unpublished M.Sc. thesis). Faculty of Earth Sciences, Shahid Beheshti University.

Abdi, N. (2003). Isotope and facies studies of the Lower Triassic Kangan Formation. In *The 22nd Symposium of Geology*. Geological Survey of Iran.

Abradat, M. (2001). Introduction of the wide out crops of Khangiran Formation of the Western Sarakhs Plain. In *The 20th Symposium of Geology*. Geological Survey of Iran.

Aftabi, A. (2003). *An introduction to sequence stratigraphy of Asmari Formation of the southeastern Kashan* (Unpublished M.Sc. thesis). Kharazmi University.

Ahia, F. (1997). Geology, geochemistry and source of Celestite-bearing horizon of the Gachasaran Formation, Likak, Behbahan, Kohgiluyeh and Boyer Ahmad Province.

Ahmadipour, H. (2003). Genetic classification of pyroclastic rocks of the upper Hazar Volcanic (Kerman Province). In *22nd Iranian Symposium of Earth Science*.

Ahmadzadeh-Heravi, M. (1971). Study of brachipods and conodonts of Alborz Central, 111.

Ain-Alahi, A. (2008). Diagenesis and its effects on the reservoir characterizations of the Asmari Formation in the Lali Oil Field, northwestern Masjed Soleyman. *Journal of Applicable Geology, 6*(2), 118–128.

Akhavan, H. (2011). *Evaluation of reservoirs characters of Shurijeh Formation (Member D1) of the Gonbadli Gas Field, Sarakhs area, Kopeh-Dagh Basin* (Unpublished M.Sc. thesis). Faculty of Sciences, University of Tehran.

Alahbakhsh-Ghiasvand, G. (2003). *Biostratigraphy and stratigraphical correlation of the Bangestan Group of the Sarkan and Malehkuh areas* (Unpublished M.Sc. thesis). Faculty of Sciences, University of Tehran.

Alavi, M., & Salehi-Ras, R. (1975). Geological Map of Damghan (scale of 1/100,000), Sheet No. 6862.

Alavi, M. (1991). Sedimentary and structural characteristics of the Paleo-Tethys remnants in North Eastern Iran. *Bulletin Geological Society of America, 103,* 983–992.

Alavi, M. (1996). Tectonostratigraphic synthesis and structural style of the Alborz Mountain system in Northern Iran. *Journal of Geodynamics, 21*(1), 1–33.

Alavi-Naeini, M., & Amidi, M. (1968). *Geology of Western Part of Takab Note, J49,* 94.

Alavi-Naini, M., Hajian, J., Amidi, A., & Bolurchi, H. (1982). *Geology of Takab-Saein Qalel: Explanatory note of 1:250,000 map of Takab quadrangl*. Geological Survey of Iran, Report 50.

M. Ghorbani, *Lithostratigraphy of Iran*, Springer Geology,
https://doi.org/10.1007/978-3-030-04963-8

Ali Nezhad-Kurdi, R. (2007). *Sedimentary petrography of the shale and marl beds of the Pabdeh Formation in the type section, and the borehole No 12, 38 of the Kupal Oil Field* (Unpublished M.Sc. thesis). Faculty of Earth Sciences, Shahid Beheshti University.

Allah Madadi, Sh. (1998). Biostratigraphical and Lithostratigraphical studies of Saran and Kand Formation. In *The 21st Iranian Symposium of Earth Science.*

Amin-Afshari, S. (2004). *Depositional environment, diagenesis and geochemistry of the carbonate deposits of the Sarvak Formation, Giskan and Mond Anticlines, Boushehr Province* (Unpublished M.Sc. thesis). Faculty of Earth Sciences, Shahid Beheshti University.

Amir Sardari, D. (1998). *Microfacies and depositional environments of the Members 3, 4 and 5 of Milla Formation, Eastern Alborz* (M.Sc. thesis). Faculty of Sciences, Kharazmi University.

Amiri-Bakhtiar, H., Sadeghi, A., Shemirani, A., Vaziri-Moghadam, H., & Safari, A. (1986). Lithostratigraphical and biostratigraphical revision of type section of Tarbur Formation, Gadivan Mountain, Shiraz, In *10th Symposium of Iranian Geological Society.*

Amiri-Bakhtiar, H., Sadeghi, A., Shemirani, A., Vaziri-Moghadam, H., & Safari, A. (1987). Introduction of Kuh-e-Chehelcheshmeh Section as the reference section of Tarbur Formation. *Journal of Sciences, Shahid Chamran University, 15,* 1–19.

Amirkhani, A. (2001). *Biostratigraphy of Tarbur Formation, southern Gandoman, based on the macrofossils* (Unpublished M.Sc. thesis). Faculty of Sciences, Kharazmi University.

Amirsardari, D. (2005). Microfaceis, depositional environment and diagenesis of Sarvak formation, Mokhtar Section, Borehole 1. In *The 21st Symposium of Earth Sciences.*

Angaji, M. (2000). Chemical evaluation and hydrocarbonic potential of the Mobarak Formation, Central and Eastern Alborz. In *The 4th Symposium of Iranian Geological Society.*

Aria-Nasab, M., Spina, A., Cirilli, S., & Daneshian, J. (2016). The palynostratigraphy of the Mississippian (middle Tournaisian–upper Viséan) Shishtu Formation from the Howz-e-Dorah section, southeast Tabas, central Iranian Basin. *Palynology, 40*(2), 247–263.

Asadi-Eskandar, A. (2006). *Permo-Triassic Boundary of South Pars Gas Field using lithological and geochemical data, and the study of its relation with reservoir quality of the related lithological units* (M.Sc. thesis). Faculty of Sciences, University of Tehran.

Asadi-Mehmandousti, E. (2005). *Geochemistry and diagensis of the Ilam Formation of the Abtimor and Mansouri Oil Fields, and Tang-e-Rashid section, Izeh area* (M.Sc. thesis). Faculty of Earth Sciences, Shahid Beheshti University.

Ashkan, F. (1982). *Early Cambrian age determination of the phosphate bearing member "Chapoghlu" of Soltanieh Formation.* Geological Survey of Iran.

Ashouri, A. (1990). *Devonian and Carboniferous conodont faunas from Iran* (Ph.D. thesis). University of Hall.

Ashouri, A. (1994). The stratigraphical position of members 1 and 6 of Khosh yeilagh Formation based on conodont fauna and introducing of three conodont zones from member 6. *Geosciences, Scientific Quarterly Journal, 4,* 64–83.

Ashouri, A. (1998). The Devonian-Carboniferous boundary in Ozbak-Kuh area. *Geosciences, Scientific Quarterly Journal, 7,* 47–53. (in Persian).

Ashouri, A. (2001). Middle Devonian-Mississippian conodont faunas from the Khoshyeilagh Formation, Alborz Mountains, north Iran, In: U. Jansen, P. Königshof, G. Plodowski, & E. Schindler (Eds.), *15th International Senckenberg Conference, Joint Meeting IGCP 421/SDS* (Abstracts, p. 6), May 2001.

Ashouri, A. (2004). Late Devonian and Middle– Late Devonian conodonts from eastern and northern Iran. *Revista Española de Micropaleontología, 3,* 355–365.

Ashouri, A., et al. (2010). Echionoids of the Tirgan Formation in the Takht-e-Arkan Section, southeastern Bojnurd and the introduction of new "*Toxaster renevieri*" for first time in Iran, In *The 12th Symposium of Geology.* Geological Survey of Iran.

Askari Pirbloufi, B. (1998). *Biostratigraphy of Upper Cretaceous strata, Tarbur Formation, in the Dorahan of the Chahar Mahal-e-Bakhtiari* (Unpublished M.Sc. thesis). Faculty of Sciences, Ferdowsi Mashhad University.

Asl-e-Shirin, F. (2011). *Nannostratigraphy of the Gurpi Formation in the Kuh-e-Sefid Section, Eastern Ramhormoz* (M.Sc. thesis). Shahid Beheshti University.

Assereto, R., & Ippolito, I. (1964). Observazioni preliminarai sul cretaceo della bassa valle del Lar (Elburz centrale, Iran).

Azimi, R. (2008). The introduction of the calpionellid biozones of the Jurassic-Cretaceous Boundary in the Shal section, southeastern Khalkhal county. *Earth Science Journal, 17*(68), 16–25.

Babakhani, A., & Rahimzadeh, F. (1988). Geological map of Ardabil (scale of 1:250,000). Supervised by: Eftekharnezhad, G, Geological Survey of Iran, Sheet No D2.

Badri-Kallou, N. (2007). *Nannostratigraphy of Gurpi Formation of the Lar area, Kuh-e-Gach and Kuh-e-Kurdeh Anticlines* (Unpublished M.Sc. thesis). Faculty of Earth Sciences, Shahid Beheshti University.

Bagheri, T. (1993). Geochemical study of carbonate deposit of Geirud Formation, southwestern Damghan. In *22nd Symposium of Geology*. Geological Survey of Iran.

Barzegar, F. (1989). The Razak Fault.Seasonal Letter in Geographical research, Tehran, Iran, 12, 72.

Bastami, & Bandpi, L. (2008). *Petrography, depositional environment and reservoir characterization of the Asmari Formation, south Gashvi Gas Field* (Unpublished M.Sc. thesis). Faculty of Earth Sciences, Shahid Beheshti University.

Berberian, M. (1973). *Two important deformational and metamorphic phases in the belt northeast of the Zagros thrust line (Iran). A brief structural review of the Sanandaj-Sirjan Belt.* Geological Survey of Iran, Intial Report 27, 35.

Berberian, M. (1984). Structural evolution and tectonics of the Iranian plateau, a plate tectonic approach, Bogdanov, N A, Tesizy, 27-Ymezhdunarodnyy geologicheskiy congress, Abstracts, 27th International Geological Congress. *Geological Survey of Iran, Tehran, 27*(3), 140.

Berberian, M. (1984b). *Recent tectonics, seismotectonics and earthquake-fault hazard study of the area West of Damghan (Amirabad), Kubankav Consulting Engineers.* Intial Report 45, 421 (in Persian).

Berberian, M., Qorashi, M., Arzhangravesh, B., & Mohajer-Ashtiani, A. (1983). *Recent tectonics, seismotectonics and earthquake-fault hazard study in the Greater at Tehran Region.* Iran: Geological Survey of Iran (in Persian).

Broumand, Z. (2008). *Palynofacies and depositional environment of the Dalichai Formation, northeastern Damghan, Talou Section* (Unpublished M.Sc. thesis). Geological Survey of Iran.

Chohen, S., & Finney, P.L. (2012). Gibbard International Commission on Stratigraphy.

Dakhili, M.T. (1995). *Petrology of magmatic and metamorphic rocks of the northeastern Fariman* (Unpublished M.Sc. thesis). Faculty of Earth Sciences, Shahid Beheshti University.

Darvishzadeh, A. (1992). *The analysis of the Ural, Iran, Madagascar and Fouron, and Hercynian Orogenic event.* Earth Science Quarterly, Geological Survey of Iran, 6.

Dayer, N. (2010). *The study of larger benthic foraminifer accumulation as the indicator of the paleoenvironment of the Asmari Formation in the Mardouk area, Northeastern Kazeroun* (Unpublished M.Sc. thesis). Payam-e-Nour University.

Dercourt, J., Zonenshain, L. P., Ricou, L. E., Kazmin, V. G., Le Pichon, X., Knipper, A. L., et al. (1986). Geological evolution of the Tethys belt from the Atlantic to the Pamirs since the Lias. *Tectonophysics, 123*, 241–315.

Diver, W., & Peat, S. C. J. (1979). On the interpretation and classification of Precambrian organic-walled microfossils. *Geology, 7*(8), 401–404.

Eftekharnezhad, J., & Stocklin, J. (1992). *Geological map of Iran sheet K8 (Birjand), scale 1: 250,000.* Tehran: Geological Survey of Iran.

Engalenc, M. (1968). *Geologie, geomorphologie, hydrogeologie de la region du Tehran (Iran)* (p. 180). Monpellier: These en sciences.

Ezzampanah, Y. (2011). *Biostratigraphy and Lithostratigraphy of the Garu Formation in the Kuzaran section and Naft borehole No 1 (southern Kermanshah)* (Unpublished M.Sc. thesis). Faculty of Earth Sciences, Shahid Beheshti University.

Feizi-Kararoudi, M. (2007). *Depositional environment and sequence stratigraphy of the Gadvan and Darian Formations of the northern Dezful Embayment (Mamatin borehole No. 10) and Izeh Zone (Ashgar anticline)* (Unpublished M.Sc. thesis). Faculty of Sciences, Kharazmi University.

Ghaderi, A., Agha, N. S. A., Hamdi, B., & Miller, J. F. (2008). Biostratigraphy of the First and Second members of type section of the Shirgesht Formation in north of Tabas with special emphasis on conodonts.

Gharabeigi, Gh., et al. (2008). Structural analysis and priority of the western Kopeh-Dagh, Subsurface Branch, Geological Office, National Iranian Oil Company, No. 2156 GR.

Ghavidel-Syooki, M. (2000). Palynostatigraphy and Palaeobiogeography of Lower Palaeozoic strata in the Ghelli area, Northeastern Alborz Range of Iran Kopeh-Dagh Region). *Journal of Sciences Islamic Republic of Iran, 11*(4), 305–318.

Ghavidel-Syooki, M. (2001). Palynostratigraphy and paleogeography of the Late Devonian in Northeastern Esfahan city, Central Iran. In D. K. Goodman, & R. T. Clarke (Eds.), *Proceeding of the IX International Palynological Congress*, Houston, Texas, USA (pp. 37–51). American Association of Stratigraphic Palynologists Foundation.

Ghavidel-Syooki, M. (2008). Palynostratigraphy and Paleogeography of upper Ordovician Gorgan Schists (Southern Caspian Sea). *Eastern Alborz Mountain Ranges, Northern Iran: Comunicocoes Geologicas, 95*, 123–155.

Ghavidel-Syooki, M., & Winchester, T. (2001). Biostratigraphy and palaeogeography of Late Ordovician chitinozoans from the Northeastern Alborz Range, Iran. *Review of Palaeobotany and Palynology, 118*, 77–99.

Ghavidel-syooki, M., & Vecoli, M. (2006). Latest Ordovician–early Silurian chitinozoans from the eastern Alborz Mountain Range, Kopet–Dagh region, Northeastern Iran: biostratigraphy and palaeobiogeography PALBO-02829, 20.

Ghazi, M., Hassanipak, A. A., Tucker, P. J., & Mobasher, K. (2001). Geochemistry and ^{40}Ar-^{39}Ar ages of the Mashhad Ophiolite, NE Iran, abstracts as: A Rare Occurrence of a 300 Ma (Paleo-Tethys) Oceanic Crust. American Geophysical Union, Fall Meeting 2001, abstract #V12C-0993.

Gheli-Netaj, M. (2005). Microfacies and depositional environment of the Elika Formation in the Bolbolkhan area, Amol. In *The 20th Symposium of Earth Science*. Geological Survey of Iran.

Gholamalian, H. (2002). *Biostratigraphy of Frasnian/Famennian Boundary in Esfahan & Tabas Areas* (p. 118) (Unpublished Ph.D. thesis). University of Isfahan.

Gohari, A. (2010). *Lithostratigraphy and nannostratigraphy of Gurpi Formation in the Siah Kuh Section, Southwestern Ilam* (Unpublished M.Sc. thesis). Faculty of Earth Sciences, Shahid Beheshti University.

Gorgij, M. N, & Boomeri, M. (1390). Two Carboniferous coral assemblage of Ozbak-Kuh and Kalmard areas, Eastern Central Iran (Zaladou and Gachal sections). *Sedimentology and Sequence Stratigraphy Journal, Isfahan University*.

Hadavand-Khani, N. (2009). *Lithostratigraphy and biostratigraphy of Pabdeh formation in the Kamestan Anticline, Northeastern Izeh, and its correlation with type section* (Unpublished M. Sc. thesis). Faculty of Earth Sciences, Shahid Beheshti University.

Hadavi, F. (2002). Ostracods of Qom Formation in the Do-Chah section. In *The 6th Symposium of Geology*. Geological Survey of Iran.

Hadi-Peykani, M. (2005). *Petrography and depostional environment of the Fahliyan Formation in the Izeh area* (Unpublished M.Sc. thesis). Faculty of Earth Sciences, Shahid Beheshti University.

Hajab, R. (1984). Stratigraphy of Member 7 of Gachsaran Formation. In *The Third Symposium of Geology*. Geological Survey of Iran.

Hajimohamadi-Shirehjini, N. (2010). *Lithostratigraphy of the Qom Formation of the Absard Area, Southern Damavand* (Unpublished M.Sc. thesis). Faculty of Earth Sciences, Shahid Beheshti University.

Hamdi, B. (1991). Stratigraphic position and correlation of Hormoz Complex in the Persian Gulf Paleontology Division. In *Canadaian Paleontology Conference I*, Vancouver, Canada.
Hamedi, M. (1988). *A revision of Shorm Beds*. Geological Survey of Iran.
Hazavehei, F. (2002). *Reservoir quality evaluation of the Asmari Formation in the Danan Oilfield* (Unpublished M.Sc. thesis). Faculty of Sciences, University of Tehran.
Hairapetian, V., Pour, M. G., Popov, L. E., Männik, P., & Miller, C. G. (2017). Silurian stratigraphy of Central Iran–an update. *Acta Geologica Polonica, 67*(2), 201–233.
Hofmann, H. J., & Jackson, G. D. (1994). Shale-facies microfossils from the Bylot Supergroup, Baffin Island, Canada. *Journal of Paleontology, 68*(4), 1–35.
Homayun-Zadeh, S. (2002). *Lithostratigraphy and biostratigraphy of the Mishan Formation in the central Dezful Embayment* (Unpublished M.Sc. thesis). Faculty of Sciences, Kharazmi University.
Hoseini-e Tabatabei, W. (2012). Lithology and depositional environment of the Lalun Formation of the Meskin and Shougolab Sections, Northwestern Urmia. In The *16th Symposium of Geology*. Geological Society of Iran.
Hoseini-Kheyrabadi, A. (2010). *The correlation of organic geochemistry data with those of sequence stratigraphy for hydrocarbon potential evaluation of the Pabdeh Formation of Ahwaz Oil Field* (Unpublished M.Sc. thesis). Faculty of Sciences, Shahid Chamran University.
Hoseinzadeh, M. (2003). *Lithostratigraphy and biostratigraphy of Shemshak Formation of the Eastern Shahmirzad, Southern Alborz, based on the ammonite fauna* (Unpublished M.Sc. thesis). Faculty of Earth Sciences, Shahid Beheshti University.
Hurford, A. J., Grunau, H. R., & Stöcklin, J. (1984). Fission track of an apatitecrystal from Hormuz Island, Iran. *Journal of Petroleum Geology, 7*(4), 365–380.
Jackson, M. P. A., Cornelius, R. R., Craig, C. H., Gansser, A., Stöcklin, J., & Talbot, C. J. (1990). Salt diapirs of the Great Kavir, central Iran. *Geological Society of America Memoir*, 177-139.
Jafarnezhad, D. (2008). *Paleoecology and palynostratigraphy of Gurpi Formation of the Tang-e-Bijar Section, Ilam Province* (Unpublished M.Sc. thesis). Faculty of Earth Sciences, Shahid Beheshti University.
Kabir, A. (2008). *Sedimentology and depositional environment of the coal-bearing sediments of Tabas* (Unpublished M.Sc. thesis). Faculty of Sciences, University of Tehran.
Kamel Abbasi, M. (2010). *Depositional environments and sequence stratigraphy of the Shurijeh Formation, Member D, in the Khangiran and Gonbadli Oil Fields* (Unpublished M.Sc. thesis). Islami Azad University, Shahroud Branch.
Kameli Azan, A., Amiri-Bakhtiar, & Vaziri-Moghadam, H. (2003). Biostratigraphy of Gurpi Formation, type section, in the northern Lali Oil field, Khuzestan. In *The 22nd Symposium of Geology*. Geological Survey of Iran.
Kapeaei, M. (2008). *Geological characterizations of the Qom Formation in the eastern and northeastern Hamedan* (Unpublished M.Sc. thesis). Bu-Ali Sina University.
Karimi-Mossadegh, Z. (2009). *Biostratgraphy, sequence stratigraphy and depositional environment of the Asmari Formation in the Central Dezful Embayment* (Unpublished Ph.D. thesis). Faculty of Earth Sciences, Shahid Beheshti University.
Karimpour, B. (2009). *Petrography and depositional environment of Jarhrom Formation in the Sarkhon Oil Field (Bandar Abbas) and type section of Jahrom Formation (Southern Jahrom)* (Unpublished M.Sc. thesis). Faculty of Earth Sciences, Shahid Beheshti University.
Kazmin, V. G., Sbortshikov, I. M., Ricou, L. E., Zonenshain, L. P., Boulin, J., & Knipper, A. L. (1986). Volcanic belts as markers of the Mesozoic-Cenozoic active margin of Eurasia. *Tectonophysics, 123*(1–4), 123–152.
Khazaei, A. (1995). *Systematic study of the Upper Cretaceous rudist-bearing limestone beds, type section of Kalat Formation, Tang-e-Neyzar, Kopeh-Dagh* (Unpublished M.Sc. thesis). Faculty of Sciences, Ferdowsi Mashhad University.
Kheradmand, A. (2002). Sedimentary facies and environment of the Middle and Upper rocks, equivalents of Bahram Formation, Kerman surroundings. In *The 6th Symposium of Geology*. Geological Society of Iran.

Kholousi, M. (1996). *Microbiostratigraphy and depositional environment of the Ziarat Formation, at the type section (southeastern Tehran) and Govije Section (Eastern Tehran and southeastern Tehran)* (Unpublished M.Sc. thesis). Faculty of Sciences, University of Tehran.

Kolodka, C., Vennin, E., Vachard, D., Trocme, V., & Goodarzi, M. H. (2012). Timing and progression of the end-Guadalupian crisis in the Fars province (Dalan Formation, Kuh-e Gakhum, Iran) constrained by foraminifers and other carbonate microfossils. *Facies, 58*(1), 131–153.

Kolosov, P. N. (1989). *Filamentous microorganism of Proterozoic terminal period*. Report at all-Union Meeting, general problems of subdivision of the USSR Precambrian, 1-47.

Kuhgard, H. (1994). *Lithostratigraphy and depositional environment of the Dahu 'Series' in the Northeastern Zarand* (Unpublished M.Sc. thesis). Faculty of Sciences, Kharazmi University.

Lasemi, Y. (1993). Depositional environments of the Geirud Formation in the Central Alborz area, Research project, Kharazmi University.

Lasemi, Y., & Raeesi, A. (2000). Microfacies, depositional environment and sequence stratigraphy of the Asmari Formation in the Southern Dezful embayment. In *The 4th Symposium of Geology*. Geological Society of Iran.

Less, G. M. (1929). Salzgletscher in Persia. *Mitteilungen Geologische, Wien, 24,* 29–34.

Lotfpour, M. (2002). Reservoir geology and sequence stratigraphy of the Dalan and Kangan Formations of the South Pars Oil Field, Persian Gulf. In The *6th Symposium of Geology*. Geological Society of Iran.

Mafi, A. (1997). *Nannostratigraphy of Chehlkaman Formation in the reference and type sections* (Unpublished M.Sc. thesis). Faculty of Sciences, Ferdowsi Mashhad University.

Mafi, A., Ghasemi-Nejad, E., Ashouri, A., & Vahidinia, M. (2014). Dinoflagellate cysts from the Upper Bajocian-Lower Oxfordian of the Dalichai Formation in Binalud Mountains (NE Iran): Their biostratigraphical and biogeographical significance. *Arabian Journal of Geosciences, 7* (9), 3683–3692.

Maghfouri-Moghaddam, A. (2000). *Lithostratigraphy, biostratigraphy and paleoecology of the Aghchegil and Apsheron Formations in the Moghan Plain* (Unpublished M.Sc. thesis). Faculty of Sciences, Kharazmi University.

Mahbubi-pour, H. (2004). *Geological and reservoir characterizations of the Kangan and Dalan Formations in the Kaman-Ghatar area, Persian Gulf* (Unpublished M.Sc. thesis). Faculty of Sciences, University of Tehran.

Mahdavipour, Z. (2003). *Sequence stratigraphy and reservoir quality of the Asmari Formation in the Shadan Oil Field* (Unpublished M.Sc. thesis). Faculty of Sciences, University of Tehran.

Majidi, B. (1981). The ultrabasic lava flows of Mashhad, Northeast Iran. *Geological Magazine, 118*(1), 49–58.

Malaki, M. (2001). *Microbiostratigraphy of Ziarat Formation in the Kalmard area* (Unpublished M.Sc. thesis). Faculty of Sciences, Kharazmi University.

Mirab-Shabestary, Gh. (1994). *Lithostratigraphy and depositional environment of the Precambrian-Lower Cambrian rocks, equivalent of Soltanieh Formation in the Sarbandan (eastern Damavand) and Shahmirzad (northern Semnan)* (Unpublished M.Sc. thesis). Faculty of Sciences, Kharazmi University.

Mirkiani, H. (1997). *Microfacies studies of the Tirgan Formation in the Mozdouran and Kalat sections, Sarakhs area, eastern Kopeh-Dagh* (Unpublished M.Sc. thesis). Faculty of Earth Sciences, Shahid Beheshti University.

Moallemi, A. (2005). *Classification of reservoir lithological groups of the Asmari Formation based on the petrophysical logs using of fuzzy logic* (Unpublished M.Sc. thesis). Faculty of Sciences, University of Tehran.

Moeinpour, M. (1998). *Sedimentological studies and recognition of porosity types* (Unpublished M.Sc. thesis). Faculty of Sciences, University of Tehran.

Mohamad-Beiknouri, M. (1997). *Provenance and thermal maturity of the organic content of the Siahoo and Sarchahan Formations (Ordovician-Silurian) of Zagros* (Unpublished M.Sc. thesis). Faculty of Earth Sciences, Shahid Beheshti University.

Mojoudi, M. (2001). *Reservoir quality evaluation of the western Ahwaz Oil Field* (Unpublished M. Sc. thesis). Faculty of Sciences, Kharazmi University.

Molki-Kheimesari, S. (1999). Lithostratigraphy and depositional environment of the Asmari Formation in the Northern Dezful Embayment. In *The 4th Symposium of Geology*. Geological Survey of Iran.

Moradi-Lir Gheshasi, N. (2002). *Foraminiferal biostratigraphy and paleoecology of the Asmari Formation* (Unpublished M.Sc. thesis). Faculty of Sciences, University of Tehran.

Moradpour, P. (2009). *Microstratigraphy of the Eocene deposits and Razak Complex in the Northern Baft, Kerman* (Unpublished M.Sc. thesis). Faculty of Earth Sciences, Shahid Beheshti University.

Mousavi, M. (2003). *Permeability of Dalan Formation in the South Pars Oil Field, based on the artificial neural network* (Unpublished M.Sc. thesis). Faculty of Sciences, University of Tehran.

Mousavi-Abnavi, N. (2009*). Biostratigraphy and lithostratigraphy of the Upper Permian deposits of the Banarizeh and Hambast Sections, based on the conodonts* (Unpublished M.Sc. thesis). Faculty of Sciences, Esfahan University.

Nabavi, M H (1978) Geological quadrangle map of Nain, 1:250,000, GSI, No. G8.

Nabavi, F. (2010). *Nannostratigraphy of the Gurpi Formation in the Kharameh (Shiraz) and Bourekh Mountain (Southern Lar)* (Unpublished M.Sc. thesis). Faculty of Earth Sciences, Shahid Beheshti University.

Naderi-Khojin, M. (2006). *Palynological studies of Kazhdomi Formation in the South Pars Oilfield* (Unpublished M.Sc. thesis). Faculty of Sciences, University of Tehran.

Naei, Z. (1995). *Microstratigraphy of the Qom Formation in the Kamarkuh and Kuh-e-Dobaradar sections, sedimentary Qom Basin* (Unpublished M.Sc. thesis). Faculty of Earth Sciences, Shahid Beheshti University.

Naeiji, M. (1997). *Depositional environment and diagenesis of the Qom Formation of the Northern Iwanaki Section* (Unpublished M.Sc. thesis). Faculty of Sciences, University of Tehran.

Najafi, M. (2010). Sequence stratigraphy of Albian-Cenomanian Aitamir Formation in the Sheikh and Bibahreh synclines, Kopeh-Dagh Basin. *Scientific Journal of Tehran University, 35*(4).

Namd-Malian, A. (1985). *Esfordi magmatic reserve and phosphate bearing horizons of Iran*. Geological Survey of Iran.

Nawrouzi, M. (2001). *Microstratigraphy of the Gurpi Formation in the Soltan Anticline, Central Lorestan, to Gurpi Anticline, type section of Gurpi Formation, Northeastern Khuzestan* (Unpublished M.Sc. thesis). Faculty of Sciences, University of Tehran.

Nazari, H, & Salamati, R. (1998). *Geological map of Rudbar, Scale 1:100000*. Geological Survey of Iran.

Nouri, N. (1997). *Microfaceis and depostional environment of the Qom Formation in the Abdol-Abad area, Semnan, and its correlation with the Kharazn of Ardestan* (Unpublished M. Sc. thesis). Faculty of Sciences, Kharazmi University.

Ogg, J. G., Ogg, G., & Gradstein, F. M. (2016). *A concise geologic time scale: 2016*. Amsterdam: Elsevier.

Okhravi, R. (2011). Ichnology analysis of the Ruteh Formation, Arou Section, Central Alborz. In *5th Symposium of Iranian*. Geological Society.

Omidpour, A. (2003). Diagenesis history of Asmari Formation, Kupal Field, and its effects on the reservoir quality.

Padiav, P. (1973). *Lithostratigraphy of Atamir Formation, based on the well log and seismic data, of the Khangiran, Gonbadli and Sangar Oil Fields* (Unpublished M.Sc. thesis). Faculty of Sciences, Kharazmi University.

Pilgrim, G. E. (1924). The geology of parts of the Persian Provinces of Fars, Kerman and Luristan. *Geological Survey of India, Memoir, 48*(2), 1–118.

Piriaei, A. (1996). *Petrography, microfacies and depositional environment of Amiran Formation of Zagros* (Unpublished M.Sc. thesis). Kharazmi University.

Piroozi, M. (2007). Geology, facies analysis, geochemistry and genesis of the fluorite ores, Pb and Br, of Shotori Formation, Kamar- e Mehdi area.

Rabbani, A. (1993). *Depositional environment and diagenesis of the lower Elika Formation of the Sarbandan Section* (Unpublished M.Sc. thesis). Faculty of Sciences, University of Tehran.

Rabbani, A. (2003). Gochemichal study of the limestone-dolomite facies of the gas reservoir of Dalan and Kangan Formations of the Salman Oil Field. In *The 22nd Symposium of Geology*. Geological survey of Iran.

Raeisi, W. (1998). *Microfacies and depositional environment of the Asmari Formation, Oligo-Miocene of the Dezful Embayment* (Unpublished M.Sc. thesis). Faculty of Earth Sciences, Shahid Beheshti University.

Rafiei-Oskuhi, N. (1992). *Palynostratography and paleobiogeography of the Padeha Formation of the Type section, Ozbak-Kuh* (Unpublished M.Sc. thesis). Faculty of Earth Sciences, Shahid Beheshti University.

Rafiei, B. (1983). *Sedimentology, depositional environment and geochemistry of Aghche gil and Apsheron Formations of the Moghan Plain, Northwestern Iran* (Unpublished Ph.D. thesis). Faculty of Sciences, Kharazmi University.

Rahim Zadeh, F. (1994). Oilgocene, Miocene and Pliocene of Iran, Geological Book Codification Project, Geological Survey of Iran, No. 12.

Rahmani Zadeh, H. (2007). *Depositional environment and geochemistry of the Ilam Formation, Surgah Mountain, and its correlation with Siahkuh Section* (Unpublished M.Sc. thesis). Faculty of Earth Sciences, Shahid Beheshti University.

Ramazan-Ghorbani, F. (2010). *Foraminiferal biostratigraphy and paleoecology of the Ziarat Formation (Paleocene-Eocene) of the Baghdareh Section, Northeastern Abhar, and its correlation with Majan Section of the northwestern Shahroud* (Unpublished M.Sc. thesis). Faculty of Sciences, Damghan University.

Ramazani, Sh. (2009). *Sedimentary facies and history of the Pesteligh Formation in the Chehel Kaman and Tang-e-Neyzar areas* (Unpublished M.Sc. thesis). Faculty of Sciences, Ferdowsi Mashhad University.

Rastegar, L. (2000). *Microfacies, depositional environment and sequence stragiraphy of the Milla Formation in the Ajabshir Section, Eastern Azarbaijan* (Unpublished M.Sc. thesis). Faculty of Sciences, Kharazmi University.

Ruttner, A. (1994). Southern borderland of Triassic Laurasia in north–east Iran. *Scientific Quarterly Journal, 12*, 28–35.

Ruttner, A. W. (1984). The Pre-Liassic Basement of the Eastern Kopeh-Dagh Range. *Neues Jahrbuch fuer Geologie und Palaeontologie Abhandlungen, 168*(2–3), 256–268.

Sadeghi, Gh. (2003). Depositional environment model of the Asmari Formation in the Bangestan Mountain, In *The 22nd Symposium of Geology*. Geological survey of Iran.

Sahabi, F. (2003). Lithostratigraphy and depositional environment of the Razak Formation of the Kavar-Firozabad. In *The 22nd Symposium of Geology*. Geological Survey of Iran.

Saheb Nouri, A. (2004). *Microfaceis study of the Aja-Gheshlaq in the Tolon Section, Moghan Plain* (Unpublished M.Sc. thesis). Faculty of Sciences, University of Tehran.

Saheb Nouri, A., Ghasemi-Nejad, E., & Amini, A. (2003). Biostratigraphy of Ajagh Gheshlaq Formation of the Tolen section, Moghan Plain. In *The 22nd Symposium of Geology*. Geological Survey of Iran.

Salehi, F. A. (2007). *Geochemistry, diagenesis and depositional environment of the Fahlian Formation in the type section, Fahlian Anticline, Northwestern Noorabad-e-Mamseni, and its correlation with Gachsaran borehole No 55 of the Gachsaran Oil Field* (Unpublished M.Sc. thesis). Faculty of Earth Sciences, Shahid Beheshti University.

Samani, B., Xoeie, Ch., Taeo, G., & Kovan, T. (1993). Precambrian geology of Central Iran; implication on the stratigraphy, magmatism and metamorphism. *Earth Science Quarterly, 10*, 40–63.

Scotese, C. R. (2005). Maps of the Phanerozoic.

Sepiani, H. (2010). *Petrography and interpretation of depositional history of the Ilam Formation if the Borehole 3 and 4 of the Malekuh Oil Field, Northwestern Andimeshk* (Unpublished M.Sc. thesis). Faculty of Sciences, Ferdowsi Mashhad University.

Seyed-Emami, K. (1996). Stratigraphy, Paleobiogeography and paleogeography of the Middle Cretaceous strata of Central Ira. *Earth Science Quarterly, 21*, 50–69.

Seyed-Emami, K., Brants, A., & Bozorgnia, F. (1971). *Stratigraphy of the Cretaceous rocks southeast of Esfahan*. Geological Survey of Iran, Report 20, 5–27.

Seyrafian, A. (2008). Depositional environment of the upper Asmari Formation, Burigialian, based on the Microfacies and foraminifer paleoecology in the southeastern Kuh-e-Laki Section, south Lordegan. *Research Journal of Esfahan University, 34*(5), 241–260.

Shafi Zadeh, M., & Seyed-Emami, K. (2005). Lithostratigraphy and Biostratigraphy of the Dalichai Formation of Western Shahroud, Eastern Alborz. *Earth Science Quarterly, Geological Survey of Iran, 44*, 98–113.

Shahbazi, H. (1999). *Petrography and petrology of Ghareh-Dash 'Series'* (Unpublished M.Sc. thesis). Faculty of Sciences, Tehran University.

Shahpasnad Zadeh, M. (1992). *Structural analysis and depositional environment of the metamorphic complex of Gorgan (Gorgan Schist)* (Unpublished M.Sc. thesis). Faculty of Earth Sciences, Shahid Beheshti University.

Shemirani, A. (2009). The new biostratigraphical findings of the Dorud and Ruteh Formations, In *The 3rd Symposium of Geology*. Geological Society of Iran.

Shemirani, A., & Amiri-Bakhtiar, H. (2000). The new lithological and biostratigraphical findings of the Darian and Kazhdomi Formations of the southeastern Iran, In *The 4th Symposium of Geology*. Geological Society of Iran.

Shemirani, A., & Arefifard, S. (1998). The new findings of the Dorud and Ruteh Formation. In: *The 3rd Symposium of Geology*. Geological Society of Iran.

Shoushtarian, F. (2003). *Geochemistry, depositional environment and digenesis of Asmari Formation in the Tang-e-Goltorsh of Asmari Anticline* (Unpublished M.Sc. thesis). Faculty of Earth Sciences, Shahid Beheshti University.

Sohrabi, S. (2009). *Microfacies and depositional environment of the Tirgan Formation in the Central Kopeh-Dagh, Chenaran-Radkan area* (Unpublished M.Sc. thesis). Faculty of Earth Sciences, Shahid Beheshti University.

Soleymani, B., & Shekhzadeh, H. (2003). Geochemical analysis and paleosalinity of the Pabdeh, Gurpi and Lower Asmari Formations in the Ahwaz Oil Field, In *The 22nd Symposium of Geology*. Geological Survey of Iran.

Soleymani, B., & Shekhzadeh, H. (2008). Geochemistry and depositional environment of the Pabdeh-Gurpi and Kazhdomi-Gadvan shaly beds, Bangestan Anticline. In *The 12th Symposium of Geology*. Geological Survey of Iran.

Soori, S. (2006). *Depositional environment and diagenesis of the Fahlian Formation in the Mangasht Section and Haftkel Borehole No 61* (Unpublished M.Sc. thesis). Faculty of Sciences, University of Tehran.

Stampfli, G. M. (1996). The Intra-Alpine terrain: A Paleotethyan remanent in the Alpine Variscides. *Ecolgae geologicae Helvetiae, 89*, 13–42.

Stampfli, G. M. (1978). Etude geologique general de l'Elbourz oriental au sud Gonbade-Qabus. Iran NE, These Geneve, 328.

Stampfli, G. M., & Borel, G. D. (2001). Plate reconstructions of the PeriTethyan Domain. In G. M. Stampfli, G. D. Borel, W. Cavazza, J. Mosar, & P. A. Ziegler (Eds.), *The Paleotectonic Atlas of the PeriTethyan Domain*. European Geophysical Society.

Stewart-Ajek, H. (1977). The study of Gurpi Formation of Iran, Research Project, University of Shiraz.

Stöcklin, J. (1968). Salt deposits of the Middle East. *Geological Society of America, 88,* 157–181. (Special Issue).
Stöcklin, J. (1968). Structural history and tectonics of Iran: A review. *American Association of Petroleum Geologists Bulletin, 52*(7), 1229–1258.
Stöcklin, J. (1977). Structural correlation of the Alpine ranges between Iran and central Asia, Anonymous, Livre a la memoire de Albert F de Lapparent (1905-1975) consacre aux Recherches geologiqes dans les chaines alpines de l`Asie du Sud-Ouest. *Soc Geol Fr, Mem Hors-Ser, 8,* 333–353.
Stöcklin, J., Eftekharnezhad, J., & Hushmandzadeh, A. (1972). *Central Lut reconnaissance, East Iran*. Geological Survey of Iran, Report 22, 62.
Tahmasebi-Sarvestani, A., & Ghavidel-Syooki, M. (2006). Microfaceis, depositional environment and sequence stratigraphy of the Asmari Formation in the Tang-e-Pabdeh section, southern flank of the Gurpi Anticline, In *The 25th Symposium of Geology*. Geological Survey of Iran.
Taifeh-Khabbazi, M., & Ghavidel-Syooki, M. (2000). The paleobiology of the Padeha and Bahram Formations in the Hootak, Abmorad and Shahzadeh Mohammad of Kerman, In *The 14th Symposium of Geology*. Geological Society of Iran, Shiraz University.
Tavakoli, W. (2005). *Electrical analysis of the Asmari Formation of the Mazoun Oil Field* (Unpublished M.Sc. thesis). Faculty of Sciences, Kharazmi University.
Tavalaei, M. (2002). The study of the source rocks of Ordovician and Silurian deposits, Sarchahan and Siahoo Formations. In *The 6th Symposium of Geology*. Geological Society of Iran.
Tavalaei, M. (2008). The study source rocks of Garo Formation of the Kalat Section, Kermanshah, In *The 12th Symposium of Geology*. Geological Society of Iran.
Teimouri, R. (2009). *Biostratography of the Geirud Formation of the Geirud Section, Northeastern Tehran, based on the Conodonts* (Unpublished M.Sc. thesis). Faculty of Earth Sciences, Shahid Beheshti University.
Thiele, O. (1968). *Explanatory text of Golpayegan Quadrangle Map1:250,000 scale*. Geological Society of Iran.
Torsvik, T. H., & Cocks, L. R. M. (2005). Norway in space and time: A Centennial cavalcade. *Norwegian Journal of Geology, 85*, 73–86. ISSN 029-196X.
Vaezjavadi, F. (2003). The introduction of plant macrofossils of the Shemshak Formation in the Narges-Chal area, Alborz. In *The 22nd Symposium of Geology*. Geological Survey of Iran.
Vahidinia, P. (2000). Echinoid studies of Abderaz Formation in the eastern Kopeh-Dagh. In *The 4th Symposium of Geology*. Geological Survey of Iran.
Walter, M. R. (1977). Interpreting stromatolites. *American Scientist, 65*, 563–571.
Wendt, J., Kaufmann, B., Belka, Z., Farsan, N., & Karimi Bavandpur, A. (2005). Devonian/Mississippian stratigraphy, facies patterns and palaeogeography of Iran, Part II: Northern and central Iran. *Acta Geologica Polonica, 55,* 31–97.
Yavari, M. (1987). *Stratigraphical and paleontological comparison of the Geirud Formation in the type and Shahmirzad section* (Unpublished M.Sc. thesis). Faculty of Earth Sciences, Shahid Beheshti University.
Zakerabbasi, Z. (2009). *Mineralogical and geological studies of the clay mineral of the marly formation of Razak and Mishan in the some outcrops of the Shiraz surroundings* (Unpublished M.Sc. thesis). Shiraz University.

MIX
Papier aus verantwortungsvollen Quellen
Paper from responsible sources
FSC® C105338
FSC
www.fsc.org

If you have any concerns about our products,
you can contact us on
ProductSafety@springernature.com

In case Publisher is established outside the EU,
the EU authorized representative is:
Springer Nature Customer Service Center GmbH
Europaplatz 3, 69115 Heidelberg, Germany

Printed by Libri Plureos GmbH
in Hamburg, Germany